玩转 Android App 开发
（微课版）

陈轶 编著

清华大学出版社
北京

内 容 简 介

Android 移动应用开发是移动应用开发领域的热点。本书介绍 Android 移动应用开发的核心技术，具体包括 Android 的开发环境、Kotlin 基础、Activity 组件、界面开发、并发处理、广播机制、Service 组件、网络应用、数据的持久化处理、ContentProvider 组件，以及 Android JetPack 的应用开发。

为了使读者更好地理解和掌握 Android 移动开发技术，本书通过翔实、丰富的项目应用实例将相应的知识点串联起来，从简单到复杂，从基础应用到综合项目开发层层推进。为了符合 Android 移动开发的发展趋势，采用 Kotlin 贯穿全书。

本书可作为普通高校本科"移动应用开发"课程的教材，也可供移动应用开发人员学习和参考。

版权所有，侵权必究。举报：010-62782989，beiqinquan@tup.tsinghua.edu.cn。

图书在版编目(CIP)数据

玩转 Android App 开发：微课版/陈轶编著. -- 北京：清华大学出版社，2025.6.
ISBN 978-7-302-69448-9

Ⅰ．TN929.53

中国国家版本馆 CIP 数据核字第 20252TF958 号

责任编辑：汪汉友　薛　阳
封面设计：常雪影
责任校对：李建庄
责任印制：杨　艳

出版发行：清华大学出版社
网　　址：https://www.tup.com.cn，https://www.wqxuetang.com
地　　址：北京清华大学学研大厦 A 座　　　　邮　编：100084
社 总 机：010-83470000　　　　　　　　　　　邮　购：010-62786544
投稿与读者服务：010-62776969，c-service@tup.tsinghua.edu.cn
质量反馈：010-62772015，zhiliang@tup.tsinghua.edu.cn
课件下载：https://www.tup.com.cn，010-83470236

印 装 者：三河市铭诚印务有限公司
经　　销：全国新华书店
开　　本：203mm×260mm　　　印　张：25　　　字　数：708 千字
版　　次：2025 年 7 月第 1 版　　　　　　　　　　印　次：2025 年 7 月第 1 次印刷
定　　价：99.00 元

产品编号：093553-01

前　言

　　近年来,移动互联网的发展非常迅猛,影响着人们生活的方方面面。作为两大主流移动操作系统平台之一的 Android 也在不断发展,新技术、新特色层出不穷,市场份额已在 2014 年超越 iOS 平台。Kotlin 具有简洁、易学、安全、快捷、开源等特点,是 Android 移动应用开发的利器;此外,Android 移动应用的架构设计为移动应用的开发奠定了基础,MVVM(Model-View-ViewModel)和 MVI(Model-View-Intent)成为移动开发的两大主流架构,2018 年,由谷歌(Google)公司推出的 Android JetPack 具有架构组件,能协助开发者快速搭建基于 MVVM 和 MVI 的 Android 移动应用。5G 技术的不断发展,为基于 Android 平台的移动终端提供了更快的网络服务,Retrofit 2.0 HTTP 网络请求框架等一系列产品让开发移动互联网应用更加方便简单,特别是 RxJava 3.0 框架在异步流的处理方面有着绝对的优势,可以更快捷地处理网络并发数据。

　　本书作者在 Android 移动应用开发的教学和科研实践过程中发现了一些更方便、更快捷,让开发流程更加清晰的方法,于是萌生了编写本书的想法。本书基于 Android 14.0 版本,采用了官方推荐的 Android Studio 开发工具,并采用 Kotlin 进行案例介绍。

　　本书分为 10 章,循序渐进地阐述了 Android 的相关知识,结合案例将相关知识点进行实践应用,通过每章的习题强化对概念的理解和掌握。作者的个人 CSDN 博客也提供了额外移动应用案例的详细介绍。

　　本书涉及的知识点包括 Android 概述、Kotlin 基础和面向对象编程、Android 的四大组件(活动、消息接收器、服务、内容提供者)、MVVM 和 MVI 开发模式在移动应用开发中的应用、基于 Compose 组件的 Android 的界面开发、Android 的并发处理、Android 的持久化处理,并将 Android JetPack 开发套件与移动应用的各个环节衔接起来。上述知识点并没有按照传统方式独立介绍,而是从实际应用出发,将知识点进行整合,采用多种方式进行介绍,突出重点知识和核心知识,避免大而全的介绍。由于 Kotlin 的知识点非常丰富,不可能用很少的篇幅介绍完整,因此本书第 2 章在介绍 Kotlin 基础时,侧重 Kotlin 在移动应用开发时常见的知识点,例如对 Kotlin 基本语法和 Kotlin 面向对象技术开展介绍,特别对移动应用开发时大量使用的函数式编程进行深入介绍。本书根据功能关联性,将相关的知识点进行融合。例如,第 4 章介绍 Android 的界面开发时,不仅介绍了 Compose 组件,还介绍了承担业务逻辑的 ViewModel 组件,并介绍了 MVVM 和 MVI 模式在移动应用开发中的应用。第 8 章介绍 Android 的网络应用时,不但介绍传统的网络应用方式(WebView 组件加载网页和 HttpURLConnection 实现基于 HTTP 进行网络处理),还引入 Retrofit 2.0 框架实现网络处理。由于网络处理经常涉及 JSON 数据解析和网络访问的异步处理,作者将 GSON 库和 RxJava 3.0 库在网络访问的异步流处理的知识点在该章进行介绍。根据功能类似性,本书将相关知识点对比介绍。例如,在介绍 Android 的并发处理时,介绍线程、Handler 消息处理机制、异步任务、Kotlin 协程和 RxJava3 处理异步流的实现。在介绍 Android 持久化处理时,介绍 DataStore、文件处理和 Room 组件处理 SQLite 数据库,使读者充分了解 3 种进行持久化处理的技术特点。本书将所讲知识点融入同一个应用实例中,采用层层推进的方式展开。例如,在同一个移动应用案例中结合 ViewModel 组件、Navigation 组件、持久化处理的 Room 组件、后台任务处理的 WorkManager 组件、分页的 Paging 组件等 Android JetPack 开发套件的组件,使读者对所学知识有更深入的理解。

本书提供了具有实际意义的移动应用案例，如掷骰子游戏、心理测试、歌词同步播放、智能聊天、调用相机和相册、播放媒体库视频、在线图片添加水印、在线视频播放等具有实践意义的移动应用案例。

本书课后练习配套答案、课件等学习资源可以在清华大学出版社的官方网站或扫描下方的二维码下载使用。

本书由**南昌大学陈轶编著**。另外，**南昌大学的武友新、江顺亮、黄伟、徐子晨教授，及刘捷、刘萍、王炜立、文喜老师在本书的编写过程中给予了支持和帮助**，在此表示深深的感谢。此外，还要衷心感谢清华大学出版社的相关编校人员，非常佩服他们的专业和敬业精神。

由于时间和编者学识有限，书中不足之处在所难免，敬请大家批评指正。

陈轶

2025 年 3 月

学习资源

目　　录

第 1 章　Android 的开发环境 ……………………………………………………………… 1
　1.1　Android 移动开发概述 ……………………………………………………………… 1
　　1.1.1　Android 的发展 ………………………………………………………………… 1
　　1.1.2　Android 平台的架构 …………………………………………………………… 2
　1.2　开发环境和开发工具 ………………………………………………………………… 4
　1.3　创建第一个 Android 项目 …………………………………………………………… 5
　　1.3.1　创建新的项目 …………………………………………………………………… 5
　　1.3.2　启动模拟器 ……………………………………………………………………… 6
　　1.3.3　运行第一个项目 ………………………………………………………………… 9
　习题 1 ……………………………………………………………………………………… 12

第 2 章　Kotlin 基础 ………………………………………………………………………… 14
　2.1　Kotlin 概述 …………………………………………………………………………… 14
　2.2　第一个 Kotlin 程序 …………………………………………………………………… 16
　2.3　函数 …………………………………………………………………………………… 18
　2.4　变量和数据类型 ……………………………………………………………………… 19
　　2.4.1　变量 ……………………………………………………………………………… 19
　　2.4.2　数据类型的种类 ………………………………………………………………… 19
　　2.4.3　数据类型的转换 ………………………………………………………………… 20
　　2.4.4　可空类型的处理 ………………………………………………………………… 21
　　2.4.5　数组和集合类型 ………………………………………………………………… 22
　2.5　操作符和表达式 ……………………………………………………………………… 24
　2.6　控制结构 ……………………………………………………………………………… 26
　2.7　Lambda 表达式 ……………………………………………………………………… 30
　　2.7.1　Lambda 表达式的应用 ………………………………………………………… 30
　　2.7.2　常见的标准函数和 Lambda 表达式 …………………………………………… 31
　2.8　面向对象编程 ………………………………………………………………………… 33
　　2.8.1　包和目录 ………………………………………………………………………… 33
　　2.8.2　类和对象 ………………………………………………………………………… 33
　　2.8.3　继承性 …………………………………………………………………………… 35
　　2.8.4　接口 ……………………………………………………………………………… 36
　2.9　异常处理 ……………………………………………………………………………… 37
　习题 2 ……………………………………………………………………………………… 38

第 3 章 Android 的 Activity 组件 … 41
3.1 Activity 的创建 … 41
3.2 Activity 和 Intent … 42
3.2.1 显式 Intent … 43
3.2.2 隐式 Intent … 46
3.3 Activity 之间的数据传递 … 50
3.3.1 传递常见数据 … 50
3.3.2 Serializable 对象的传递 … 54
3.3.3 Parcelable 对象的传递 … 56
3.3.4 数据的返回 … 60
3.4 Activity 的生命周期 … 64
3.4.1 Activity 的返回栈 … 64
3.4.2 Activity 的启动方式 … 64
3.4.3 Activity 的生命周期 … 70
3.5 掷骰子游戏 … 76
习题 3 … 82

第 4 章 Android 的界面开发 … 86
4.1 JetPack Compose 组件 … 86
4.1.1 可组合函数和预览函数 … 86
4.1.2 Modifier 修饰符 … 88
4.1.3 Compose 常见的 UI 组件 … 92
4.1.4 ConstraintLayout … 109
4.2 搭建 Scaffold … 111
4.3 Compose 组件的状态管理和重组 … 124
4.3.1 可组合项的状态 … 124
4.3.2 无状态的可组合函数和有状态的可组合函数 … 126
4.3.3 状态提升 … 127
4.3.4 状态丢失和状态保留 … 129
4.3.5 状态容器 … 132
4.4 ViewModel 组件 … 133
4.4.1 MVVM 模式 … 135
4.4.2 MVI 模式 … 138
4.5 Navigation 组件 … 143
4.5.1 页面导航的实现 … 144
4.5.2 在目的地之间安全传递数据 … 147
4.6 心理测试移动应用实例 … 150
4.6.1 项目说明 … 150
4.6.2 心理测试移动应用的功能实现 … 150
习题 4 … 168

第 5 章　Android 的并发处理 ····· 169
5.1　多线程 ····· 169
5.2　Handler 机制 ····· 171
5.3　协程 ····· 174
5.4　Compose 的附带效应 ····· 180
5.4.1　附带效应概述 ····· 180
5.4.2　LaunchedEffect 和 rememberUpdatedState ····· 182
5.4.3　DisposableEffect 和 Lifecycle ····· 185
5.4.4　SideEffect ····· 190
5.4.5　produceState 和 derivedStateOf ····· 193
5.5　RxJava 库实现异步操作 ····· 198
5.5.1　Observer 模式 ····· 198
5.5.2　RxJava 的相关概念 ····· 200
5.6　歌词同步播放 ····· 204
习题 5 ····· 209

第 6 章　Android 的广播机制 ····· 212
6.1　BroadcastReceiver 组件 ····· 212
6.2　发送广播 ····· 215
6.2.1　标准广播 ····· 215
6.2.2　有序广播 ····· 216
习题 6 ····· 222

第 7 章　执行后台任务 ····· 224
7.1　Service 组件 ····· 224
7.2　Service 的生命周期 ····· 229
7.3　Activity 和 Service 的通信 ····· 230
7.4　Notification 通知和前台服务 ····· 234
7.4.1　Notification ····· 234
7.4.2　前台服务 ····· 237
7.5　WorkManager 组件 ····· 244
7.5.1　WorkManager 的基本使用方法 ····· 245
7.5.2　任务链 ····· 251
习题 7 ····· 257

第 8 章　Android 的网络应用 ····· 259
8.1　网络访问相关配置 ····· 259
8.2　WebView 组件 ····· 260
8.3　使用 HttpURLConnection 访问网络资源 ····· 263
8.4　JSON 数据的解析 ····· 269

	8.4.1 JSON 格式	269
	8.4.2 JSONObject 解析 JSON 数据	270
	8.4.3 GSON 解析 JSON 数据	274
8.5	使用 Retrofit 库访问网络资源	277
8.6	智能聊天移动应用实例	289
	8.6.1 功能需求分析和设计	289
	8.6.2 系统的实现	291
习题 8		302

第 9 章 数据的持久化处理和 ContentProvider 组件 — 305

9.1	DataStore 存储处理	305
9.2	文件处理	320
9.3	Room 组件	332
	9.3.1 用 Room 实现数据库的基本操作	333
	9.3.2 用 Room 实现迁移数据库	341
9.4	ContentProvider 组件	347
	9.4.1 创建 ContentProvider 组件	348
	9.4.2 使用 ContentProvider 组件	354
9.5	调用相机和媒体库	359
	9.5.1 运行时权限	359
	9.5.2 拍照和显示媒体库的图片	365
	9.5.3 访问媒体库中的视频	369
习题 9		374

第 10 章 Paging 组件 — 376

10.1	分页组件概述	376
10.2	分页处理	378
习题 10		391

参考文献 — 392

第 1 章　Android 的开发环境

Android 是美国的谷歌(Google)公司为手机等移动终端开发的一款基于 Linux 内核的操作系统和编程平台。目前,有超过八成的手机应用是基于 Android 平台的,这使得它成为移动终端中应用最广泛的平台。基于 Android 的移动应用与人们的生活密切相关。学习和掌握 Android 技术已经成为很多程序员的选择。下面开始介绍基于 Android 平台移动应用开发的相关知识。

1.1　Android 移动开发概述

1.1.1　Android 的发展

Android 平台主要服务于移动终端,是基于 Linux 内核的自由开源的操作系统和编程平台。2003 年,Android 操作系统由 Andy Robin 开发并发布,2005 年被谷歌公司收购。它以 Apache 开源许可证的授权方式,发布了 Android 的源代码。在免费开源以及 Android 平台不断优化更新的驱动下,Android 操作系统发展迅速。2007 年,第一台基于 Android 操作系统的手机问世;2014 年,Android 手机市场份额首次超越 iOS 手机市场份额。自此,Android 的市场份额不断扩展。国际数据公司(IDC)于 2019 年发布的研究数据表明,Android 平台的手机已经占据 87% 的市场份额。这意味着有近九成的手机使用了 Android 系统。因此,基于 Android 平台的移动终端成为主导,在移动应用开发中,Android 处于非常重要的地位。

Android 之所以能获得成功,与其本身的不断发展有关。表 1-1 中列出了 Android 的各阶段的版本。

表 1-1　Android 的历史版本

代 码 名 称	版　　本	Linux 内核版本	最初发布时间	Android SDK
KitKat	4.4～4.4.4	3.10	2013 年 10 月	19～20
Lollipop	5.0～5.1.1	3.16	2014 年 11 月	21～22
Marshmallow	6.0～6.0.1	3.18	2015 年 10 月	23
Nougat	7.0～7.1.2	4.4	2016 年 8 月	24～25
Oreo	8.0～8.1	4.10	2017 年 8 月	26～27
Pie	9.0	4.4.107、4.9.84 和 4.14.42	2018 年 8 月	28
Android 10/Android Q	10.0	4.9q、4.14-q 和 4.19q	2019 年 5 月	29
Android 11/Android R	11.0	4.14-stable 和 4.19-stable	2020 年 9 月	30
Android 12/Android S	12.0	5.4 和 5.10	2021 年 10 月	31
Android 12L/Android Sv2	12L	5.10.66	2021 年 9 月	32
Android 13/Tiramisu	13	5.15.41	2022 年 7 月	33
Android 14/UpsideDownCake	14	6.1.23	2023 年 2 月	34
Android 15/Vanilla Ice Cream	15	6.30	2024 年 2 月	35

从表 1-1 可以发现，从 2008 年至今，Android 的版本推进时间非常短。几乎每年都推出一个新的版本，每次新版本的问世也意味着新功能和新特性的出现。例如，Android V（Vanilla Ice Cream）即 Android 15 已经发布正式版本，可以对曲面屏幕有更好的支持，为 5G 网络的应用进行提供技术保障，支持分区存储等功能。在新版本中，一些新的特性也会出现，例如新增动态性能框架、增强多任务处理能力。

1.1.2 Android 平台的架构

Android 平台架构定义软件层次分为应用层、应用框架层、安卓原生库和安卓运行时硬件抽象层（HAL）、Linux 内核层。Android 软件架构可以由图 1-1 所示的软件堆栈表示。

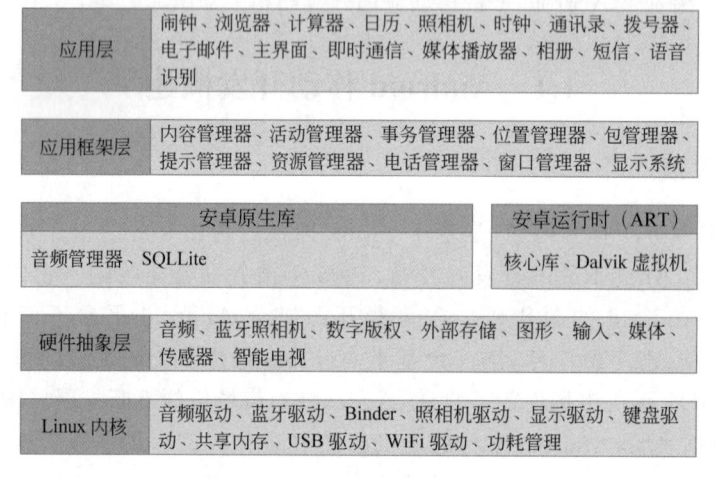

应用层	闹钟、浏览器、计算器、日历、照相机、时钟、通讯录、拨号器、电子邮件、主界面、即时通信、媒体播放器、相册、短信、语音识别
应用框架层	内容管理器、活动管理器、事务管理器、位置管理器、包管理器、提示管理器、资源管理器、电话管理器、窗口管理器、显示系统
安卓原生库	音频管理器、SQLLite
安卓运行时（ART）	核心库、Dalvik 虚拟机
硬件抽象层	音频、蓝牙照相机、数字版权、外部存储、图形、输入、媒体、传感器、智能电视
Linux 内核	音频驱动、蓝牙驱动、Binder、照相机驱动、显示驱动、键盘驱动、共享内存、USB 驱动、WiFi 驱动、功耗管理

图 1-1　Android 软件堆栈

1. 应用层

Android 中内置了很多移动应用，这些移动应用包括联系人、电话、电子邮件、计算器、照相机等，用户也可以自行安装。这些移动应用包含在应用层（Application Layer）中。如果将一个 Android 移动应用的 APK 文件解压，可以发现一个移动应用的结构如图 1-2 所示，主要包括如下内容。

图 1-2　应用层的结构

（1）AndroidManifest.xml：移动应用的系统配置清单，设置了移动应用的名称、版本、访问权限等。

（2）assets 目录：其中包括应用中没有编译的资源。

（3）res 目录：其中包括应用中已经编译的资源，例如字符串、布局、尺寸大小等。

（4）META-INF 目录：其中包括 CERT.RSA 和 CERT.SF，应用的签名文件，说明移动应用的数字签名。MANIFEST.MF 说明移动应用的包和扩展的信息。

（5）resources.arsc：为已经编译的资源 res 目录下的资源指定资源编号。

（6）classes.dex：在 Dalvik 虚拟机中运行的压缩的字节码代码。

（7）lib 目录：保存了运行移动应用的本地库。

注意：在 2018 年的 Google I/O 大会上，Android App Bundle（Android 动态化框架）被正式提出，该技术支持 Google Play 的动态交付（Dynamic Delivery）。这使得在用户下载移动应用时，动态框架会根据每个用户的设备信息和开发者上传的 App Bundle 生成与之对应且经过优化的 APK，并将其提供给设备进行安装。这使得下载的移动应用不再臃肿。

2. 应用框架

应用框架提供了大量的应用程序接口（API），供开发者开发 Android 应用。移动应用开发者通过调用这些 API，开发实现移动应用，如图 1-3 所示。应用框架包括 Android SDK API 和 Android SOFT API。Android SDK API 就是开发需要的库，其中包括 Java API 库。Android SOFT API 包括了权限配置、构建参数、意图配置等应用配置的相关内容。

3. Android 原生库和安卓运行时

Android 的原生库是 C/C++ 函数库的集合，包括 Surface Manager、Media Framework、SQLite、OpenGL ES、FreeType、Webkit、SGL 和 SSL。这些 C/C++ 库支持 Android 的应用框架，为开发基于 Android 平台的移动应用提供帮助。

图 1-3 Android 应用调用应用框架

安卓运行时（Android Runtime，ART）主要负责运行 Android 应用程序。ART 引入了预编译机制。从 Android 7.0 开始，结合使用 AOT（Ahead Of Time）、JIT（Just In Time）编译和配置文件引导编译，不需要重新编译 APK 安装包就可以直接加载和运行它。

ART 内置了 AOT 编译器，可结合配置文件对常使用的代码进行 AOT 编译，在 APK 运行之前，就对其包含的 dex 字节码进行翻译，生成对应的本地机器指令。因此，运行时就可直接执行这些机器指令。如图 1-4 所示，Android 的源代码编译生成 DEX 文件，这些编译后的文件与资源以及一些必要的本地代码一起压缩生成 APK 文件。将 APK 文件安装到移动终端时，通过 AOT 技术，可利用 dex2oat（Dalvik executable file to optimized art file）将 DEX 文件编译优化生成可执行、可链接的文件格式（Executable and Linkable Format，ELF）的机器码 OAT 可执行文件。最后通过 ART 运行 OAT 文件。经过 AOT 编译器编译后，可使 ART 反复运行编译后的本地机器指令，从而提高运行效率。

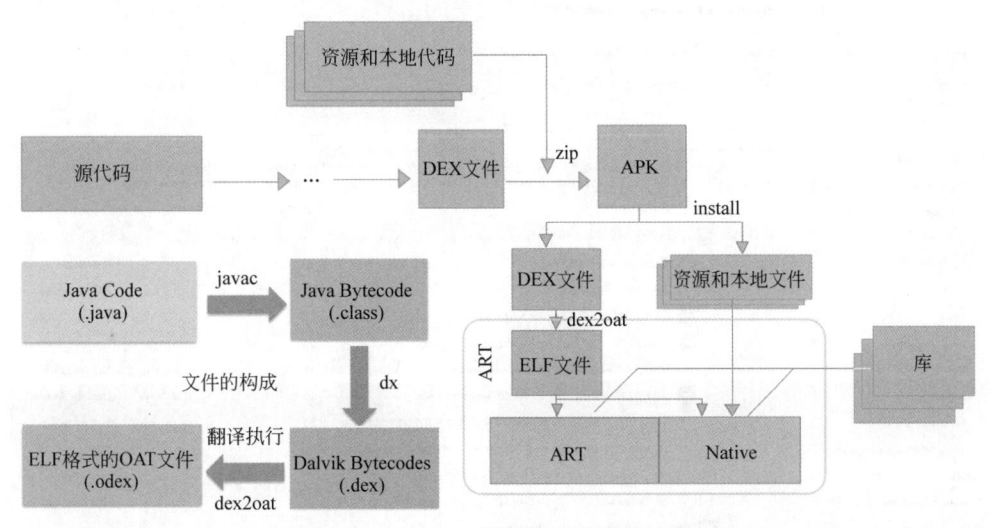

图 1-4 Android 运行时编译执行过程

ART 包含了具有代码分析功能的 JIT 编译器。JIT 编译器可对 AOT 编译进行补充。当用户启动应用时，ART 加载 DEX 文件，如果有 OAT 可执行文件，ART 会直接执行该文件。如果 OAT 文件中不含经过编译的代码，ART 会通过 JIT 和解释器执行。JIT 会将经常执行的方法添加到配置文件，使得下一次重新启动通过配置文件引导执行代码。通过这种方式，JIT 可以节省存储空间，加快应用和系统更新速度。

4. 硬件抽象层

硬件抽象层（Hardware Abstract Level，HAL）用于为第三方硬件厂商提供标准的接口。通过这些标准接口，可使更高级别的 Java API 框架调用硬件。HAL 包含多个库模块，其中每个模块都为特定类型的硬件组件实现一个接口，例如照相机或者蓝牙模块。当框架 API 要求访问硬件时，Android 系统将为该硬件加载库模块。HAL 采用硬件接口定义语言（Hardware Interface Definition Language，HIDL）描述了 HAL 和它的用户之间的接口。

5. Linux 内核

Android 平台基于 Linux 内核。Linux 内核提供了底层的硬件驱动，包括 Audio Driver（音频驱动）、Display Driver（显示驱动）、KeyPad Driver（按键驱动）、Bluetooth API（蓝牙 API）、Camera API（照相机 API）、Shared Memory Driver（共享存储驱动）、USB Driver（USB 驱动）、WiFi Driver（WiFi 驱动）、Power Management（电源管理）等。

1.2 开发环境和开发工具

下面介绍搭建 Android 开发环境的流程以及所用的开发工具。

开发 Android 应用需要进行以下准备。

（1）JDK。它是 Java 开发软件库，包括 Java 的运行环境、工具集合、基础库等支持。

（2）Android SDK。它是开发 Android 应用的 Android 软件开发工具包，包括软件包、软件框架、硬件平台、操作系统等建立应用软件的开发工具的集合。Android SDK 可直接下载，无须安装，在 Android Studio 中配置后即可使用，如图 1-5 所示。

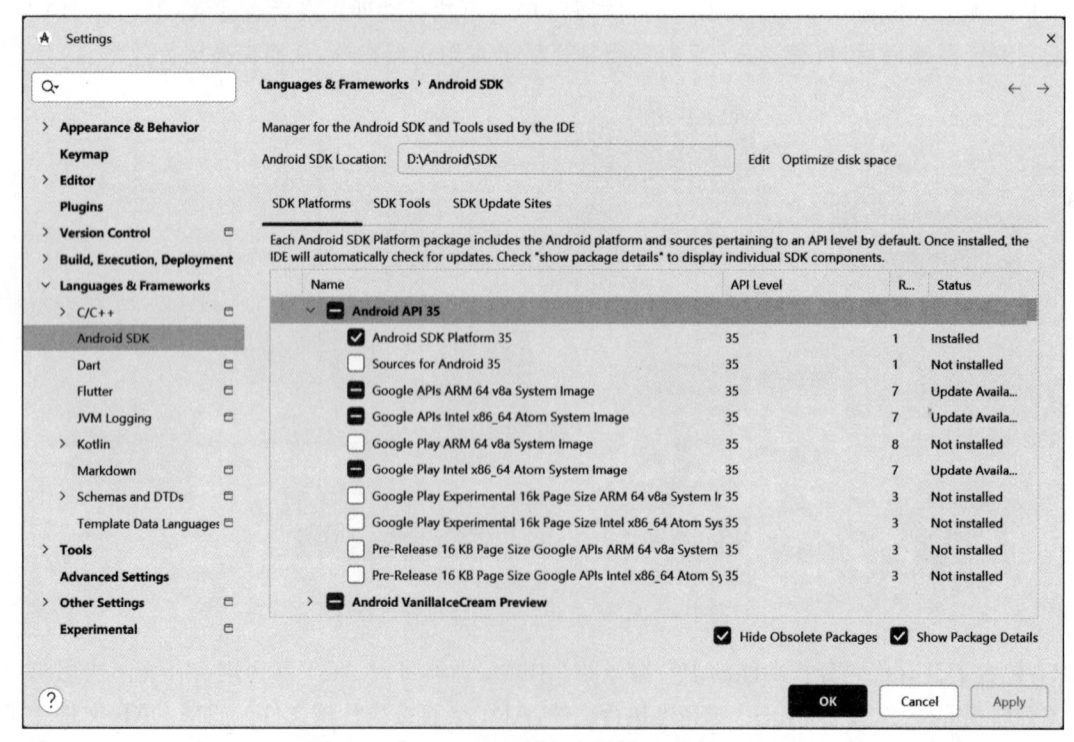

图 1-5　在 Android Studio 中配置 Android SDK

(3) Android Studio 是官方推出的 Android 应用开发工具。本书采用的是 Android Studio Koala｜20241.1 Patch2。当然，更新版本的 Android Studio 也会不断地推出。其他开发工具如 IntelliJ IDEA 也可以用来开发 Android 应用。

Android Studio 可以从 Android 开发者网站 https://developer.android.google.cn/studio?hl＝zh-cn 直接下载安装。

安装 Android Studio 非常容易，根据安装提示依次执行操作即可，本书将不再介绍。

1.3 创建第一个 Android 项目

本节通过一个简单的 Hello World 项目介绍 Android 项目的开发流程，并对项目的结构层次进行介绍和说明。

1.3.1 创建新的项目

首先，启动 Android Studio，在启动的界面中选中 New Project，新建一个项目，如图 1-6 所示。

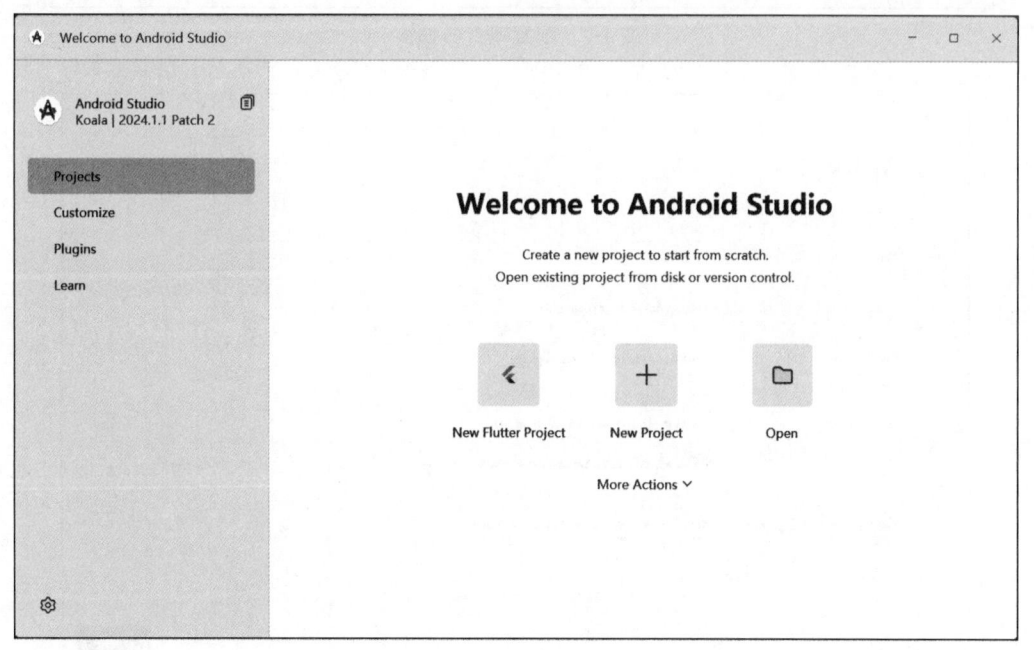

图 1-6　启动界面

选中 New Project 选项，进入新项目配置的模板界面，选中 Phone and Tablet｜Empty Activity 选项，如图 1-7 所示。

单击 Next 按钮，进入项目的配置界面，开始配置项目的 Name(名称)、Package name(包名)、Save location(保存的位置)、Build configuration language(开发语言)和 Minium SDK(最小的 SDK)，如图 1-8 所示。配置完毕，单击 Finish 按钮，结束项目的创建。

项目配置完毕后，就进入 Android Studio 的编辑界面，如图 1-9 所示。

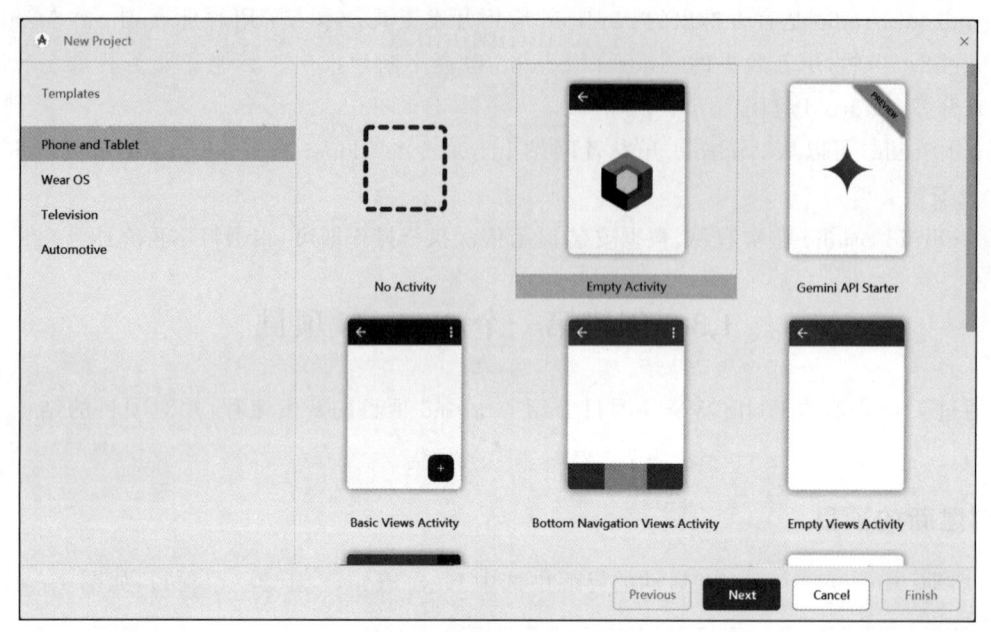

图 1-7　配置项目的模板

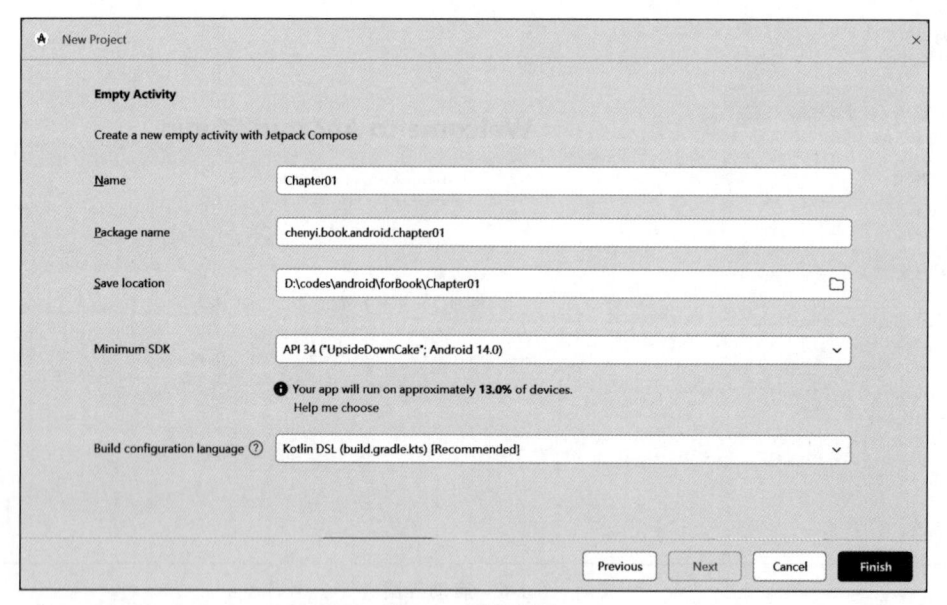

图 1-8　配置项目

1.3.2　启动模拟器

运行 Android 移动应用项目需要在移动终端设备真机或模拟器上进行。通过 Android Studio 的 Android 虚拟设备管理器(Android Virtual Device Manager,AVD Manager)可以启动模拟器。方法是单击 Android Studio 右侧工具栏中的图标,可启动 AVD Manager,如图 1-10 所示。

单击窗口左上角的 Create Virtual Device(创建虚拟设备)按钮,可创建模拟器并配置硬件,如图 1-11 所示。

图 1-9　编辑界面

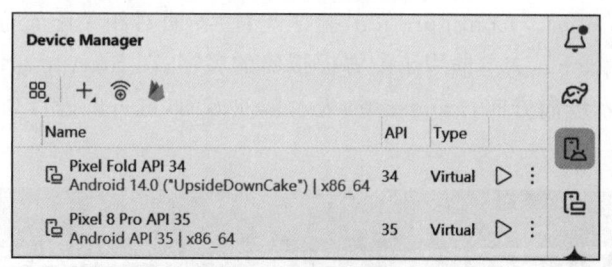

图 1-10　AVD Manager

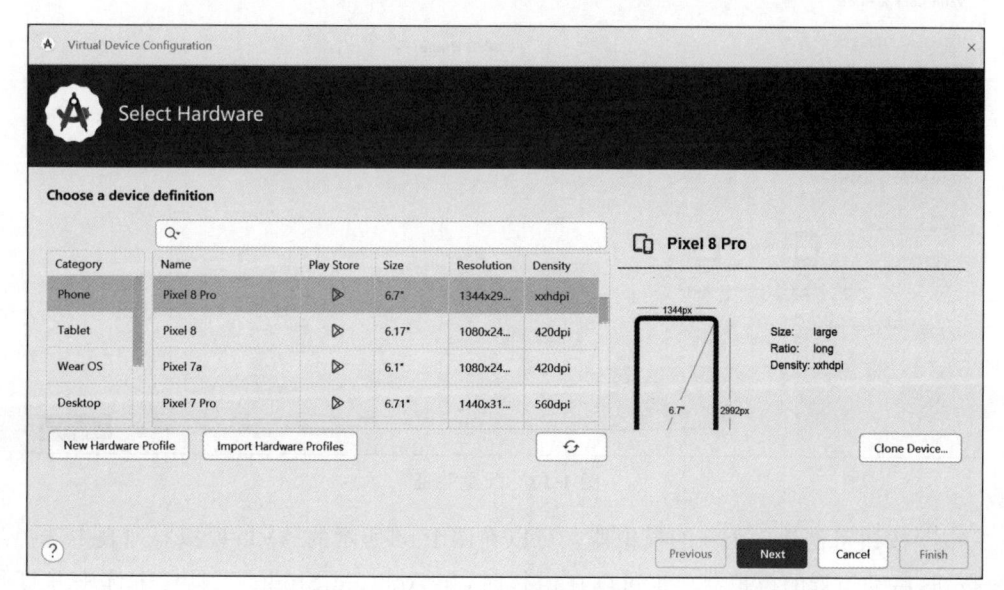

图 1-11　配置模拟器

在 Phone(手机)选项中选择对应的手机模拟器。根据 API 级别选择 System Image(系统映像)，如图 1-12 所示。

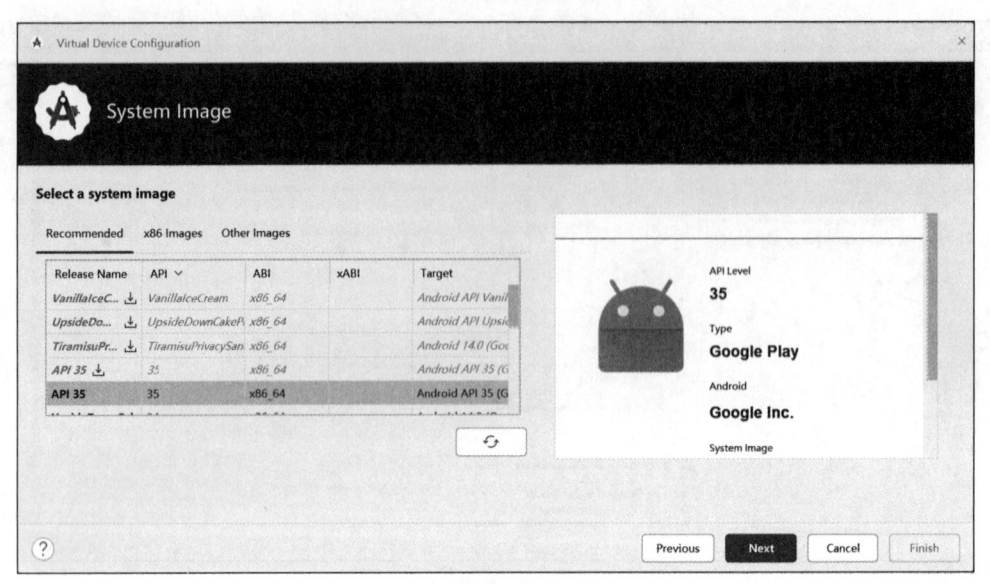

图 1-12　配置系统映像

单击 Next 按钮,进入 Verify Configuration(验证配置)界面对相关配置进行设置,如图 1-13 所示。在验证配置界面中,AVD Name 文本框用于配置模拟器的名字;在 Startup orientation 栏中可指定屏幕的方向,其中,Portrait 表示纵向方向,Landscape 表示横向方向,也可以进行高级配置。此处不再介绍。

图 1-13　配置验证

单击 Finish 按钮可创建一个新的模拟器。可以在图 1-14 所示的 AVD 模拟器列表中选择某个模拟器,单击▶按钮,启动所选的模拟器。也可以在运行时,在 Android Studio 工具栏中选中并打开对应的模拟器运行应用。

1.3.3 运行第一个项目

项目创建成功后,选中 Android Studio 中合适的模拟器,然后单击工具栏中的 ▶ 按钮,运行已经创建的项目,运行结果如图 1-15 所示。

图 1-14　选择模拟器　　　　图 1-15　第一个项目的运行结果

在第一个项目成功运行后,一些重要的文件让项目正常运行发挥了关键作用。在 Android 模式下,项目的基本结构如图 1-16 所示,Android 模式下项目的基本结构有利于方便、快速地开发移动应用。

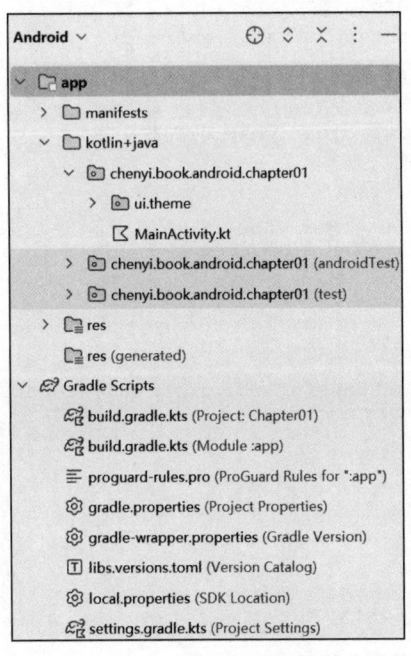

图 1-16　Android 模式下项目的基本结构

1. app 模块

项目中的 app 目录又称 app 模块，它是在创建项目时自动生成的。当前项目的所有的源代码和资源都放置在这个模块中。

在 app|manifest 目录中定义的 AndroidManifest.xml 是项目的配置清单文件，对移动应用的应用、活动、权限等进行配置。

在 app|kotlin+java 目录中可以放置 Kotlin 或 Java 源代码。为了维护管理，在 java 目录下的源代码文件被放置在对应的包中。本项目中自动生成了 MainActivity 源代码，它定义了如图 1-15 所示效果的代码内容。

在 app|res 目录中保存了应用中需要的各种资源。它的下级目录 values 中保存了 strings.xml 定义字符串、colors.xml 定义需要的各种颜色、dimens.xml 定义应用需要的各种尺寸数据等各种取值的资源；下级目录 mipmap 中放置图片资源，这些资源可以根据屏幕进行大小适配。下级目录 drawable 中虽然也放置了图片资源，但是图片资源不可以按照大小适配。

2. build.gradle.kts 和 libs.versions.toml

在项目中有两个 build.gradle.kts 文件，第一个是针对整个项目的顶层 build.gradle.kts 构建的配置脚本文件。具体如下：

```
plugins {                                                           //定义插件
    alias(libs.plugins.android.application) apply false
    alias(libs.plugins.jetbrains.kotlin.android) apply false
}
```

另一个是针对应用模块的 build.gradle.kts 构建的脚本文件。它实现对模块（在这里是针对 app 模块）提供依赖、插件、编译器等内容的配置。

```
plugins {                                                           //引入 Gradle 插件库的插件
    alias(libs.plugins.android.application)
    alias(libs.plugins.jetbrains.kotlin.android)
}
android {
    namespace="chenyi.book.android.chapter01"
    compileSdk=34                                                   //编译 SDK 版本
    defaultConfig {                                                 //默认配置
        applicationId="chenyi.book.android.chapter01"               //移动应用编号
        minSdk=34                                                   //最小 SDK 版本
        targetSdk=34                                                //目标 SDK 版本
        versionCode=1
        versionName="1.0"
        testInstrumentationRunner="androidx.test.runner.AndroidJUnitRunner"
        vectorDrawables {
            useSupportLibrary=true                                  //允许使用系统支持库
        }
    }
    buildTypes {                                                    //构建类型
        release {
            isMinifyEnabled=false                                   //关闭混淆
```

```
            proguardFiles(
                getDefaultProguardFile("proguard-android-optimize.txt"),
                "proguard-rules.pro"                              //指定混淆文件
            )
        }
    }
    compileOptions {                                              //编译选项
        sourceCompatibility=JavaVersion.VERSION_1_8               //源代码兼容
        targetCompatibility=JavaVersion.VERSION_1_8               //目标兼容
    }
    kotlinOptions {                                               //Kotlin 选项设置
        jvmTarget="1.8"
    }
    buildFeatures {                                               //构建特性
        compose=true                                              //使用 compose
    }
    composeOptions {                                              //compose 选项
        kotlinCompilerExtensionVersion="1.5.1"                    //kotlin 编译扩展版本
    }
    packaging {
        resources {
            excludes+="/META-INF/{AL2.0,LGPL2.1}"
        }
    }
}
dependencies {                                                    //依赖库
    implementation(libs.androidx.core.ktx)
    implementation(libs.androidx.lifecycle.runtime.ktx)
    implementation(libs.androidx.activity.compose)
    implementation(platform(libs.androidx.compose.bom))
    implementation(libs.androidx.ui)
    implementation(libs.androidx.ui.graphics)
    implementation(libs.androidx.ui.tooling.preview)
    implementation(libs.androidx.material3)
    testImplementation(libs.junit)
    androidTestImplementation(libs.androidx.junit)
    androidTestImplementation(libs.androidx.espresso.core)
    androidTestImplementation(platform(libs.androidx.compose.bom))
    androidTestImplementation(libs.androidx.ui.test.junit4)
    debugImplementation(libs.androidx.ui.tooling)
    debugImplementation(libs.androidx.ui.test.manifest)
}
```

  上述两种不同的 build.gradle.kts 构建文件配置的插件名和依赖名都与 libs.versions.toml 文件定义的版本、插件和依赖项的别名一致。从 Gradle 7.0 开始引入了版本目录功能,即在项目的 libs.versions.toml 文件中统一定义插件和依赖项的版本和别名,使得这个项目的不同模块可以使用相同的依赖项名和插件名,以方便 Gradle 在 libs.versions.toml 文件查找目录,实现统一管理依赖项和插件,克服早期 Gradle 版本针对单独的模块中硬编码定义的插件和依赖项,不能共享配置的缺点。这使得管理应用项目的所有模块共享插件和依赖库更加容易。lib.versions.toml 的文件结构如下所示:

```
[versions]
[libraries]
[bundles]
[plugins]
```

其中，versions 用于定义插件或依赖项的引用的版本。libraries 用于定义依赖库，bundles 用于定义依赖包，plugins 用于定义插件。

3. gradle-wrapper.properties

gradle-wrapper.properties 文件是 Gradle 的安装和配置的文件。

4. proguard-rules.pro

proguard-rules.pro 定义了代码混淆规则。代码编写成功后，若不希望被破解，可对代码进行混淆，增加破解难度。

5. gradle.properties

gradel.properties 文件是项目的 Gradle 全局配置文件。这个文件用于配置全局键值对的数据。例如：

```
org.gradle.jvmargs=-Xmx2048m -Dfile.encoding=UTF-8
```

表示 jvm 参数设置分配最大内存为 2048MB，文件的字符集为 UTF-8。

6. settings.gradle.kts

settings.gradle.kts 用于指定项目中引入的模块。在项目中，Android Studio 会默认创建 app 模块。例如：

```
include ':app'                           //包含的模块 app
rootProject.name="Chapter01"             //项目名称
```

也可以通过在 Android Studio 中选中 New|New Module 菜单选项新建模块，这些模块的名称会记录在 settings.gradle.kts 中进行保存。如果要删除某个模块，需先在 settings.gradle.kts 文件中把包含模块的内容删除，然后才能对模块删除。

7. local.properties

这个文件非常重要，它指定了 Android SDK 的路径。该路径通常是自动生成的。当 Android SDK 位置发生变化时，可以修改这个文件来实现重新配置。

项目中还有一些其他文件和目录，在 Android 开发中，了解上述项目结构中的文件有助于理解和掌握 Android 应用的开发和运行。

至此，学习 Android 移动开发已经迈开了第一步。希望在后续的学习中，能更好地了解和掌握所讲内容。

习 题 1

一、选择题

1. 在 Android 架构中，＿＿＿＿＿＿＿通过 Webkit 库提供了浏览器内核支持。
   A. 应用层　　　　B. Linux 内核　　　C. 应用框架层　　　D. Android 核心库
2. ＿＿＿＿＿＿＿提供了大量的 API 供 Android 应用开发者使用。
   A. 应用层　　　　B. 应用框架层　　　C. 硬件抽象层　　　D. Linux 内核

3. _____负责运行 Android 应用程序。
   A. Android 运行时   B. 应用框架层   C. 硬件抽象层   D. Linux 内核
4. 提供了底层硬件驱动的是_____。
   A. Android 运行时   B. 应用框架层   C. 硬件抽象层   D. Linux 内核
5. HAL 硬件抽象层可以使用_____来定义 HAL 与用户的接口。
   A. Kotlin   B. Java   C. HIDL   D. OAT
6. 2003 年，Android 操作系统由_____开发并发布。
   A. Andy Robin   B. 谷歌   C. JetBrain   D. IDC
7. 在 Android Studio 中，使用_____可创建模拟器。
   A. SDK Manager   B. AVD Manager
   C. Resource Manager   D. LayoutManager
8. 对移动应用的应用、活动、权限等进行配置的信息保存在 Android 移动项目的_____中。
   A. AndroidManifest.xml   B. classes.dex
   C. resources.arsc   D. CERT.RSA

二、填空题
1. Android 平台是自由开源的_____和_____。
2. 谷歌公司发布的 Android 源代码是以_____方式授权的。
3. Android 平台的软件框架分为_____、_____、_____、_____和_____。
4. Android 系统架构包括_____、_____、_____、_____和_____组件。
5. 利用 dex2oat 可将_____文件编译优化后生成机器能识别的_____可执行文件。

三、上机实践
1. 搭建 Android 移动应用开发环境，并安装 Android Studio。
2. 开发一个 Android 移动应用，输出"Hello World!"，并在模拟器中运行。
3. 从网络下载一个 APK 文件，并分析其内部结构。
4. 从网络下载一个 APK 文件，并安装到手机模拟器中运行。

第 2 章 Kotlin 基础

Kotlin 是由 JetBrains 公司在 2010 年创建的一种基于 Java 虚拟机(Java Virtual Machine,JVM)的计算机语言,并于 2016 年发布了第一个版本。谷歌公司于 2017 年将其定为 Android 开发的官方语言。这使得 Kotlin 成功进入 Android 开发移动应用领域。由于 Kotlin 具有简洁、易学、安全性好、实用性强、Java 框架兼容性好等优点,所以用其开发 Android 移动应用已成为趋势。本章主要对 Kotlin 的基本语法进行介绍。

2.1 Kotlin 概述

Kotlin 是一种高级的静态强类型语言,可以在 JVM 上运行。虽然 Kotlin 受到 Java、Scala、Groovy 等语言的影响,但是仍具有自己鲜明的特色,与 Java 的语法存在明显区别。

(1) Kotlin 是一种简洁的编程语言,代码量少、易读易懂、运行更快、效率更高。下面分别使用 Java 和 Kotlin 定义一个 Student 类,对两种语言加以比较。

用 Java 定义 Student 类,代码如下:

```java
class Student {                                              //Java 定义的 Student 类
    private String name;
    private String gender;
    public String getName() {
        return name;
    }
    public void setName(String name) {
        this.name=name;
    }
    public String getGender() {
        return gender;
    }
    public void setGender(String gender) {
        this.gender=gender;
    }
    @Override
    public boolean equals(@Nullable Object obj) {
        return super.equals(obj);
    }
    @Override
    public String toString() {
        return super.toString();
    }
}
```

用 Kotlin 定义 Student 类,代码如下:

```
data class Student(val name:String,val gender:String)          //Kotlin定义的Student类
```

对比后发现,同样定义一个实体类,Kotlin只用一行代码即可实现。由此可见,Kotlin的表达方式简单明了、代码量更少。Kotlin只需通过data class说明Student类是数据类,Kotlin编译器便会自动生成对应的toString、equals等函数。

(2) Kotlin是一种安全性好的编程语言。Kotlin的很多特性都能避免运行时崩溃。例如,自动检测对象是否为null,为null直接抛出NullPointerException。在数据类型中,明确区分了可以为null的数据,例如:

```
var intValue:Int?
```

则表示变量intValue可以为null;若定义形式如下:

```
var intValue:Int
```

则表示intValue变量不能取值为null。在语法层面上避免了可能对null的误判。这与Java有明显不同。

Kotlin是一种基于JVM的语言,它可以与Java代码交互。Kotlin可以方便地利用Java API以及第三方基于Java开发的库。上述Kotlin例子也可以使用Java的java.lang.String来定义字符串对象。

Kotlin可以开发任何基于Java生态的应用。无论是服务器端的应用,还是Android移动端的应用,都可以涉足其中。Kotlin不但能开发基于Java的应用,还能开发跨平台的应用。例如,通过应用Intel Multi-OS Engine,可以让Kotlin代码运行在iOS移动终端。

与Java语言一样,Kotlin是一种静态强类型的语言,所有表达式的类型在编译期已经确定。Kotlin不需要在代码中显式地声明每个变量的类型,变量类型可以根据上下文进行自动判断。例如,通过代码

```
var intValue=2
```

可以直接推断出intValue的数据类型为Int。Kotlin的静态特性可使代码性能更好。因为不需要在运行时判断调用的具体方法,所以方法调用速度更快。Kotlin是可靠的,在通过编译器验证程序的正确性后,运行时崩溃的概率更低。因为可以直接从代码中知道对象的类型,所以代码更容易维护。此外,静态类型使IDE能提供可靠的重构、精确的代码补全以及其他特性。

(3) Kotlin支持函数式编程。Kotlin支持函数类型,允许函数把其他函数作为参数或者返回其他函数。Kotlin支持Lambda表达式,使用最少的样板代码就能方便地传递代码块。对于数据类,Kotlin提供了创建不可变值对象的简明语法。可以方便地使用函数式编程风格操作对象和集合。例如:

```
val lst=listOf("hello","welcome","this","is","kotlin","world")    //定义列表
lst.forEach{                                                       //遍历列表中的所有元素
    print(it)
}
```

通过Lambda表达式lst.forEach{}实现了对列表中所有元素的依次遍历。

(4) Kotlin的代码可以编译成Java的字节码或JavaScript,以便在没有Java虚拟机的设备上运行。编译Kotlin的命令行的形式如下:

```
kotlinc <源代码或目录> -include-runtime -d <jar 名>
```

例如,已知有一个名为 hello.kt 的文件,它在命令行的编译命令可以写成如下形式:

```
kotlinc hello.kt -include-runtime -d hello.jar
```

编译 Kotlin 与 Java 代码的编译过程如图 2-1 所示。

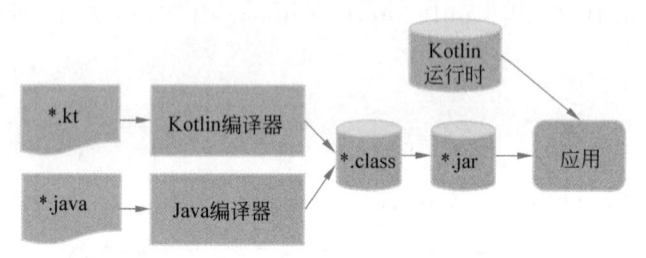

图 2-1　Kotlin 代码与 Java 代码的编译

Kotlin 的代码会保存在扩展名为 .kt 的文件中,通过 Kotlin 编译器进行编译会生成 class 字节码文件。打包后可生成 .jar 文件。应用可以直接打开 .jar 文件,使用 Kotlin 代码定义的功能。开发 Kotlin 应用可以使用 JetBrains 的 IntelliJ IDEA 平台,也可以使用 Android Studio。

2.2　第一个 Kotlin 程序

下面先观察一下代码 Ch02_01.kt。通过这段代码对 Kotlin 编码做一个初步的了解。该代码的完整内容如下:

```kotlin
/**Kotlin代码 Ch02_01.kt */
fun main(){
    val reader=Scanner(System.`in`)        //定义 Scanner 对象
    var name=reader.next()                  //输入
    print("你好: $name")                    //输出
}
```

运行结果如图 2-2 所示。

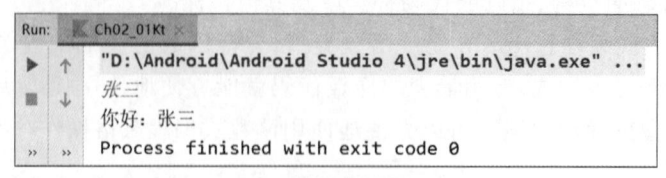

图 2-2　运行结果

上述代码需要注意以下内容。
(1) 在 Kotlin 中,可以使用注释来标注信息和解释代码。上述代码中定义了 3 种不同的注释:

```
/** 文档注释 */
/* 多行注释 */
//单行注释
```

这些注释的定义与 Java 表述的注释是一致的。其中，文档注释可以生成注释的文档 HTML 文件。在 Android Studio 的 Settings 对话框中设置 Plugins 项，增加 BugKotlinDocument 插件，可以自动生成注释。BugKotlinDocument 插件的安装如图 2-3 所示。

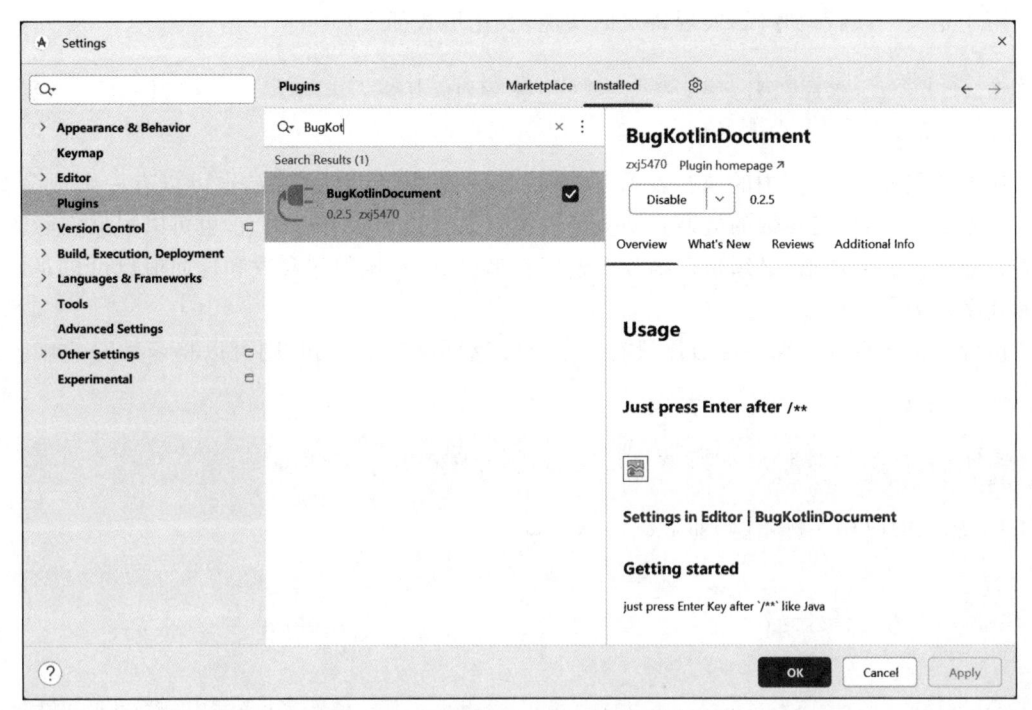

图 2-3　BugKotlinDocument 插件安装成功

安装成功后，在 Kotlin 文件中直接使用"/＊＊"并按回车键，就会生成对应的文档注释以及自动加入相应的参数配置内容。另外，可以通过 Dokka 插件（https://github.com/Kotlin/dokka）生成文档注释对应的 HTML 帮助文件。

（2）fun 表示定义函数，fun 后面跟函数的名称。其中，main 表示主函数，也是程序的执行入口。Kotlin 定义的主函数不需要定义在类中，这与 Java 明显不同。

（3）主函数 main 中定义了两种不同类型变量，通过 val 定义了只读变量 reader，表示一旦定义之后，reader 就是固定对象，不能变化。var 定义了变量，可以多次赋值，修改变量的取值。

（4）上述代码中调用 Scanner 对象 reader 的 next 函数，表示输入一个字符串，并赋值给可变量 name。

（5）调用函数

```
print("你好：$name")
```

本质上就是调用 Java 标准库的标准输出对象的 print 函数，对应的 Java 的完整表示是 System.out.print 函数。由于 Kotlin 标准库重新包装了 Java 的标准库函数，提供了更简洁的包装，所以表达形式更简洁。

（6）字符串 $name 是字符串模板，用于对字符串格式化，在字符串中可使用"$"引用局部量和表达式，也可以直接嵌套在""""内。

通过上例，可以发现编写 Kotlin 代码并不复杂。

· 17 ·

2.3 函　　数

函数用于定义一段代码片段,实现特定的功能,表达形式如下:

```
fun 函数名([参数:参数类型,参数:参数类型…])[:返回值类型]{
}
```

在上述形式定义中,"[]"中的内容表示可选项。说明函数可以无参数,也可以有一个或多个参数。一旦定义了参数,必须指定参数的类型。函数可以执行执行操作没有返回值,也可以根据实际需要返回一个取值。如果要返回取值,必须指定函数的返回类型。其中,函数的参数和返回值也可以是函数。具有这种形式函数为高阶函数。

"{}"包含的是函数体。如果函数体只有一个直接返回结果的语句,则可以将函数写成函数表达式的形式:

```
fun 函数名([参数:参数类型,参数:参数类型 …])[:返回值类型]=表达式
```

例 2-1　函数的应用实例,代码如下:

```kotlin
//例 2-1 Ch02_02.kt
fun add(x:Int,y:Int):Int{                              //定义函数
    return x+y
}
fun sub(x:Int,y:Int):Int=x-y                           //函数表达式,指定函数的返回值类型
fun mul(x:Int,y:Int)=x * y                             //函数表达式,不指定函数的返回值类型
fun operate(op:String):(Int,Int)->Int=when(op){        //高阶函数,函数的返回值为函数对象
    "+"->::add
    "-"->::sub
    "*"->::mul
    else ->throw RuntimeException("操作符有误!")
}
fun display(x:Int,op:String,y:Int,result:(Int,Int)->Int){   //高阶函数,函数的参数是函数对象
    print("输出: ${x}${op}${y}=${result(x,y)}\n")
}
fun main(){                                            //定义主函数
    val num1=23
    val num2=44
    display(num1,"+",num2,operate("+"))                //调用函数
    display(num1,"-",num2,operate("-"))                //调用函数
}
```

运行结果如图 2-4 所示。

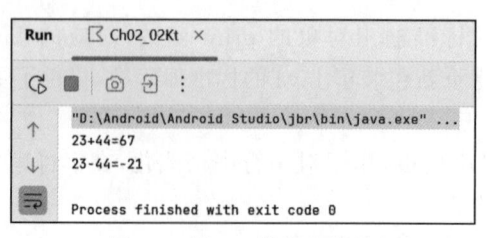

图 2-4　例 2-1 的运行结果

注意：在上述代码中，"＄{}"是字符串格式符,因为代码中调用函数的表达式要按照约定的格式进行输出,因此需要将函数调用的表达式写入"＄{}"中。

2.4 变量和数据类型

2.4.1 变量

Kotlin 分别使用 var(variable,变量)和 val(value,取值)来定义变量。但是二者有明显不同。var 定义的是可变量,可通过多次赋值修改 var 修饰的变量。val 定义的是只读变量,只进行一次初始化,一旦赋值,定义就不会再发生变化了。例如：

```
var a: Int=23
a=45                                        //成立
val b: Int=23
b=45                                        //编译错误;因为b已经初始化了
```

无论是何种变量,都必须有固定的数据类型。在 Kotlin 中,数据类型有两种情况：非 null 值的数据类型和可以 null 值的数据类型。例如：

```
var a: Int?=null
```

这时,变量 a 的初始值为 null。
如果定义成如下形式：

```
var a: Int
a=null                                      //编译错误
```

因为变量 a 定义的是 Int 整数类型,因此不能赋值于一个 null 值。

在变量的定义时,如果已经赋予初始值,则可以不写变量的数据类型,编译器根据对象的取值及上下文推断出数据对象的具体的类型。例如：

```
var a=2
```

根据初始值 2,编译器推断出 a 的数据类型为 Int,变量 a 就是一个 Int 类型的变量。例如：

```
var b=3.4
```

根据初始值 3.4,编译器推断出 b 的数据类型是 Double,变量 b 就是一个 Double 类型的变量。

2.4.2 数据类型的种类

Kotlin 中所有的数据都是对象。即使是一个整数 12,在 Kotlin 中表达的也是一个 Int 对象。这使得 Kotlin 的表示形式与其他语言有所不同。Kotlin 常见的数据类型如表 2-1 所示。

和 Java 不一样,Kotlin 并不区分基本数据类型和包装类型。所有的非空类型都是 Any 类的子类,所有的可空(null)类型都是 Any?类的子类。图 2-5 中展示了 Any 和 Any?与其他数据类型如 String 字符串类的相互关系。

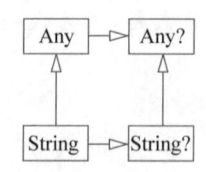

图 2-5 Kotlin 的 Any 和 Any？类之间的关系

表 2-1 常见的数据类型

数据类型	说明	数据类型	说明
Short	短整数(16 位)	Short?	可取 null 的短整数(16 位)
Int	整数(32 位)	Int?	可取值 null 的整数(32 位)
Long	长整数(64 位)	Long?	可取 null 的长整数(64 位)
Double	双精度实数(64 位)	Double?	可取 null 的双精度实数(64 位)
Float	单精度实数(32 位)	Float?	可取 null 的单精度实数(32 位)
Byte	字节类型(8 位)	Byte?	可取 null 的字节类型(8 位)
Boolean	布尔类型，取值 true 或 false	Boolean?	可取 null 的布尔类型，取 true 或 false

Any？不但是所有可空类型的根类型，而且是 Kotlin 所有数据类型的根类型。它的取值包括了 null 及具体的对象。Any 是 Any？的子类，二者之间构成了继承关系。而 Any 是所有非空类型的根类型。因此，通过这些关系的确定，带有"？"的数据类型的变量可以取值 null，而没有"？"的数据类型变量必须表示具体的对象。

Kotlin 中还定义了两种特殊类型：Unit 和 Nothing。Unit 对应 Java 语言的 void，表示没有返回值。在大多数情况下，代码并不需要显式地返回 Unit，或者申明一个函数的返回类型为 Unit。编译器会通过上下文推断它。至于 Nothing 类型，则是永远不返回。

2.4.3 数据类型的转换

Kotlin 支持类型转换，即将不同类型的数据转换成其他类型的数据。Kotlin 不支持数字类型的隐式转换，只支持显式转换。在表 2-2 中展示了数字类型相互转换的函数。

表 2-2 数字类型转换函数

函数	说明	函数	说明
toByte	转换成字节类型	toFloat	转换成单精度实数
toShort	转换成短整数	toDouble	转换成双精度实数
toInt	转换成整数	toChar	转换成字符
toLong	转换成长整数		

例如，将整数 23 转换成长整数的代码如下：

```
var a=23
var b: Long=a                                                //编译错误
```

要实现上述的转换，可以直接调用对象的转换函数，写成以下形式：

```
var b: Long=a.toLong()
```

在 Kotlin 中还提供了 as 运算符实现类型转换,例如:

```
var c=23
var d : Long=c as Long                    //编译通过,但是运行抛出 ClassCastException 异常
```

采用 as 直接进行转换存在安全隐患。如果 c 的数据类型不是 Long,即使编译通过,也会抛出 java.lang.ClassCastException,这是因为 c 的数据类型为 Int 不是 Long。因此,可以写成如下形式:

```
var f: Long=23 as? Long
```

这种表达方式是安全转换,如果 23 是 Long 数据类型,则 23 可以赋值给 f,如果 23 不是 Long 数据类型,则返回 null 给 f。在此,f 的取值最后为 null,因为 23 是 Int。这种转换方式可以确保转换可以正常执行。

也可以通过 is 运算符来判断取值是否为特定的数据类型的实例。而!is 表示不是某个数据类型的实例,例如:

```
var a=23
a is Int                                   //表达式为 true
a !is Int                                  //表达式为 false
```

2.4.4 可空类型的处理

Kotlin 定义的可空类型对象存在两种情况:一种取值为 null,另一种取值为非空。在对可空类型的对象进行处理时,需要对 null 进行判断和处理。这种方式略显复杂,因此 Kotlin 针对这种可空类型的对象提出了几种处理方式。

1. 安全调用

Kotlin 定义了安全调用符"?."。这种调用方式允许把一次 null 检查和一次方法调用合并成一个操作,例如:

```
var s: String?=null                        //定义可空类型,并初始化 s 为 null
...
s?.length                                  //安全调用求 s 的长度
```

此处的安全调用符表示先对可空类型对象 s 进行判断,如果 s!=null,则表达式的结果就是对象的长度 s.length;如果 s 判断为 null,则整个表达式的结果为 null。

2. Elvis 运算符

Elvis 运算符"?:"。Elvis 是双目运算符,如果第一个运算数不为 null,则运算结果就是第一个运算数,如果第一个运算数为 null,则运算结果解释第二个运算数。例如:

```
var tyre: String?=null
...
tyre?:"报废"
```

上述表达式首先对 tyre 对象进行判断,如果 tyre!=null,则表达式的结果就为 tyre 本身;如果 tyre

==null,则表达式的结果为"报废"。

3. 非空断言运算符

非空断言运算符"!!"是在假设可空类型的对象是非空的情况下调用对象或对象的函数。非空断言可将任何值转换成非空类型,如果值本身为 null 时,做非空断言,则会抛出 NullPointerException 异常。例如:

```
var instance:Stirng?=null
…
instance!!
```

若不清楚 instance 是否为空 null,则可使用非空断言。如果 instance!=null,则为 instance 本身;如果 instance==null,则抛出 KotlinNullPointerException 空指针异常。

2.4.5 数组和集合类型

程序中往往需要处理一组数据。这组数据可用数组或集合框架的列表、数学集等存储类型表示。

Kotlin 的数组可通过 ByteArray(字节数组)、IntArray(整数数组)、DoubleArray(双精度实数数组)等具体类型的数组定义,也可以通过 arrayOf 函数创建数组。如果创建的数组中包含 null 空值,可以使用 arrayOfNulls 创建包含 null 元素的数组。例如:

```
val array=arrayOf("Hello","Android")
```

array 是数组对象,其中包含了两个字符串"Hello"和"Android"。

```
var arrayNull: Array<String?>=arrayOfNulls<String?>(3)
```

arryNull 数组对象包含了 3 个 null 值元素的数组对象。实际上,通过数组来批量处理数据在 Kotlin 的开发中比较少见。对于一组数据存储,多利用集合类型来实现。Kotlin 支持集合类型。集合类型主要有 List(列表)、Set(集合)和 Map(映射)。Kotlin 将集合类型分成两种类型:一种为可变的类型,即存储结构包含的元素可以变化,可以增加元素,也可以删除元素;另一种为不可变的只读类型,一旦定义之后,存储结构包含的元素就是固定的,是不可变的,只能读取元素。表 2-3 罗列出了常见集合类型创建的方式。

表 2-3 列表 List、集合 Set 和映射 Map 创建

集合类型	创建只读集合类型函数	创建可变集合类型函数	说 明
List(列表)	listOf	mutableListOf	列表中的元素都有对应的顺序索引进行访问
		arrayListOf	
Set(集合)	setOf	mutableSetOf	此处的集合表示数学集的概念,即 Set 对象中不能包含重复的元素
		hashSetOf	
		linkedSetOf	
		sortedSetOf	

续表

集合类型	创建只读集合类型函数	创建可变集合类型函数	说　明
Map(映射)	mapOf	mutableMapOf hashMapOf linkedMapOf sortedMapOf	映射存储的是一组组键值对,通过关键字来获得对应的取值

例 2-2　集合类型的应用实例,代码如下:

```
//例 2-2 Ch02_03.kt
fun main(){
    val lst1: List<String>=listOf("Hello","Android")              //创建不可变的列表
    val lst2: MutableList<String>=mutableListOf("Hello","Android") //创建可变的列表
    lst2.add("World")
    lst2.addAll(lst1)
    println("列表: $lst2")
    val set1=setOf("Java","Scala")                                //创建不可变的集合
    val set2=mutableSetOf("Kotlin","Java","Scala")                //创建可变的集合
    set2.removeAll(set1)                                          //从 set2 中删除 set1
    println("集合: $set2")
    val map1=mapOf("Kotlin" to 1,"Java" to 2)                     //创建不可变的映射
    val map2=mutableMapOf("Scala" to 3,"Groovy" to 4)             //创建可变的映射
    map2.putAll(map1)                                             //将映射 map1 加入到 map2
    println("映射: $map2")
}
```

运行结果如图 2-6 所示。

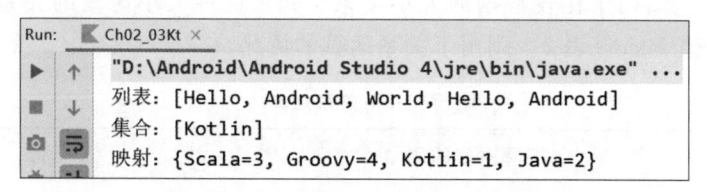

```
"D:\Android\Android Studio 4\jre\bin\java.exe" ...
列表: [Hello, Android, World, Hello, Android]
集合: [Kotlin]
映射: {Scala=3, Groovy=4, Kotlin=1, Java=2}
```

图 2-6　集合类型应用实例的运行结果

在上例的代码中,所有对象的类型并没有明确指定,这些只读变量的数据类型是由编译器通过上下文推断。lst1 被推断为 List<String>类型;set1 对象被推断为 Set<String>类型;map1 被推断为 Map<String,Int>类型,其中 String 表示关键字对应的类型,Int 是对应的取值的类型。这些对象均为不可变只读类型。一旦初始化完成,对于包含的元素对象永远不会发生变化,因此它们也没有对应地变化自身的函数。lst2 被推断为 MutableList<String>类型,set2 被推断为 MultableSet<String>类型,map2 被推断为 MultableMap<String,Int>类型。它们属于可变集合类型,可以增加或者删除包含的元素。此外,映射的键值对定义中采用"关键字 to 取值"的方式,形式简洁。代码如下:

```
val lst: List<String>=multableListOf("Hello","Android")
```

注意：lst 已经明确指定为 List<String>不可变的集合类型，而它只能识别不可变的只读集合类型，因此 lst 是无法增加或删除元素的。

2.5 操作符和表达式

操作符用于实现特定的运算。Kotlin 包括的运算有算术运算、条件运算、逻辑运算、范围运算和赋值运算。Kotlin 与 Java 的运算非常类似。

1. 算术运算

算术运算符是用于执行基本算术计算（加、减、乘、除、求余）的运算符。在 Kotlin 中，简单的 a＋b 被解释为 a.plus(b)作为函数调用。表 2-4 列出了算术运算的基本情况。

表 2-4 算术运算

运　　算	解释对应的函数	说　　明
a＋b	a.plus(b)	加法运算，a 加上 b
a－b	a.minus(b)	减法运算，a 减去 b
a＊b	a.times(b)	乘法运算，a 乘以 b
a/b	a.div(b)	除法运算，a 除以 b
a％b	a.rem(b)	求余运算，a 余 b
a＋＋或＋＋a	a.inc()	自增运算
a－－或－－a	a.dec()	自减运算

2. 关系运算

关系运算即比较运算，用于比较数据的大小关系。如果这种大小比较的关系成立，返回布尔真值 true，否则返回布尔假值 false。表 2-5 列出了关系运算的情况。

表 2-5 关系运算

运　　算	解释对应的函数	说　　明
a＞b	a.compareTo(b)＞0	大于
a＞＝b	a.comparetTo(b)＞＝0	大于或等于
a＜b	a.compareTo(b)＜0	小于
a＜＝b	a.compareTo(b)	小于或等于
a＝＝b	a?.equals(b) ?：(b＝＝＝null)	相等
a!＝b	!(a?.equals(b) ?：(b＝＝＝null))	不等于

3. 逻辑运算

逻辑运算又称布尔运算。Kotlin 支持 &&、||和! 这 3 种逻辑运算。表 2-6 列出了逻辑运算的情况。

表 2-6 逻辑运算

运 算 符	说　　明
a&&b	逻辑与运算,若二者同时为 true,则表达式为 true,否则为 false
a||b	逻辑或运算,若二者同时为 false,则表达式为 false,否则为 true
!a	逻辑非运算,若 a 取值为 true,则!a 为 false;若 a 取值为 false,则!a 取值为 true

4. 范围运算和 in 运算

Kotlin 使用".."表示区间,用于定义范围运算,用两个数字表示范围,一个数字是起始值,另一个是结束值。例如:

```
val oneToTen=1..10
```

与范围运算相关的 in 运算,表示某个数据是否包含在指定范围中,例如:

```
var a=1..10
var b=2
```

则 b in a 表示 b 变量是否在 a 变量指定的范围内中。当然,in 运算符也可以表示在数组和集合中是否包含特定元素的判断,例如:

```
var a=2
var lst=listOf(2,3,4,5)
```

则表达式 a in lst 表示判断 a 是否包含在 lst 列表中。如果是,则返回 true;否则返回 false。在此处,表达式返回 true。

5. 赋值运算

赋值运算将取值赋值给变量的操作。Kotlin 支持赋值运算,也支持赋值运算符"＝"与其他二元运算符组合构成的复合赋值运算符。表 2-7 列出了常见的复合赋值运算。

表 2-7 常见的复合赋值运算

运 算 符	说　　明	运 算 符	说　　明
a＋＝b	a＝a＋b	a/＝b	a＝a/b
a－＝b	a＝a－b	a%＝b	a＝a%b
a*＝b	a＝a*b		

例 2-3 运算符的应用实例,代码如下:

```
//例 2-3 Ch02_04.kt
fun main(){                                          //Kotlin 的运算符
    var a: Int
    var b: Int
    val sc=Scanner(System.`in`)
    a=sc.nextInt()
    b=sc.nextInt()
```

```
        var c=a+b>++a && (a-b)<--b
        println("a=$a,b=$b,c=$c")
}
```

运行结果如图 2-7 所示。

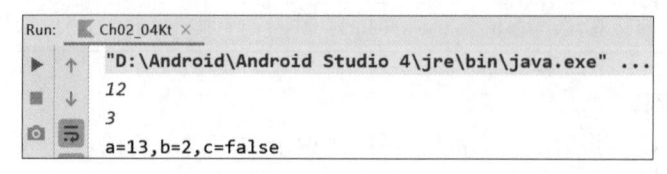

图 2-7　运行结果

在进行运算时,运算符有不同的优先级,一般算术运算优先级大于关系运算的优先级,关系运算优先级大于逻辑运算的优先级。在优先级相等的情况下,需要考虑运算符的结合方向,例如 a+b 的"+"运算结合方向是从左到右的。在运算时,表达式"a+b>++a && (a-b)<--b"先执行++a 运算,再执行 a+b 运算。比较 a+b 是否比增 1 后的 a 大的结果为 true,后面(a-b)<--b 的运算继续,故 b 的取值减 1。整个表达式的结果为 false。

2.6　控 制 结 构

Kotlin 支持 3 种控制结构:顺序结构、选择结构和循环结构。在顺序结构中,语句从前到后依次执行。本书在此之前的代码都是顺序结构的,本节不再赘述。

1. 选择结构

选择结构是对给定的条件进行判断,并根据判断结果决定程序执行指令的方向。Kotlin 提供了两种形式实现选择结构:if 语句和 when 语句。

if 语句的使用与 Java 语言相似,形式如下:

```
if (条件)
    语句块 1
[else
    语句块 2]
```

该语句表示,如果条件为真,则执行"语句块 1"。else 子句是可选项,如果没有 else 子句,条件判断不满足,则返回逻辑 false,直接退出 if 语句,执行后续的其他语句。如果有 else 子句,则当条件判断返回 false 时,会执行 else 子句的"语句块 2",然后再退出 if 语句,执行后续的命令。

例 2-4　输入年份,判断是否是闰年。如果是,则输出是闰年的信息,否则输出"不是闰年",代码如下:

```
//例 2-4 Ch02_05.kt
/* *定义函数表达式*/
fun isLeapYear(year: Int)=if(year%4==0 && year%100==0 || year%400==0)"$year 闰年" else
    "$year 不是闰年"
fun main(){
    val sc=Scanner(System.`in`)
```

```
    println("输入年份：")
    val year=sc.nextInt()                              //输入数据
    println(isLeapYear(year))                          //调用年份判断并输出结果
}
```

运行结果如图 2-8 所示。

图 2-8 闰年判断的运行结果

在例 2-4 所示的 isLeapYear 函数中，由于语句只有一条，因此可将 isLeapYear 函数写成函数表达式的形式。它的表达形式与下列代码片段中采用传统的 if 条件判断的形式是一致的。

```
fun isLeapYear(year: Int): String{
    var result: String
    if (year%4==0 && year%100==0 || year%400==0)
        result="$year 是闰年"
    else
        result="$year 不是闰年"
return result
}
```

when 语句是 Kotlin 自己定义的多条件判断。虽然 if 语句可以实现多条件判断，但是通过多层次的嵌套 if 语句结构较为复杂。when 语句的使用可使程序的结构变得清晰，形式如下：

```
when (条件参数){
    条件表达式 1 ->语句块 1
    条件表达式 2 ->语句块 2
    …
    [ else->语句块 n ]
}
```

对条件参数取值进行判断，如果条件表达式 1 返回逻辑 true 值，则执行语句块 1，否则执行条件表达式 2；如果条件表达式 2 返回逻辑 true 值，则执行语句块 2，否则继续对后续的条件表达式进行判断；如果后续的条件表达式的结果均为 false，则当没有 else 子句时从 when 语句退出，否则执行 else 子句的语句块 n，然后再从 when 语句退出。

在带有参数的 when 语句中，参数在所有的条件表达式可用的情况下均有效。但是，并不是所有的参数都是必需的，这时可使用不带参数的 when 语句，形式如下：

```
when {
    条件表达式 1 ->语句
    …
    else ->表达式 n
}
```

例 2-5 成绩等级的判断。输入一个学生的百分制成绩，判断成绩等级（优秀：90～100 分，良好：

80~89分,中等:70~79分,通过:60~69分,不及格:0~59分),代码如下:

```kotlin
//例2-5 Ch02_06.kt
fun judgeGrade(score: Int)=when (score/10){
    10,9 ->"$score 成绩优秀"
    8 ->"$score 成绩良好"
    7 ->"$score 成绩中等"
    6 ->"$score 成绩通过"
    else ->"$score 不及格"
}
fun main(){
    val sc=Scanner(System.`in`)
    print("输入成绩: ")
    var score=sc.nextInt()
    print(judgeGrade(score))
}
```

运行结果如图2-9所示。

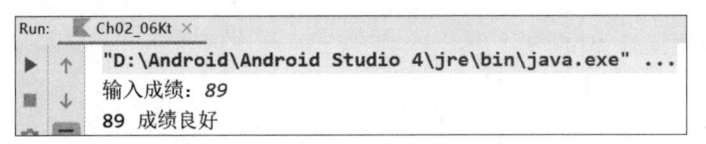

```
Run:   Ch02_06Kt ×
       "D:\Android\Android Studio 4\jre\bin\java.exe" ...
       输入成绩: 89
       89 成绩良好
```

图2-9 成绩等级判断的运行结果

在例2-5中,judgeGrade函数用带参数的when语句实现多条件的判断。当然,也可以将judgeGrade函数修改成用不带参数的when语句判断,表达的含义完全一致,但是表示的结构会更加清晰,代码如下:

```kotlin
fun judgeGrade(score: Int)=when{
    score in 90..100 ->"$score 成绩优秀"
    score in 80..89 ->"$score 成绩良好"
    score in 70..79 ->"$score 成绩中等"
    score in 60..69 ->"$score 成绩通过"
    else ->"$score 不及格"
}
```

上述代码还可以优化成如下形式:

```kotlin
fun judgeScore(score: Int): String=when (score) {
    in 90..100 ->"$score 成绩优秀"
    in 80..89 ->"$score 成绩良好"
    in 70..79 ->"$score 成绩中等"
    in 60..69 ->"$score 成绩通过"
    else ->"$score 不及格"
}
```

2. 循环结构

循环结构用于执行反复相同操作。Kotlin提供了while循环、do…while循环和for循环。while循环在执行时,先对条件进行判断,如果条件为逻辑true(即满足循环条件),则执行语句块;否则,退出

while 语句,执行后续的语句,形式如下:

```
while (条件){
    语句块
}
```

do…while 循环与 while 语句的区别在于,先执行循环体中的语句块,然后再对循环条件进行判断。如果条件为逻辑 true,则继续执行循环;否则,退出循环,执行后续的语句。

```
do{
    语句块
} while (条件)
```

Kotlin 提供的 for 语句的表达形式与 Java 的 for 循环有所不同,形式如下:

```
for (循环变量 in 范围表达式) {
    语句块
}
```

在 for 语句的范围表达式中包含了多种含义,既可以表示为数据的区间范围,又可以是迭代器,还可以是列表、数组、Stream 流等。例如:

```
for (i in 1..10)                                            //表示区间范围
    print("$i")
val lst=listOf("Python","Java","Kotlin")                    //表示列表
for (x in lst)
    print("$x")
```

例 2-6 对一个班 10 名学生 Kotlin 课程的考试成绩进行等级判断。代码如下:

```
//例 2-6 Ch02_07.kt
fun judge(score: Int)=when (score) {
    in 90..100 ->"$score 成绩优秀"
    in 80..89 ->"$score 成绩良好"
    in 70..79 ->"$score 成绩中等"
    in 60..69 ->"$score 成绩通过"
    else ->"$score 不及格"
}
fun main(){                             //模拟生成 10 名学生的成绩,成绩的范围是 50～100 分
    val rand=Random()                   //创建随机对象
    val scores=rand.ints(10L,50,100)    //创建 IntStream 对象
    for (score in scores)
        println(judge(score))
}
```

运行结果如图 2-10 所示。

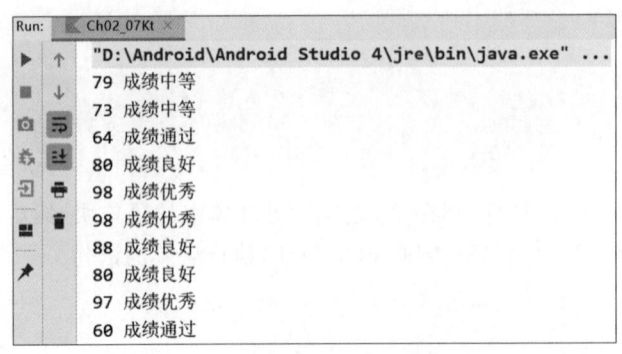

图 2-10　10名学生考试成绩的等级

2.7　Lambda 表达式

2.7.1　Lambda 表达式的应用

Lambda 表达式本质是代码片段，可以传递给其他函数。Kotlin 大量使用了 Lambda 表达式，这样使得代码简洁、清晰。Kotlin 的 Lambda 表达式是闭包代码块，可以认为是函数的一个变体，因此所有的函数都可以转换为 Lambda 表达式，所有的 Lambda 表达式也都可以转换为函数。注意，Lambda 表达式不是匿名函数，形式如下：

```
{参数变量：类型[,参数变量：类型,…,参数变量：类型]->表达式}
```

例如，实现加法运算的形式如下：

```
{x: Int,y: Int->x+y}
```

也可以将 Lambda 表达式赋值给变量，例如：

```
val sum={ x: Int,y: Int ->x+y }
print(sum(3,4))
```

Lambda 表达式往往与集合类型结合得非常紧密。集合类型对象的批量化处理可以使用 Lambda 表达式实现，下列的代码片段展示了列表中所有元素的遍历：

```
val lst=listOf("Hello","Android","World")
lst.forEach{
    it: String->print(it)
}
```

有时，Lambda 表达式中的参数可以省略，编译器通过上下文推断出 Lambda 表达式的参数和参数类型，因此对列表遍历可以修改成如下形式：

```
val lst=listOf("Hello","Android","World")
lst.forEach{
```

· 30 ·

```
    print(it)
}
```

Lambda 表达式形式灵活,可以作为参数传递给函数。例如,已知定义实体数据类 Student,具有两个属性 name 和 gender:

```
data class Student(val name: String,val gender: String)
val lst=listOf(Student("张三","男"),Student("李四","男"))         //定义列表
//将列表中 Student 对象的 name 属性相连接,分隔符为空格
val names: String=lst.joinToString(separator=" ",transform={it: Student->it.name })
                                                               //Lambda 作为参数
```

2.7.2 常见的标准函数和 Lambda 表达式

Kotlin 中在 Standard.kt 中定义了标准函数,这些标准函数中往往可以使用 Lambda 表达式作为函数的参数。常见的标准函数有 let 函数、with 函数、run 函数和 apply 函数。

1. let 函数

let 函数与安全调用操作符"?."结合,可实现对可空对象的判断。let 函数与安全调用运算符一起,允许对表达式求值,检查求值结果是否为 null,并把结果保存为一个变量。例如:

```
var message: String?=null
...
message?.let{
    message ->println(message)
}
```

若 message!=null,则执行 Lambda 表达式;若 message==null,则什么也不执行。

2. with 函数

with 函数用于对某个对象连续多次调用相关的方法。with 函数的表达式形式如下:

```
with (对象){
    ...
}
```

with 函数实际接收了两个参数:第一个参数是任意对象,第二个参数是 Lambda 表达式。第一个参数的任意对象作为第二个参数的上下文;第二个参数 Lambda 表达式的最后一行代码为返回值。例如:

```
var name="张三"
var message:String=with(name){
    println(this.length)
    "你好!$this"
}
```

上述的 with 函数中,this 对象就是传递的第一个参数 name 对象,执行输出"张三"字符串长度后,返回一个字符串信息"你好! 张三",正是 Lambda 表达式调用的结果。

3. run 函数

run 函数用于对某个对象的多次不同调用,并返回结果。它接收一个 Lambda 表达式作为参数。

并将当前对象作为 Lambda 表达式的上下文。在调用 run 函数时,返回值为 Lambda 函数块最后一行或者指定返回表达式。run 函数的表达式形式如下:

```
对象.run{
    ...
    返回值
}
```

下列代码片段是 run 函数调用的示例:

```
var name="张三"
var message=name.run{
    "$this,你好"
}
```

在上述 run 函数中,利用 name 对象生成了一个新的字符串"张三,你好",并将这个字符串返回并赋值给变量 message。

4. apply 函数

apply 函数的主要作用是简化初始化对象,用其调用某个对象的形式如下:

```
对象.apply{
    ...
}
```

apply 函数接收一个 Lambda 表达式,而这个 Lambda 表达式的上下文就是这个对象本身。apply 函数的返回值也是对象本身。例如:

```
var name=StringBuilder("张三")
name.apply {
    name.append(",你好")
}
```

在上述的 apply 函数中修改了对象中保存的值,并将当前对象 this 作为 apply 函数的返回值。执行上述语句后,name 动态字符串中包含的内容为"张三,你好"。

5. also 函数

also 函数适用于 let 函数的任何场景,一般可用于多个扩展函数链式调用。例如:

```
对象.also {
    ...
}
```

also 函数与 let 函数有些相似,但又有自己的特点:一方面 also 函数也需要判断可空对象是否为空,并将当前对象作为 Lambda 表达式的上下文;另一方面,also 函数的返回值就是传入的当前对象。例如:

```
var name: StringBuilder?=StringBuilder("张三")
name.also{
    it?.append(",你好!")
}
```

在上述的代码片段中，name 是一个可空的动态字符串对象。在 also 函数中，传递的是 name 这个可空对象本身。因此，需要通过安全调用判断 name 是否为 null，如果非 null，才可以执行相关的操作，此处是调用 name 的 append 函数。执行 also 函数后，name 的字符串变为"张三，你好！"。

2.8 面向对象编程

Kotlin 是一种面向对象的语言，可以实现面向对象开发应用。Kotlin 的面向对象编程可以实现面向对象的四大特性：抽象性、封装性、继承性与多态性。与 Java 相比，Kotlin 增加了许多新特点，在实现面向对象编程方面具有更大的灵活性和便利性。

2.8.1 包和目录

为了便于项目代码的组织和管理，往往会将功能类似的代码归于同一个包。在 Kotlin 中，不但可以将类和接口定义在包中，而且可以将函数和变量定义在包中。与 Java 定义包的概念相比，范围有所扩大。Kotlin 定义包的形式如下：

```
package 包名
```

包的名称（简称包名）总是小写且不使用下画线，仍然按照从大到小的原则命名，包名之间通过"."进行分隔，例如 com.example.kotlin.code。Kotlin 包的层次结构与文件所在的目录结构无关。Kotlin 中包的定义，只是让类、接口、函数或属性之间不产生冲突，并不在意它们在磁盘的存储位置。导入包的形式如下：

```
import 包名
```

Kotlin 中不允许类、接口、函数或属性重名，因此，一旦出现重名现象，必须使用 as 实现自定义的导入名称，例如：

```
import java.util.Date
import java.sql.Date
```

以上的类名均为 Date，会产生歧义，从而导致编译错误，因此可以将其中的一个导入修改成自定义的导入名，修改为如下形式：

```
import java.util.Date
import java.sql.Date as SqlDate
```

2.8.2 类和对象

Kotlin 仍使用 class 进行类的定义。在 Kotlin 中，一个类可以有一个主构造函数以及一个或多个次构造函数。主构造函数是类头的一部分，表示形式如下：

```
class 类名[constructor(参数名表)]{
    类体
}
```

其中，"[]"中的内容是可选项。如果在 Kotlin 定义类时所用的构造方法是无参的默认构造方法，则可直接表示如下：

```
class Test{
}
```

如果类只需要一个主构造函数实现初始化的功能，就需要在类名的后面增加参数列表实现主构造函数初始化的功能。假设定义带有两个参数类 Test，就可以写成如下形式：

```
class Test constructor(val message: String,val date: LocalDate){
}
```

如果在类的定义主构造函数时没有任何注解或者可见性修饰符，可以将 constructor 省略。这时，类 Test 的定义可以写成如下形式：

```
class Test (val message: String,val date: LocalDate){
}
```

因为在 Kotlin 的主构造函数中并不能定义代码，所以可以使用 init 代码片段来实现对初始化的操作。这时，可以通过增加 init 片段的方法实现初始化的处理，例如：

```
class Test(val message: String,val date: LocalDate){
    init{
        print("$date 发出: $message")
    }
}
```

如果类中需要多种初始化的方式构造对象，就需要使用 constructor 定义多个辅助构造函数。例如：

```
class Test(val message: String,val date: LocalDate){
    init{                                                           //初始化处理
        println("$date 发出: $message")
    }
    constructor(message: String): this(message,LocalDate.now()){    //定义辅助构造函数
        println("$message")
    }
    constructor(date: LocalDate): this("",date){                    //定义辅助构造函数
        println("$date")
    }
}
```

Kotlin 并没有 new 关键字创建类的对象实例。创建对象实例的形式如下：

```
类名([实参列表])
```

针对上述 Test 类创建 3 个不同的对象的代码如下：

```
val t1=Test("消息 1",LocalDate.now())        //利用主构造函数创建对象实例
```

```
val t2=Test("消息 2")                          //利用辅助构造函数创建对象实例
val t3=Test(LocalDate.now())                   //利用辅助构造函数创建对象实例
```

Kotlin 中还定义了一些特殊形式的类,满足一些特定的要求。

1. 数据类

数据类的定义形式其实并不陌生。本章第一个代码定义的就是一个数据类。数据类采用 data 修饰,往往用来保存数据。编译器会自动从主要构造函数中声明的所有属性扩展出 equals、hashCode、toString 等函数。例如:

```
data class Student(val name: String,val birthday: LocalDate)
```

2. 密封类

密封类表示受限的类继承结构。当一个值为有限的几种类型,而不能有任何其他类型时,采用密封类。密封类的定义形式如下:

```
sealed class 类名
```

密封类是一个抽象类,并不能实例化一个密封类,它用于扩展使用。例如:

```
sealed class Express{
    data class StringExpress(val str: String): Express()
    data class NumberExpress(val num1: Int,val num2: Int): Express()
}
```

3. 开放类

只有开放类才能被继承和扩展。这是因为 Kotlin 定义类和类的方法默认是 final,不可修改,以确保子类不能对基类进行修改。若使用 open 修饰类,则说明该类可以修改,但是子类只能用 open 修饰符指定的函数对继承的父类进行修改。

```
open class 类名[constructor(参数名表)]{
    类体
}
```

2.8.3 继承性

Kotlin 中所有类都可继承 Any 类。Kotlin 中的 Any 类不是 java.lang.Object,它只有 hashCode、equals 和 toString 这 3 个函数。对于没有超类型声明的类均默认为超类 Any。要定义一个类是另外一个类的子类,其父类或基类必须是开放类,否则无法继承。例如:

```
open class Person(val name: String,val birthday: LocalDate)         //定义父类,为开放类
class Teacher(name: String,birthday: LocalDate,val no: String): Person(name,birthday){
                                                                    //定义子类
}
```

如果父类定义了主构造函数,则子类中必须在主构造函数中提供参数对父类的主构造函数进行初始化操作。注意,子类的主构造函数初始化父类的主构造函数时,参数不需要增加 val 和 var 的限定。

如果子类无主构造函数,则必须在每个二级构造函数中用 super 关键字初始化父类,或者再代理另一个构造函数。初始化父类时,可以调用父类的不同构造方法。例如:

```kotlin
open class Person(val name: String,val birthday: LocalDate)          //定义父类
class Teacher: Person{                                               //定义无参子类
    constructor(name: String,birthday: LocalDate) : super(name,birthday){
    }
    constructor(name: String,birthday: LocalDate,no: String) : super(name,birthday){
    }
}
```

子类不但可以继承父类,而且具有重新定义父类的能力,这使得子类具有了自身的特性。这种能力称为"重写"(Overriding)。例如:

```kotlin
open class Person(val name: String,val birthday: LocalDate){
    open fun eat(){
        println("吃饭")
    }
}
class Teacher: Person{
    constructor(name: String,birthday: LocalDate) : super(name,birthday){
    }
    constructor(name: String,birthday: LocalDate,no: String) : super(name,birthday){
    }
    override fun eat(){                                              //重写 eat 函数
        println("到食堂吃饭")
    }
}
fun main(){
    val teacher=Teacher("张老师",LocalDate.of(1990,1,1))
    teacher.eat()
}
```

子类重新定义父类的函数时有一个前提条件就是重写的函数必须是 open 类型的。并且在子类重写这个函数时,必须使用 override 关键字修饰,说明为子类重写的函数。

2.8.4 接口

接口用于定义一组函数。Kotlin 使用 interface 来定义接口。形式如下:

```kotlin
interface 接口名{
    fun 函数名 1(参数列表)
    ...
    fun 函数名 n(参数列表)
}
```

定义接口的函数可以是默认函数,默认函数可以有函数体。形式如下:

```kotlin
interface 接口名{
    fun 函数名 1(参数列表){                                           //默认函数
        函数体
```

```
    }
    ...
    fun 函数名 n(参数列表)
}
```

要实现接口,可以写成如下形式:

```
class 类名: 接口名表{
    override fun 函数名 1(参数列表){…}
    ...
    override fun 函数名 n(参数列表){…}
}
```

下列代码展示了接口的定义和实现的应用:

```
interface Shape{
    fun display(){                              //定义默认函数
        println("显示形状")
    }
    fun reconstruct()                           //定义抽象函数
}
//定义类 Rect 实现接口 Shape
class Rect: Shape{
    override fun reconstruct() {                //对 Shape 的抽象函数 reconstruct()重写
        println("构建一个正方形")
    }
}
fun main(){
    val shape=Rect()                            //创建 Rect 对象
    //调用
    shape.display()
    shape.reconstruct()
}
```

2.9 异常处理

Kotlin 提供了异常处理机制。当执行代码出现异常时,会做出相应的处理,避免程序非法中断。Kotlin 中所有的异常都是 Throwable 的子类。可以根据异常对象的堆栈跟踪获取错误的相关信息。Kotlin 不区分已检查异常和未检查异常,Kotlin 的函数不需要 throws 声明异常,也没有类似 Java 中 try-with-resources 这样的异常处理方式。

Kotlin 抛出异常对象是采用 throw 子句来实现的。具体的形式如下:

```
throw 异常类型对象
```

一旦抛出异常,就需要捕获异常对象并进行处理,具体的处理形式如下:

```
try {
    ...
} catch [(异常列表)]{
    ...
} finally {
    ...
}
```

例 2-7　异常处理机制的应用实例。

```kotlin
//例 2-7 Ch02_08.kt
fun div(x: Int,y: Int): Int{
    if (y==0){
        throw ArithmeticException("除零错误")
    }
    return x/y
}
fun main(){
    val sc=Scanner(System.`in`)
    println("输入两个整数: ")
    var xx=sc.nextInt()
    var yy=sc.nextInt()
    try {
        println("$xx / $yy=${div(xx, yy)}")
    } catch (e: Exception) {
        e.printStackTrace()
    }
}
```

运行结果如图 2-11 所示。

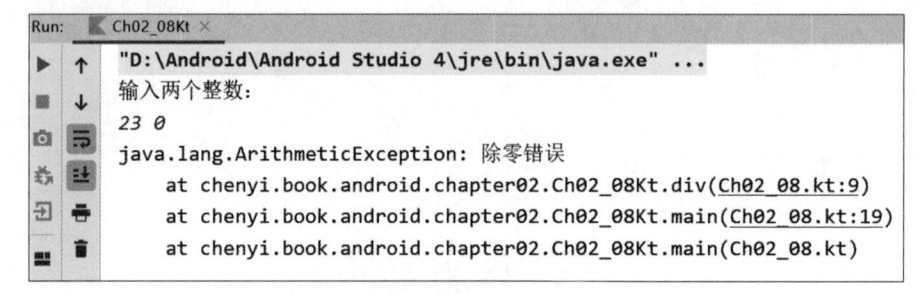

图 2-11　异常处理的运行结果

习　题　2

一、选择题

1. 已知有 Kotlin 代码片段 val a=23，则下列表达式表示正确的是_____。

　　A. a=a+1　　　　　　　　　　　　B. ++a

　　C. a*=3　　　　　　　　　　　　　D. 以上表达式均不正确

2. 已知有 Kotlin 代码片段 var a=23，则下列表达式正确的是_____。

　　A. a is Integer　　　　　　　　　　B. a is Int

C. a is int   D. 以上表达式均不正确

3. 已知有如下代码片段：

```
var a=23
val b=24
```

则下列表达式正确的是_____。

A. a＝b

B. b＝a

C. a 和 b 对应于 Java 的整型数值，只是一个数字

D. 以上表达式均不正确

4. 下列关于 Kotlin 的 main 函数说法正确的是_____。

A. main 函数定义形式只能是 fun main(args：Array＜String＞){}

B. main 函数定义形式只能是 fun main(){}

C. main 函数必须定义在 Kotlin 的类中

D. 以上说法均不正确

5. 下列 Kotlin 代码片段表示正确的是_____。

A. val s：String?＝null   B. val s：String?
　　s＝"hello"   s＝null

C. val s：String?   D. 以上定义均不正确
　　s＝null
　　s＝"hello"

6. 已知有如下代码片段：

```
val lst=listOf("hello","welcome")
```

则 lst is List＜String＞的执行结果是_____。

A. true   B. false

C. 无法通过编译   D. 编译通过，运行抛出异常

7. 已知有如下代码片段：

```
var lst: List<String?>?=mutableListOf<String>("hello","Kotlin")
```

在执行 lstList.add(null)之后，lst.size()的结果是_____。

A. 3   B. 2

C. 编译无法通过   D. 编译通过，运行抛出运行时异常

8. 已知有如下代码片段：

```
val lst=mutableListOf<String>("hello","kotlin")
```

执行 lst.add(null)后，lst.size()的结果是_____。

A. 2   B. 3

C. 编译无法通过   D. 编译正常，运行会抛出运行时异常

9. 有如下 Kotlin 代码片段：

```
val lst=listOf("hello","welcome","this","is","kotlin","world")
lst.forEach{e->print(" $e")}
```

则运行结果是_____。

  A. hello welcome this is kotlin world   B. 编译错误

  C. 通过编译，运行抛出异常   D. 以上说法均不正确

10. 有如下 Kotlin 代码片段：

```
val lst=listOf("hello","welcome","this","is","kotlin","world")
lst.forEach{e->print(" $e.size()")}
```

则运行结果是_____。

  A. 5 7 4 2 6 5   B. 编译无法通过

  C. 通过编译，运行错误   D. 以上说法均不正确

二、填空题

1. Kotlin 的编译命令行形式为 _____。
2. 使用 Kotlin 编写的代码保存在扩展名为 _____ 的文件中。
3. Kotlin 中用定义字符串模板的方式实现字符串的格式化。假设已知变量 val name="张三"，且要求利用 name 输出"你好，张三"，则需要执行 print(_____)。
4. 已知 var type：String?="非空"，则表达式 type? "空"的运算结果是_____。则执行 type!!.toString()的结果是_____。
5. 已知 val map=mapOf("hello" to 1,"welcome" to 2)，则 map 的数据类型是_____。
6. 已知 val map=mutableMapOf("hello" to 1,"welcome" to 2)，则 map 的数据类型是_____。
7. 要导入包，可以使用_____。
8. 用 apply 标准函数接收 Lambda 表达式时，apply 函数的返回值是_____。
9. run 标准函数接收 Lambda 表达式作为参数时，run 函数的返回值是_____。
10. let 标准函数与操作符_____结合，可实现对可空对象的判断。

三、上机实践题

1. 编写程序，实现从键盘输入一个年份，并将该年的所有月历打印输出。
2. 编写程序，实现从键盘输入一个学生的百分制成绩，并输出该名学生成绩的等级。其中，优：90～100 分；良：80～89 分；中：70～79 分；及格：60～69 分；不及格：0～59 分。
3. 编写程序，用列表结构存放学生姓名，要求向指定班级学生发送通知。通知内容如下：

```
×××同学：
    今天下午 2:00—4:00 在信工楼 B104 室召开班级会议，请准时到场。
                                                        班委会
```

4. 编写程序，从键盘输入一段英文，统计其中每个单词的出现次数。
5. 编写程序，用面向对象的方法实现发红包的应用。

要求：群主可以发红包，普通用户只能接收红包，最后按照每个人接收红包的金额升序排序并输出。

第 3 章 Android 的 Activity 组件

Activity(活动组件)是 Android 的四大组件之一,是 Android 移动应用的基本组件,用于设计移动应用界面的屏幕显示。Android 移动应用往往需要多个界面,因此移动应用中必须定义一个或多个 Activity,以实现界面的交互。自从 Compose 组件问世以来,便采用 Compose 组件定义 Android 应用界面。目前,采用单个 Activity 和多个 Compose 的可组合界面方式已成为移动应用界面的首选。本章将开始介绍嵌入 Compose 可组合界面的 Activity。

3.1 Activity 的创建

Activity 用于创建移动界面的屏幕。它的主要作用就是实现移动界面和用户之间的交互。Activity 类用于创建和管理用户界面,一个应用程序可以有多个 Activity,但在同一个时间内只有一个 Activity 处于激活状态。在移动应用中,Activity 之间的依赖关系低。

在创建 Activity 时,需要完成如下步骤。
(1) 定义 Activity 类。
(2) 在配置清单文件 AndroidManifest.xml 中声明并配置 Activity 的相应属性。

1. 定义 Activity 类

Activity 必须是自己的子类或扩展子类的子类。因为 Activity 的一个子类为 ComponentActivity,用于定义 Material Design 风格的界面,所以在当前 Android 中创建 Activity 时一般是定义为 ComponentActivity 的子类。形式如下:

```
class MainActivity: ComponentActivity() {
    override fun onCreate(savedInstanceState: Bundle?) {
        super.onCreate(savedInstanceState)
        enableEdgeToEdge()
        setContent {
            Chapter02Theme {
                Box(modifier=Modifier.fillMaxSize(),
                    contentAlignment=Alignment.Center){
                    Text(text="Hello Android!",fontSize=40.sp)
                }
            }
        }
    }
}
```

在 MainActivity(主活动)中定义了 onCreate()函数,该函数是必须定义的回调函数,它会在系统创建 Activity 时被调用。可在 MainActivity 的扩展函数 setContent 内调用 Compose 组件定义的可组合函数(如上例的 Box 函数和 Text 函数)生成可组合视图树,实现界面的定义,如图 3-1 所示。

2. 在配置清单文件 AndroidManifest.xml 中注册 Activity

要让创建的 Activity 被移动应用识别和调用,就必须在 AndroidManifest.xml 系统配置清单文件中

图 3-1 Activity 和可组合函数的关系示意

进行声明和配置。AndroidManifest.xml 文件保存在项目的 manifests 目录中。代码如下：

```xml
<?xml version="1.0" encoding="utf-8"?>
<manifest xmlns: android="http://schemas.android.com/apk/res/android"
    package="chenyi.book.android.Ch03_01">
    <application
        android: allowBackup="true"
        android: icon="@mipmap/ic_launcher"
        android: label="@string/app_name"
        android: roundIcon="@mipmap/ic_launcher_round"
        android: supportsRtl="true"
        android: theme="@style/Theme.Chapter03">
        <activity
            android: name=".MainActivity"
            android: exported="true">
            <intent-filter>
                <action android: name="android.intent.action.MAIN" />
                <category android: name="android.intent.category.LAUNCHER" />
            </intent-filter>
        </activity>
    </application>
</manifest>
```

在 application 元素（表示移动应用）中增加下级元素 activity，并通过 android：name 属性指定活动的类名。在 manifest 元素的 package 属性中指定了包名，所以在 activity 元素的 android：name 中只需要定义成

```
android: name=".MainActivity"
```

即可。在 activity 元素下定义了 intent-filter 意图过滤器。intent-filter 定义的意图过滤器非常强大，通过它可以根据显式请求启动 Activity，也可以根据隐式请求启动 Activity。下面代码用于指定 MainActivity（主活动），它是整个移动应用的入口活动，表示应用程序可以显示在程序列表中。

```
action android: name="android.intent.action.MAIN"
category android: name="android.intent.category.LAUNCHER"
```

注意：在 Android 12 以及以上版本中使用 intent-filter 元素配置时，必须使 Activity 增加配置 android：exported 属性。

3.2 Activity 和 Intent

一个应用可以定义多个 Activity，这些 Activity 可以相互切换和跳转。要实现这些交互功能，就必须了解什么是 Intent（意图）。Intent 表示封装执行操作的意图，是应用程序启动其他组件的 Intent 对象启动组件。这些组件可以是 Activity，也可以是其他基本组件，如 Service（服务）组件、BroadcastReceiver（广播接收

器)组件。从一个基本组件导航到另一个组件。通过 Intent 实现组件之间的跳转和关联。Intent 分为显式 Intent 和隐式 Intent 两种。

3.2.1 显式 Intent

显式 Intent 就是在 Intent 函数启动组件时,需要明确指定的激活组件名称,例如:

```
Intent(Context packageContext, Class<?>c)
```

其中,packageContext 用于提供启动 Activity 的上下文,c 用于指定想要启动的目标组件类。

例 3-1 显式 Intent 的应用。在本应用中定义 3 个 Activity 类:MainActivity、FirstActivity 和 SecondActivity。其中,MainActivity 用于定义 3 个按钮:"第一个活动"按钮用于跳转到 FirstActivity,"第二个活动"按钮用于跳转到 SecondActivity,"退出"按钮用于从移动应用中退出。

在新建移动应用模块时,因为本例涉及按钮的应用和字体的控制,所以需要先定义相关的资源文件。

(1) 定义相关资源。

移动应用往往需要大量的字符串来配置相关的组件,因此可以将字符串统一定义在 res/values 目录下的 strings.xml 配置文件中,通过设置 string 元素配置字符串。string 元素的 name 属性表示字符串的名字,string 元素的内容表示字符串的取值,代码如下:

```xml
<!--模块 Ch03_02 定义字符串资源文件 res/values/strings.xml-->
<resources>
    <string name="app_name">Ch03_02 </string>
    <string name="title_first_activity">第一个活动 </string>
    <string name="title_second_activity">第二个活动 </string>
    <string name="title_exit_app">退出 </string>
</resources>
```

(2) 定义 MainActivity。MainActivity 是当前应用的入口。它定义了 3 个按钮,以实现不同活动的跳转,代码如下:

```kotlin
//模块 Ch03_02 主活动的定义文件 MainActivity.kt
class MainActivity : ComponentActivity() {
    override fun onCreate(savedInstanceState: Bundle?) {
        super.onCreate(savedInstanceState)
        enableEdgeToEdge()
        setContent {
            Chapter03Theme {
                Scaffold(modifier=Modifier.fillMaxSize()) {
                    innerPadding ->
                        Column(modifier=Modifier.padding(innerPadding).fillMaxSize(),
                                                                        //占据全屏
                            horizontalAlignment=Alignment.CenterHorizontally,
                                                                        //内容水平居中
```

```
                    verticalArrangement=Arrangement.Center){        //内容垂直居中
                    Button(onClick={                                 //定义按钮动作
                        turnTo(FirstActivity::class.java)            //执行意图
                    }){                                              //定义按钮文本
                        Text(stringResource(id=R.string.title_first_activity))
                    }
                    Button(onClick={
                        turnTo(SecondActivity::class.java)
                    }){
                        Text(stringResource(id=R.string.title_second_activity))
                    }
                    Button(onClick={
                        exitProcess(0)                               //退出应用
                    }){
                        Text(stringResource(id=R.string.title_exit_app))
                    }
                }
            }
        }
    }
    private fun <T>turnTo(activityClass:Class<T>){
        val intent=Intent(MainActivity@this,activityClass)
        startActivity(intent)
    }
}
```

turnTo 函数是自定义的函数,在这个函数中使用了泛型,即定义了类型变量 T,将类型变量传递给 Class<T>中,表示各种 Class 类型。在函数体中,通过创建显式 Intent 对象,并启动 startActivity 函数数实现从 MainActivity 跳转到指定类的 Activity 中。如果调用的是 turnTo(FirstActivity::class.java)函数,则等价执行的代码如下:

```
//创建从当前的 MainActivity 跳转到 FirstActivity 活动的意图对象
val intent=Intent(MainActivity@this, FirstActivity::class.java)
//在当前 Activity 中启动 Intent
MainActivity@this.startActivity(intent)
```

其中,MainActivity@this 表示 MainActivity 的当前对象。

(3) 定义其他的 Activity。移动应用模块中还定义了 FirstActivity 和 SecondActivity,它们都定义了一个文本标签中显示字符串。

```
//模块 Ch03_02 第一个活动的定义文件 FirstActivity.kt
class FirstActivity : ComponentActivity() {
    override fun onCreate(savedInstanceState: Bundle?) {
        super.onCreate(savedInstanceState)
        enableEdgeToEdge()
        setContent {
            Chapter03Theme {
                Scaffold(modifier=Modifier.fillMaxSize()) {
```

```kotlin
                innerPadding ->Box(modifier=Modifier.padding(innerPadding),
                    contentAlignment=Alignment.Center){
                    Text(stringResource(id=R.string.title_first_activity),fontSize=40.sp)
                }
            }
        }
    }
}
//模块 Ch03_02 第二个活动的定义文件 SecondActivity.kt
class SecondActivity : ComponentActivity() {
    override fun onCreate(savedInstanceState: Bundle?) {
        super.onCreate(savedInstanceState)
        enableEdgeToEdge()
        setContent {
            Chapter03Theme {
                Scaffold(modifier=Modifier.fillMaxSize()) {
                    innerPadding->Box(modifier=Modifier.padding(innerPadding),
                        contentAlignment=Alignment.Center){
                        Text(stringResource(id=R.string.title_second_activity),fontSize=40.sp)
                    }
                }
            }
        }
    }
}
```

指定的尺寸采用了 sp 单位,表示比例无关的像素单位,常常表示字体的大小。表示布局的尺寸往往使用单位 dp,单位 dp 代表与密度无关的像素,是基于屏幕的物理密度的抽象或虚拟单元。

(4) 配置清单文件中注册 Activity,代码如下:

```xml
<!--模块 Ch03_02 配置清单文件 AndroidManifest.xml -->
<?xml version="1.0" encoding="utf-8"?>
<manifest xmlns: android="http://schemas.android.com/apk/res/android"
    package="chenyi.book.android.ch03_02">
    <application .......>      <!--其他配置省略 -->
        <!--配置 SecondActivity -->
        <activity
            android: name=".SecondActivity"
            android: label="@string/title_second_activity"/>
        <!--配置 FirstActivity -->
        <activity
            android: name=".FirstActivity"
            android: label="@string/title_first_activity"/>
        <!--配置 MainActivity -->
        <activity
            android: name=".MainActivity"
            android: exported="true">
            <intent-filter>
                <action android: name="android.intent.action.MAIN" />
                <category
                    android: name="android.intent.category.LAUNCHER" />
```

```
            </intent-filter>
        </activity>
    </application>
</manifest>
```

在配置清单文件中,配置了 FirstActivity、SecondActivity 和 MainActivity。其中,MainActivity 作为入口,在程序加载时首先启动和显示。

运行结果如图 3-2 所示。

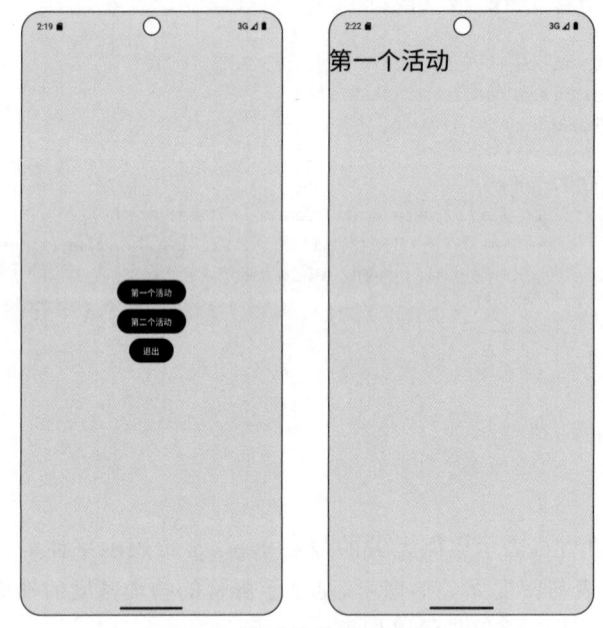

图 3-2 活动跳转的运行结果

3.2.2 隐式 Intent

隐式 Intent 没有明确地指定要启动哪个 Activity,而是通过 Android 系统分析 Intent,并根据分析结果来启动对应的 Activity。执行过程如图 3-3 所示,即某个 Activity A 通过 startActivity 函数操作某个隐式 Intent 调用另外的 Activity。Android 系统先分析应用配置清单 AndroidManifest.xml 中声明的 IntentFilter 内容,再搜索应用中与 Intent 相匹配的 IntentFilter 内容,若匹配成功后,则 Android 系统调用匹配的 Activity,用 Activity B 的 onCreate 函数启动新的 Activity,实现从 Activity A 跳转到 Activity B。

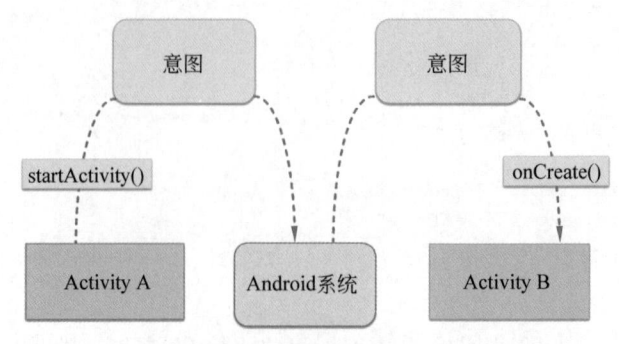

图 3-3 通过隐式 Intent 通过 Android 系统启动活动

一般情况下，隐式Intent需要在配置清单文件AndroidManifest.xml中指定IntentFilter的action、category、data属性。Android系统会对IntentFilter的3个属性进行分析和匹配，以搜索对应的Activity或其他组件。

(1) action：表示该Intent所要完成的一个Activity对应的"动作"。

(2) category：用于为action增加额外的附加类别信息。

(3) data：向action提供操作的数据。

但是这3个属性并不是必须全部配置，可以根据具体的需要进行组合配置。例如：

```xml
<activity
    android: name=".CustomedActivity"
    android: exported="true">
    <intent-filter>
        <action android: name="chenyi.book.android.ch03_03.ACTION" />
        <!--自定义动作的名称-->
        <category android: name="android.intent.category.DEFAULT" />
        <!--指定默认类别-->
        <category android: name="chenyi.book.android.ch03_03.MyCategory" />
        <!--指定自定义的类别-->
    </intent-filter>
</activity>
```

例3-2 使用隐式Intent的应用实例。MainActivity可以分别调用自定义的Activity和拨打电话。

本应用中定义了两个活动：一个是自定义的CustomizedActivity；另一个表示MainActivity，用于移动应用的入口。在MainActivity中提供了两个按钮，单击可以实现跳转CustomizedActivity和拨打电话的功能。在系统的配置清单文件中，首先配置并注册这些Activity。

(1) 在配置清单AndroidManifest.xml中注册活动，代码如下：

```xml
<?xml version="1.0" encoding="utf-8"?>
<manifest
xmlns:android="http://schemas.android.com/apk/res/android">
    <application …>          <!--其他配置省略-->
        <!--配置CustomizedActivity-->
        <activity
            android:name=".CustomizedActivity"
            android:exported="false"
            android:label="@string/title_activity_customized"
            android:theme="@style/Theme.Chapter03">
            <intent-filter>
                <action android:name="chenyi.book.android.ch03_03.ACTION" />
                <category android:name="android.intent.category.DEFAULT" />
                <category android:name="chenyi.book.android.ch03_03.MyCategory" />
            </intent-filter>
        </activity>
        <!--配置MainActivity-->
        <activity
            android:name=".MainActivity"
            android:exported="true"
            android:label="@string/app_name"
            android:theme="@style/Theme.Chapter03">
            <intent-filter>
                <action android:name="android.intent.action.MAIN" />
```

```
                <category android:name="android.intent.category.LAUNCHER" />
            </intent-filter>
        </activity>
    </application>
</manifest>
```

(2) 定义 CutomizedActivity,代码如下:

```
//模块 Ch03_03 定义的 CustomizedActivity 的文件 CustomizedActivity.kt
class CustomizedActivity : ComponentActivity() {
    override fun onCreate(savedInstanceState: Bundle?) {
        super.onCreate(savedInstanceState)
        enableEdgeToEdge()
        setContent {
            Chapter03Theme {
                Scaffold(modifier=Modifier.fillMaxSize()) {
                    innerPadding->Text(text="自定义的 Customer 组件",
                        fontSize=30.sp,modifier=Modifier.padding(innerPadding))
                }
            }
        }
    }
}
```

(3) 定义 MainActivity。在 MainActivity 布局文件中定义了两个按钮,目的是实现不同 Activity 的跳转的接口。代码如下:

```
//模块 Ch03_03 定义的 MainActivity 的文件 MainActivity.kt
class MainActivity : ComponentActivity() {
    override fun onCreate(savedInstanceState: Bundle?) {
        super.onCreate(savedInstanceState)
        enableEdgeToEdge()
        setContent {
            Chapter03Theme {
                Scaffold(modifier=Modifier.fillMaxSize()) {
                    innerPadding->Column(modifier=Modifier.padding(innerPadding)
                        .fillMaxSize(),
                        horizontalAlignment=Alignment.CenterHorizontally,
                        verticalArrangement=Arrangement.Center) {
                        Button(onClick={
                            //利用 action 定义 intent
                            val myIntent=Intent("chenyi.book.android.ch03_03.ACTION")
                            //增加类别
                            myIntent.addCategory("chenyi.book.android.ch03_03.
                                MyCategory")                              //指定所在包名
                            myIntent.`package`="chenyi.book.android.ch03_03"
```

```
                startActivity(myIntent)
            }){
                Text("跳转到CustomizedActivity",fontSize=28.sp)
            }
            Spacer(modifier=Modifier.padding(bottom=20.dp))
            Button(onClick={                                         //拨打电话
                val myIntent=Intent(Intent.ACTION_DIAL)
                startActivity(myIntent)
            }){
                Text("拨打电话",fontSize=28.sp)
            }
        }
      }
     }
    }
   }
 }
}
```

运行结果如图 3-4 所示。

(a) MainActivity　　(b) CustomizedActivity　　(c) 系统的拨号界面

图 3-4　隐式 Intent 实例的运行结果

MainActivity 是移动应用中第一个调用和加载的界面,在配置清单文件 AndroidManifest.xml 中也注册了 CustomizedActivity,在 activity 元素定义了下级标签 intent-filter,指定了 CustomizedActivity 的动作名称,添加了两个类别:android.intent.category.DEFAULT 和 chenyi.book.android.ch03_03.MyCategory。android.intent.category.DEFAULT 是默认的类别。在调用 startActivity 函数时,Android 系统会自动将 android.intent.category.DEFAULT 添加到 Intent 中。chenyi.book.android.ch03_03.MyCategory 是自定义的类别,因此需要调用 addCategory 函数将这个自定义的类别添加到 Intent 中,即单击第一个按钮可通过 myIntent 跳转到 CustomizedActivity。

注意:单击第二个按钮,实现拨打电话。Intent.ACTION_DIAL 内置的系统动作,对应 android

.intent.action.DIAL。Android 系统中定义常见的内置动作如表 3-1 所示。

表 3-1 Intent 内置的动作

动　作	说　明
ACTION_ANSWER	处理来电
ACTION_CALL	拨打电话,使用 Intent 的号码,需要设置 android.permission.CALL_PHONE
ACTION_DIAL	调用拨打电话的程序,使用 Intent 的号码,没有直接打出
ACTION_EDIT	编辑 Intent 中提供的数据
ACTION_VIEW	查看动作,可以浏览网页、短信、地图路肩规划,根据 Intent 的数据类型和数据值决定
ACTION_SENDTO	发送短信、电子邮件
ACTION_SEND	发送信息、电子邮件
ACTION_SEARCH	搜索,通过 Intent 的数据类型和数据判断搜索的动作

调用 Intent 的 setData 函数可以设置 Intent 的数据,用 Kotlin 的表达是 intent.data;调用 Intent 的 setType 函数可以设置 Intent 数据的类型;调用 setDataAndType 函数可以设置数据和数据的类型。

3.3　Activity 之间的数据传递

Activity 之间往往需要传递数据。数据的传递可以借助 Intent 来实现。Intent 提供了两种实现数据传递的方式:一种是通过 Intent 的 putExtra 传递数据;另一种是通过 Bundle 传递多类型数据。

3.3.1　传递常见数据

在这里常见的数据类型包括 Int(整型)、Short(短整型)、Long(长整型)、Float(单精度实型)、Double(双精度实型)、数组、ArrayList(数组列表)等多种类型。在 Activity 之间,可以将这些类型的数据进行传递。

首先,Intent 提供了一系列的 putExtra 函数实现数据传递。在数据的发送方,通过 putExtra 函数种指定键值对,然后启动 startActivity 将意图从当前的 Activity 跳转到 intent 指定的活动,代码如下:

```
val intent=Intent(this,OtherActivity：：class.java)    //定义跳转到 OtherActivity 的 Intent
intent.putExtra("intValue",23)                        //配置传递数据的键值对
startActivity(intent)                                 //根据 Intent 启动 Activity
```

数据接收方(设为 OtherActivity)调用 getIntent 函数获得启动该 Activity 的 Intent,代码如下:

```
val intent=getIntent()
```

在 Kotlin 中也可以直接表示成

```
val intent=intent
```

然后,根据数据的关键字,调用接收数据类型对应 get×××Extra 函数来获得对应的取值。例如:

```
val received=intent.getIntExtra("intValue")          //根据关键字 intValue 获取对应的 Int 数据
```

第二种方式结合 Bundle 数据包和 putExtras 函数实现数据的传递。具体的执行过程与第一种方式类似。数据的发送方创建 Bundle 数据包,代码如下:

```
val bundle: Bundle=Bundle()
```

然后,调用 Bundle 对象的对应的 put×××函数,设置不同类型的数据,例如:

```
bundle.putInt("bundleIntValue",1000)
bundle.putString("bundleStringValue","来自 MainActivity 的问候!")
```

在发送方通过调用 Intent 的 putExtras 函数,设置要传递的数据,再调用 startActivity 实现 Activity 的跳转和数据的发送,代码如下:

```
val intent=Intent(this,OtherActivity::class.java)
intent.putExtras(bundle)
startActivity(intent)
```

在数据的接收方,例如 OtherActivity 有两种方式接收数据:一种是通过 Bundle 对象的对应的 get×××函数来获得;另一种是直接通过调用启动该 Activity 的 Intent 的 get×××Extra 函数,根据关键字获得对应的取值,代码如下:

```
val intent=intent
val bundle: Bundle?=intent.extras
//根据 Bundle 来获得数据
val intValue=bundle?.getInt("bundleIntValue",0)              //0 为默认值,取值失败时会赋值
val strValue=bundle?.getString ("bundleStringValue")
//或根据 Intent 直接获得数据
val intValue=intent.getIntExtra("bundleIntValue")
val strValue=intent.getStringExtra("bundleStringValue")
```

例 3-3 常见数据类型数据的传递的应用实例 MainActivity 分别使用两种不同的方式传递数据到其他两个 Activity,代码如下:

```
//模块 Ch03_04 定义数据的发送方的 MainActivity.kt
class MainActivity: ComponentActivity() {
    override fun onCreate(savedInstanceState: Bundle?) {
        super.onCreate(savedInstanceState)
        enableEdgeToEdge()
        setContent {
            Chapter03Theme {
                Scaffold(modifier=Modifier.fillMaxSize()) {
                    innerPadding->Column(modifier=Modifier.padding(innerPadding)
                        .fillMaxSize(),
                        horizontalAlignment=Alignment.CenterHorizontally,
                        verticalArrangement=Arrangement.Center) {
```

```kotlin
                    Button(onClick={
                        sendData()
                    }){
                        Text(stringResource(id=R.string.title_send_base),fontSize=30.sp)
                    }
                    Spacer(modifier=Modifier.padding(bottom=10.dp))
                    Button(onClick={
                        sendArrayData()
                    }){
                        Text(stringResource(id=R.string.title_send_array),fontSize=30.sp)
                    }
                }
            }
        }
    }
    private fun sendData(){
        val intent=Intent(this,FirstActivity::class.java)
        intent.putExtra("intData",23)
        intent.putExtra("strData","来自MainActivity的问候!")
        startActivity(intent)
    }
    private fun sendArrayData(){
        val bundle=Bundle()                                        //创建Bundle对象
        bundle.putString("str","来自MainActivity的问候!!")
        val intArr=intArrayOf(23,45,67,220,23)                     //创建只读的整型数组
        val strArr=arrayOf("hello ","android ","world")            //创建只读的字符串数组
        bundle.putIntArray("intArray",intArr)
        bundle.putStringArray("strArray",strArr)
        val intent=Intent(this,SecondActivity::class.java)
        intent.putExtras(bundle)
        startActivity(intent)
    }
}
```

在MainActivity中定义了两个按钮,并分别为这两个按钮定义了相应的处理动作。作为数据的发送方MainActivity,提供了两种数据发送的方法。第一种方法是直接利用Intent的putExtra函数将数据传递出去;第二种是将数据通过键值对放置在Bundle对象中,然后再利用Intent将Bundle数据包整体传递出去。有数据发送,必须有一个接收方来接收数据,否则发送数据没有任何意义。在此处,定义了FirstActivity和SecondActivity分别接收数据,代码如下:

```kotlin
//模块Ch03_04 定义数据接收方FirstActivity.kt
class FirstActivity: ComponentActivity() {
    override fun onCreate(savedInstanceState: Bundle?) {
        super.onCreate(savedInstanceState)
        enableEdgeToEdge()
        val receivedInt=intent.getIntExtra("intData",0)            //按照关键字接收数据,0为默认值
        val receivedStr=intent.getStringExtra("strData")           //按照关键字获取字符串
```

```
            setContent {
                Chapter03Theme {
                    Scaffold(modifier=Modifier.fillMaxSize()) {
                        innerPadding ->Box(modifier=Modifier.padding(innerPadding).fillMaxSize()
                            .padding(20.dp),contentAlignment=Alignment.Center){
                            Text(text="接收的整数是$receivedInt,接收的字符串是 $receivedStr !",
                                fontSize =24.sp)
                        }
                    }
                }
            }
        }
```

FristActivity 中接收的数据是利用 Intent 调用 get×××Extra 函数,通过关键字来获得对应的数据。在代码中,分别接收了整型数组和字符串数组,调用数组对象的 contentToString 函数,获取数组内容的字符串的表达形式。代码如下：

```
//模块 Ch03_04 定义数据接收方 SecondActivity.kt
class SecondActivity : ComponentActivity() {
    override fun onCreate(savedInstanceState: Bundle?) {
        super.onCreate(savedInstanceState)
        enableEdgeToEdge()
        val bundle:Bundle?=intent.extras          //intent 获得意图,extras 获得 Bundle 对象
        val receivedStr=bundle?.getString("str")                //接收字符串
        val receivedIntArr=bundle?.getIntArray("intArray")      //接收整型数组
        val receivedStrArr=bundle?.getStringArray("strArray")   //接收字符串数组
        setContent {
            Chapter03Theme {
                Scaffold(modifier=Modifier.fillMaxSize()) {
                    innerPadding->Box(modifier=Modifier.padding(innerPadding).fillMaxSize(),
                        contentAlignment=Alignment.Center){        //显示数据
                        Text(text ="接收的字符串是${receivedStr} \n " +
                            "接收的整数数组是 ${receivedIntArr.contentToString()}\n" +
                            "接收的字符串数组是 ${receivedStrArr.contentToString()} !",
                            fontSize=24.sp)                        //将数组的内容换成字符串
                    }
                }
            }
        }
    }
}
```

SecondActivity 接收数据的方式有两种：一种方式是启动该活动的意图获得 Bundle 对象,然后从 Bundle 对象中获得关键字对应的数据；另一种方式是通过调用 Intent 的 get×××Extra 函数根据传递数据的关键字获得对应的数据。

运行结果如图 3-5 所示。

(a) 数据的发送方　　　　(b) 数据的接收方一　　　　(c) 数据的接收方二

图 3-5　活动传递数据实例的运行结果

3.3.2　Serializable 对象的传递

在 Activity 之间传递数据有一种特殊情况，就是传递自定义类的对象或者自定义类的对象数组或集合类型中包含自定义类的对象。需要对这些自定义类进行可序列化处理。可序列化处理，就是将对象处理成可以传递或存储的状态，使得对象可以从临时瞬间状态保存持久状态，即存储到文件、数据库等中。要在 Java 中实现可序列化，就需要用类实现 java.io.Serializable 接口。作为基于 JVM 的语言，Kotlin 也需要实现 java.io.Serializable 接口，这个类的对象就可以进行序列化的处理。形式如下：

```
class DataType: Serializable{
    …
}
```

通过 Serializable 传递对象的过程，如图 3-6 所示。发送方将数据序列化成字节流发送，在接收方，将字节流重组成对应的对象。

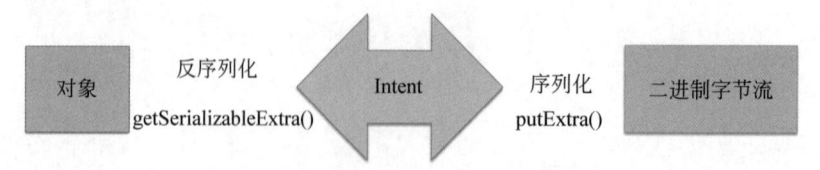

图 3-6　Serializable 方式传递对象的原理

要在 Activity 之间发送 Serializable 对象，仍需要通过 Intent 实现。Intent 通过调用 putExtra 函数，设置"关键字"和对象之间的键值对，形式如下：

```
val intent=Intent(this,OtherActivity: : class.java)
val object:Type= …
```

· 54 ·

```
intent.putExtra("key",object)
startActivity(intent)
```

Activity 接收方 OtherActivity 在接收数据时,首先启动该活动的意图,再通过意图的 getSerializableExtra 函数通过关键字获得对应的 Serializable 对象,第二个参数 Class<T>是指数据对象的类型对象,实现了根据接收的状态数据反序列化生成对应的对象。形式如下:

```
val object=intent.getSerializableExtra("key",Class<T>)
```

例 3-4 在 Activity 之间传递 Serializable 对象应用实例。实现将学生的信息(学号、姓名、出生日期)从一个 Activity 传递到另外一个 Activity。

(1) 定义实体类 Student,代码如下:

```
//模块 Ch03_05 定义数据实体类 Student
data class Student(val no: String,val name: String,
    val birthday: LocalDate) : Serializable
```

(2) 定义数据的发送方 MainActivity。MainActivity 发送 Student 对象到 FirstActivity。调用 Intent 的 putExtra 函数设置发送数据的键值对。关键字是 student,取值是 Student 对象 student。然后通过 startActivity(intent)启动 Intent 封装活动,实现数据通过 Intent 发送到 FirstActivity,代码如下:

```
//模块 Ch03_05 发送方的主活动定义 MainActivity.kt
class MainActivity : ComponentActivity() {
    override fun onCreate(savedInstanceState: Bundle?) {
        super.onCreate(savedInstanceState)
        enableEdgeToEdge()
        setContent {
            Chapter03Theme {
                Scaffold(modifier=Modifier.fillMaxSize()) {
                    innerPadding ->Column(modifier=Modifier.padding(innerPadding)
                    .fillMaxSize(),
                    horizontalAlignment=Alignment.CenterHorizontally,
                    verticalArrangement=Arrangement.Center) {
                        Button(onClick={
                            sendData()                //调用发送数据的函数
                        }) {
                            Text("发送数据",fontSize =30.sp)
                        }
                    }
                }
            }
        }
    }
    private fun sendData(){
        val intent=Intent(this,FirstActivity::class.java)
        val student=Student("910023","张三", LocalDate.of(2001,5,12))
```

```
        intent.putExtra("student",student)
        startActivity(intent)
    }
}
```

（3）定义数据接收方的 FirstActivity，代码如下：

```
//模块 Ch03_05 接收方定义 FirstActivity.kt
class FirstActivity: ComponentActivity() {
    override fun onCreate(savedInstanceState: Bundle?) {
        super.onCreate(savedInstanceState)
        enableEdgeToEdge()
        //获取可序列化对象
        val data:Student?=intent.getSerializableExtra("student",Student::class.java)
        setContent {
            Chapter03Theme {
                Scaffold(modifier=Modifier.fillMaxSize()) {
                    innerPadding ->Box(modifier=Modifier.padding(innerPadding)
                    .fillMaxSize().padding(20.dp),contentAlignment=Alignment
                    .Center) {
                        Text(text="接收的学生的信息是：$data！", fontSize=24.sp)
                    }
                }
            }
        }
    }
}
```

运行结果如图 3-7 所示。

(a) 对象发送方　　　　(b) 对象接收方

图 3-7　活动传递 Serializable 对象的运行结果

3.3.3　Parcelable 对象的传递

通过 Java 的 java.io.Serializable 方式传递对象,将对象整体数据序列化处理,将对象转换为可传递状态进行发送。接收方将接收的状态数据重新反序列化重新生成对象。由于这种处理方式效率低下,因此 Android 提供了自带的 Parcelable 方式实现对象的传递。

以 Parcelable 方式传递对象的原理是,通过一套机制,将一个完整的对象进行分解,分解后的每一部分数据都属于 Intent 支持的数据类型。可以将分解后的可序列化的数据写入一个共享内存中,其他进程通过 Parcel 从这块共享内存中读出字节流,并反序列化成对象,通过这种方式来实现传递对象的功能,如图 3-8 所示。

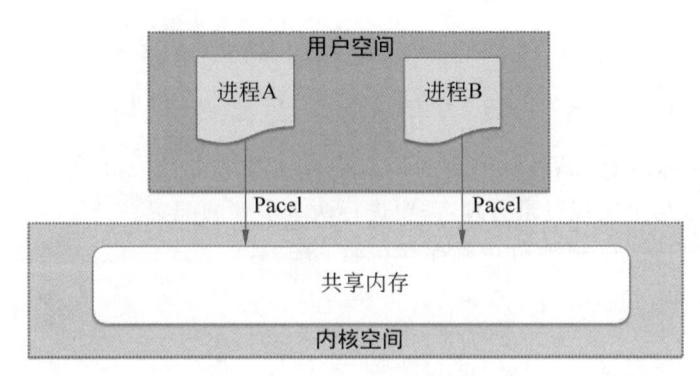

图 3-8　Parcelable 方式传递对象

要实现 Parcelable 方式传递对象,需要定义的类必须实现 Parcelable 接口。必须对其中的方法进行覆盖重写,具体如下。

(1) writeToParcel:实现将对象的数据进行分解写入。

(2) describeContent:内容接口的描述,默认只需要返回为 0。

(3) Parcelable.Creator:需要定义静态内部对象实现 Parcelable.Createor 接口的匿名类。该接口提供了两个函数:一个是 createFromParcelable 函数实现从 Parcel 容器中读取传递数据值,读取数据的顺序与写入数据的顺序必须保持一致,然后封装成 Parcel 对象返回逻辑层;另一个是 newArray 函数创建一个指定类型指定长度的数组,供外部类反序列化本类数组使用。下面定义了实现 Parcelable 接口的 Student 类,代码如下:

```kotlin
data class Student(val no: String,val name: String,val birthday: LocalDate): Parcelable {
    constructor(parcel: Parcel):
        this(parcel.readString()!!,parcel.readString()!!,
        parcel.readSerializable() as LocalDate)
    override fun writeToParcel(parcel: Parcel, flags: Int) {           //分解数据
        parcel.writeString(no)
        parcel.writeString(name)
        parcel.writeSerializable(birthday)
    }
    override fun describeContents(): Int=0
        companion object CREATOR : Parcelable.Creator<Student>{        //重组生成对象
```

```
            override fun createFromParcel(parcel: Parcel): Student {
                return Student(parcel)
            }
            override fun newArray(size: Int): Array<Student?>{
                return arrayOfNulls(size)
            }
        }
    }
```

但是,这种表示方式非常复杂,在 Kotlin 中提供了一种更简单的表示方式,即利用@Parcelize 标注实体类。形式如下:

```
@Parcelize
data class Student(val no: String,val name: String,val birthday: LocalDate): Parcelable
```

在使用@Parcelize 标注时,需要对模块的构建配置文件 build.gradle.kts 增加插件 org.jetbrains.kotlin.plugin.parcelize,为了实现目录管理,需要进行两个步骤的配置。

第 1 步,在 libs.versions.toml 文件中确保存在如下配置:

```
[versons]
kotlin="1.9.25"
[plugins]
kotlin-parcelize={id ="org.jetbrains.kotlin.plugin.parcelize", version.ref="kotlin" }
```

第 2 步,在模块的构建配置文件 build.gradle.kt 中增加如下配置:

```
plugins {
    …
    id(libs.plugins.kotlin.parcelize.get().pluginId)
}
```

一旦数据实体类定义完毕,就可以在发送方的 Activity 中执行如下代码:

```
val intent=Intent(MainActivity@this,FirstActivity: : class.java)
intent.putExtra("student", Student("6001232","李四", LocalDate.of(2005,3,20)))
startActivity(intent)
```

接收方的 Activity 可以根据关键字接收数据,形式如下:

```
val data:Student?=intent.getParcelableExtra("student",Student::class.java)
```

例 3-5 Parcelable 方式传递对象的应用实例。
(1) 定义数据实体类 Student,代码如下:

```
//模块 Ch03_06 定义数据实体类
@Parcelize
data class Student(val no: String,val name: String,val birthday: LocalDate): Parcelable
```

· 58 ·

(2) 定义数据的发送方的 MainActivity,代码如下:

```kotlin
//模块 Ch03_06 定义数据发送方 MainActivity.kt
class MainActivity : ComponentActivity() {
    override fun onCreate(savedInstanceState: Bundle?) {
        super.onCreate(savedInstanceState)
        enableEdgeToEdge()
        setContent {
            Chapter03Theme {
                Scaffold(modifier=Modifier.fillMaxSize()) {
                    innerPadding ->Column(modifier=Modifier.padding(innerPadding)
                        .fillMaxSize(),
                        horizontalAlignment=Alignment.CenterHorizontally,
                        verticalArrangement=Arrangement.Center) {
                        Button(onClick={
                            sendData()                          //调用发送数据的函数
                        }) {
                            Text("发送数据",fontSize =30.sp)
                        }
                    }
                }
            }
        }
    }
    private fun sendData(){
        val intent=Intent(this,FirstActivity::class.java)
        val student=Student("910023","张三", LocalDate.of(2001,5,12))
        intent.putExtra("student",student)
        startActivity(intent)
    }
}
```

(3) 定义接收数据的 FirstActivity,代码如下:

```kotlin
//模块 Ch03_06 定义数据接收方的 FirstActivity.kt
class FirstActivity: ComponentActivity() {
    override fun onCreate(savedInstanceState: Bundle?) {
        super.onCreate(savedInstanceState)
        enableEdgeToEdge()
        //获取可序列化对象
        val data:Student?=intent.getParcelableExtra("student",Student::class.java)
        setContent {
            Chapter03Theme {
                Scaffold(modifier=Modifier.fillMaxSize()) {
                    innerPadding ->Box(modifier=Modifier.padding(innerPadding)
                        .fillMaxSize().padding(20.dp),contentAlignment=Alignment
                        .Center){
                        Text(
                            text="接收的学生的信息是: $data !",
                            fontSize=24.sp
```

```
                    )
                }
            }
        }
    }
}
```

运行结果如图3-9所示。

(a) 数据发送方　　　　(b) 数据接收方

图3-9　用Parcelable方式传递数据实例的运行结果

3.3.4　数据的返回

在Activity之间可以传递数据。当从一个Activity跳转到另外一个Activity时,默认情况下,按移动终端的Back键,可以实现从已跳转的Activity返回前一个Activity。如果返回前一个Activity,要求同时返回传递的数据。当然,也可以通过Intent直接传递数据,但是这种传递数据的方式无法区分数据是从哪个Activity返回的。按Back键直接返回上一个Activity时并不是将数据返回前一个Activity,因此在实际中若需要将数据返回前一个Activity,就需要做如下处理。

(1) 从前一个Activity跳转到其他Activity时,可以利用ActivityResultLauncher对象约定从其他Activity返回的相关操作。ActivityResultLauncher对象是通过调用Activity的registerForActivityResult函数来创建的。registerForActivityResult有两个参数。

第一个参数是ActivityResultContract,表示返回Activity返回结果的约定类型,可以设置为ActivityResultContracts.StartActivityForResult,表示启动活动并要求返回的约定。其中,将原始的Intent作为输入,活动返回结果的ActivityResult作为输出。

第二个参数是ActivityResultCallback,表示返回Activity时传递处理的具体操作。创建ActivityResultLauncher对象的代码如下:

```kotlin
val resultLauncher=registerForActivityResult(
    ActivityResultContracts.StartActivityForResult()) {
    it: ActivityResult ->if (it.resultCode==Activity.RESULT_OK) {
        //从其他活动返回的操作
    }
}
```

(2) 对于前一个 Activity,还需要定义跳转并传递数据到其他活动的操作,代码如下:

```kotlin
val intent=Intent(this,OtherActivity::class.java)    //定义跳转到其他活动的意图
intent.putExtra("DATA",info)                          //设置发送的数据
resultLauncher.launch(intent)                         //加载运行意图
```

(3) 对于返回的 Activity,需要做一些处理,让它在返回前一个 Activity 时可以携带数据。为了返回前一个 Activity,通过调用 setResult 函数实现。setResult 函数中传递两个实参,其中一个是用整数表示结果码,表示返回前一个活动的处理结果。结果值由两个值构成:Activity.RESULT_OK 表示处理成功;Activity.RESULT_CANCELED 表示处理取消。如果需要返回数据,必须通过第二个实参的 Intent 对象来实现。最后通过调用 finish()函数来结束这个返回 Activity。例如:

```kotlin
val intent=Intent()
intent.putExtra("data","数据")                        //设置返回的数据的键值对
setResult(Activity.RESULT_OK,intent)                  //设置结果码
finish()                                              //结束 Activity
```

如果只是单纯地返回前一个 Activity,并没有携带返回数据,也可以写成

```kotlin
setResult(RESULT_OK)
finish()
```

例 3-6 返回数据的应用实例。实现从一个 MainActivity 分别跳转到其他两个 Activity,其他两个 Activity 通过按 Back 键的方式返回时,可以分别返回字符串给 MainActivity。

(1) 定义 MainActivity,代码如下:

```kotlin
//模块 Ch03_07 MainActivity.kt
class MainActivity : ComponentActivity() {
    private lateinit var resultLauncher:ActivityResultLauncher<Intent>
    override fun onCreate(savedInstanceState: Bundle?) {
        super.onCreate(savedInstanceState)
        enableEdgeToEdge()
        resultLauncher=registerForActivityResult(
            ActivityResultContracts.StartActivityForResult()) {
            it: ActivityResult ->if (it.resultCode==Activity.RESULT_OK) {
                val returnData=it.data?.getStringExtra("RESULT")
                Toast.makeText(this, "$returnData", Toast.LENGTH_LONG).show()
            }
        }                                             //配置从其他活动返回的约定操作
        setContent {
            Chapter03Theme {
```

```
                Scaffold(modifier=Modifier.fillMaxSize()) {
                    innerPadding ->Column(modifier=Modifier.padding(innerPadding)
                        .fillMaxSize(),
                        horizontalAlignment=Alignment.CenterHorizontally,
                        verticalArrangement=Arrangement.Center) {
                        Button(onClick={
                            turnTo(FirstActivity::class.java,"从 MainActivity 跳转到
                                FirstActivity")
                        }) {
                            Text(text="跳转第一个活动",fontSize=24.sp)
                        }
                        Spacer(modifier=Modifier.padding(bottom=20.dp))
                        Button(onClick={
                            turnTo(SecondActivity::class.java,"从 MainActivity 跳转到
                                SecondActivity")
                        }) {
                            Text(text="跳转第二个活动",fontSize=24.sp)
                        }
                    }
                }
            }
        }
    }
    private fun <T>turnTo(activityType:Class<T>,info:String){
        val intent=Intent(this,activityType)                    //定义意图
        intent.putExtra("DATA",info)                            //设置发送的数据
        resultLauncher.launch(intent)                           //加载意图
    }
}
```

在 MainActivity 中定义了两个按钮,单击后,分别跳转到 FirstActivity 和 SecondActivity。在此处定义了一个通用方法 turnTo()函数,表示跳转到参数指定的类对应的 Activity。如果从其他活动返回 MainActivity,则执行 ActivityResultLauncher 对象约定的返回处理。在 MainActivity 中,如果判断是从 FirstActivity 返回,则要求接收从 FirstActivity 返回的数据,并通过 Toast 显示相应的消息提示;如果从 SecondActivity 返回 MainActivity,则调用 Toast 显示文本提示消息。

(2) 定义 FirstActivity,代码如下:

```
//模块 Ch03_07 FirstActivity.kt
class FirstActivity : ComponentActivity() {
    override fun onCreate(savedInstanceState: Bundle?) {
        super.onCreate(savedInstanceState)
        enableEdgeToEdge()
        val data=intent.getStringExtra("DATA")
        setContent {
            Chapter03Theme {
                Scaffold(modifier=Modifier.fillMaxSize()) {
                    innerPadding ->Column(modifier=Modifier.padding(innerPadding)
                        .fillMaxSize().padding(20.dp),
```

```
                    horizontalAlignment=Alignment.CenterHorizontally,
                    verticalArrangement=Arrangement.Center){
                    Text(text="接收的信息是$data !", fontSize =24.sp)
                    Button(onClick={
                        val intent=Intent()
                        intent.putExtra("RESULT","FirstActivity返回MainActivity")
                                                                    //返回数据
                        setResult(Activity.RESULT_OK,intent)
                        finish()
                    }){
                        Text("返回")
                    }
                }
            }
        }
    }
}
```

定义SecondActivity,代码如下：

```
//模块 Ch03_07SecondActivity.kt
class SecondActivity: ComponentActivity() {
    override fun onCreate(savedInstanceState: Bundle?) {
        super.onCreate(savedInstanceState)
        enableEdgeToEdge()
        val data=intent.getStringExtra("DATA")
        setContent {
            Chapter03Theme {
                Scaffold(modifier=Modifier.fillMaxSize()) {
                    innerPadding ->Column(modifier=Modifier.padding(innerPadding)
                        .fillMaxSize().padding(20.dp),horizontalAlignment=Alignment
                        .CenterHorizontally,
                        verticalArrangement=Arrangement.Center) {
                        Text(
                            text="接收的信息是$data !",
                            fontSize=24.sp
                        )
                        Button(onClick={
                            setResult(Activity.RESULT_OK)          //没有数据返回
                            finish()
                        }){
                            Text("返回")
                        }
                    }
                }
            }
        }
    }
}
```

在 FirstActivity.kt 和 SecondActivity.kt 分别处理了带数据的返回和没有带数据的返回。

运行结果如图 3-10 所示。

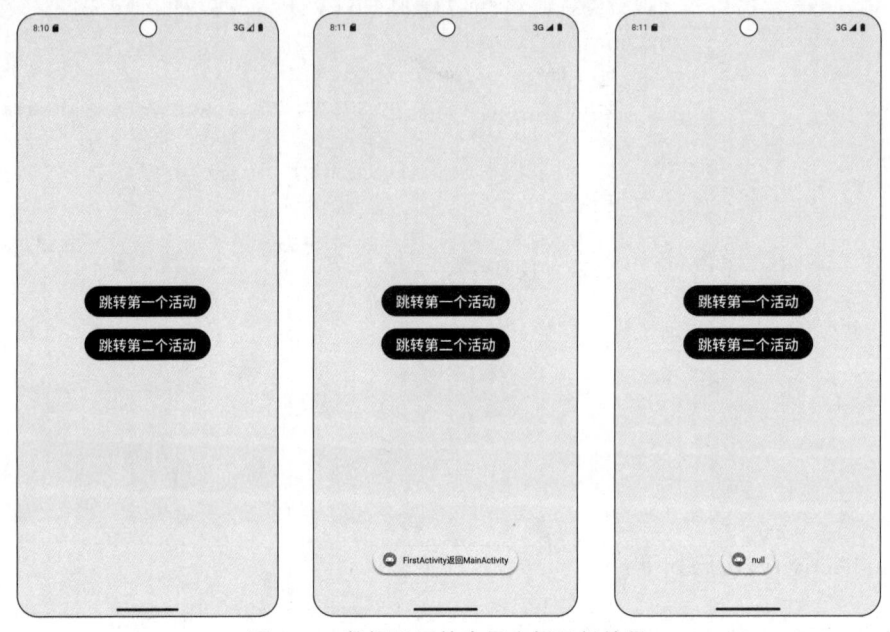

图 3-10 数据返回的应用实例运行结果

3.4 Activity 的生命周期

Activity 的生命过程中多种状态相互转换。Activity 不同状态的转换过程中涉及多种相关概念。本节会对这些相关概念进行介绍。

3.4.1 Activity 的返回栈

活动的返回栈(Back Stack)又称任务栈,用于实现对 Activity 的管理。这与活动的执行情况相关联。当启动一个 Activity,使得 Activity 显示在屏幕,就意味着这个 Activity 进入返回栈,并处于栈顶的位置。当执行 finish()或异常或其他的函数使得该 Activity 退出时,该 Activity 从返回栈的栈顶移除,使得返回栈的下一个 Activity 成为新的栈顶活动,成为屏幕的当前界面。当新的 Activity 启动入栈时,原有的 Activity 会被压入栈的下一层。一个 Activity 在栈中的位置变化反映了它在不同状态间的转换。

这种返回栈的执行过程如图 3-11 所示。

3.4.2 Activity 的启动方式

可以通过 Activity 的启动方式对返回栈管理 Activity 的方式有更清楚的认识。在配置清单文件 AndroidManifest.xml 文件中对 Activity 对象配置 android:launchMode 属性,用来指定 Activity 的加载模式,支持 5 种加载模式:standard、singleTop、singleTask、singleInstance 和 singleInstancePerTask。通过 Activity 的交互,了解 Activity 加载方式的作用。定义 MainActivity 和 OtherActivity,分别提供按键实现相互的调用。定义 MainActivity 和 OtherActivity 的 Activity 类的代码如下:

```kotlin
//模块 Ch03_08 MainActivity.kt
class MainActivity: ComponentActivity() {
    override fun onCreate(savedInstanceState: Bundle?) {
        super.onCreate(savedInstanceState)
        enableEdgeToEdge()
        setContent {
            Chapter03Theme {
                Scaffold(modifier=Modifier.fillMaxSize()) {
                    innerPadding ->Box(modifier=Modifier.padding(innerPadding).fillMaxSize(),
                        contentAlignment=Alignment.Center){
                        Button(onClick={                                            //跳转按钮
                            turnTo()                                                //跳转处理
                        }) {
                            Text("MainActivity->OtherActivity",fontSize=24.sp)
                        }
                    }
                }
            }
        }
    }
    private fun turnTo(){
        val intent=Intent(this,OtherActivity::class.java)
        startActivity(intent)
    }
}
```

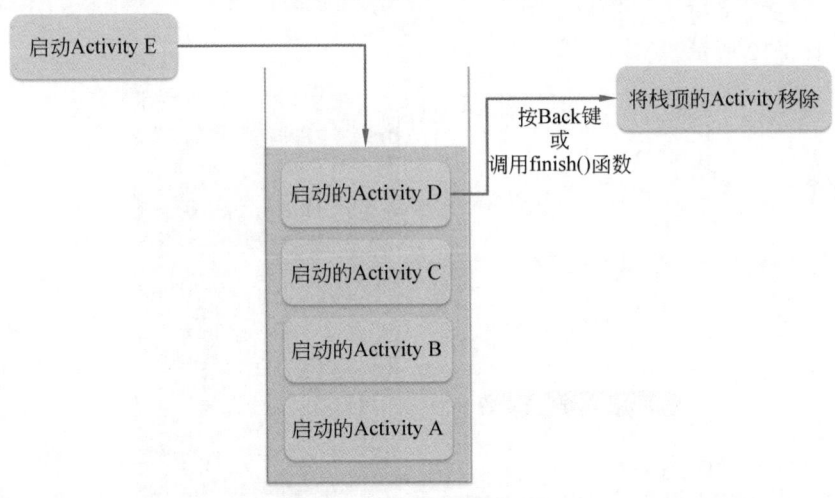

图 3-11 返回栈的执行过程

在 MainActivity 中定义了一个按钮,单击该按钮可从 MainActivity 跳转到 OtherActivity 表示的界面。OtherActivity 定义了一个 FloatingActionButton 按钮。单击该按钮,可以返回 MainActivity。OtherActivity 的代码定义如下:

```kotlin
//模块 Ch03_08 OtherActivity.kt
class OtherActivity : ComponentActivity() {
    override fun onCreate(savedInstanceState: Bundle?) {
```

```
        super.onCreate(savedInstanceState)
        enableEdgeToEdge()
        setContent {
            Chapter03Theme {
                Scaffold(
                    modifier=Modifier.fillMaxSize(),
                    floatingActionButton={
                    FloatingActionButton(onClick={
                        backToMain()                              //返回 MainActivity 界面
                    }) {
                        Icon(Icons.AutoMirrored.Filled.ArrowBack,
                            contentDescription="back")
                     }
                }) {
                        innerPadding ->Box(modifier=Modifier.padding(innerPadding)
                            .fillMaxSize().background(colorResource(id=R.color.teal_200)),
                            contentAlignment=Alignment.Center){
                            Text("OtherActivity",fontSize=30.sp,color=Color.White)
                        }
                    }
                }
            }
    }
    private fun backToMain(){
        val intent=Intent(this,MainActivity::class.java)
        startActivity(intent)
    }
}
```

运行结果如图 3-12 所示。

图 3-12　Activity 的运行效果

1. standard 方式

Activity 的加载默认方式就是 standard 方式。如果 AndroidManifest.xml 对 Activity 没有任何加载方式的设定，则默认指定 Activity 的加载方式为 standard 方式。当然，也可以在 Activity 设置中增加 android：launchMode 属性并定义为 standard。

standard 方式启动一个 Activity 会创建一个新 Activity，并将这个新的 Activity 推进返回栈中。对上述的 MainActivity 和 OtherActivity 在 AndroidManifest.xml 配置，代码如下：

```xml
<?xml version="1.0" encoding="utf-8"?>
<manifest xmlns: android="http://schemas.android.com/apk/res/android"
    package="chenyi.book.android.ch03_10">
    <application ...><!--application 的配置省略 -->
        <activity
            android: name=".OtherActivity"
            android: label="@string/title_other_activity"
            android: launchMode="standard" /><!--配置 OtherActivity -->
        <activity
            android: name=".MainActivity"
            android: label="@string/title_main_activity"
            android: launchMode="standard"
            android: exported="true"><!--配置 MainActivity -->
            <intent-filter>
                <action android: name="android.intent.action.MAIN" />
                <category android: name="android.intent.category.LAUNCHER" />
            </intent-filter>
        </activity>
    </application>
</manifest>
```

在这种配置下，MainActivity 和 OtherActivity 启动均为 standard 方式运行。首先要显示 MainActivity，将 MainActivity 实例先压入栈中；单击 MainActivity 按钮，进入 OtherActivity，此时是将 OtherActivity 实例推入栈中，然后通过 OtherActivity 的 FloatingActionButton 按钮启动 MainActivity。这时会创建一个新的 MainActivity 实例并压入返回栈中。这一执行过程返回栈如图 3-13 所示。

当按 Back 键 3 次，退出该应用时，可以观察到显示顺序是 MainActivity→OtherActivity→MainActivity，然后退出应用。

2. singleTop 方式

singleTop 方式表示在这种模式下，当启动目标 Activity 已经在返回栈顶时，系统会直接复用已有的 Activity 实例，不会创建新的 Activity 实例。但是，当启动目标不再返回栈顶时，Android 会为目标 Activity 创建一个新的实例，并将 Activity 添加到当前返回栈中，成为栈顶新 Activity。这时将系统配置清单 AndroidManifest.xml 文件配置，代码如下：

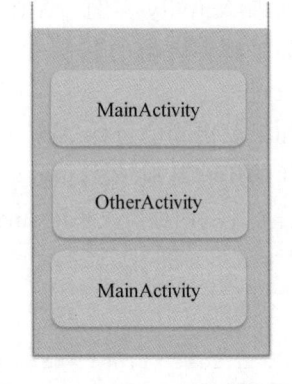

图 3-13 standard 方式下返回栈

```xml
<?xml version="1.0" encoding="utf-8"?>
<manifest xmlns: android="http://schemas.android.com/apk/res/android"
    package="chenyi.book.android.ch03_10">
    <application ...><!--application 的配置省略 -->
```

```xml
        <!--配置 OtherActivity,加载模式是 singleTop -->
        <activity
            android: name=".OtherActivity"
            android: label="@string/title_other_activity"
            android: launchMode="singleTop" />
        <!--配置 MainActivity,加载模式是 singleTop -->
        <activity
            android: name=".MainActivity"
            android: label="@string/title_main_activity"
            android: launchMode="singleTop"
            android: exported="true">
            <intent-filter>
                <action android: name="android.intent.action.MAIN" />
                <category android: name="android.intent.category.LAUNCHER" />
            </intent-filter>
        </activity>
    </application>
</manifest>
```

在这种配置下,MainActivity 和 OtherActivity 的启动均为 singleTop 方式。将 MainActivity 实例压入栈中,显示 MainActivity 界面;单击 MainActivity 按钮,进入 OtherActivity,此时是将 OtherActivity 实例推入栈中,然后通过 OtherActivity 的 FloatingActionButton 按钮启动 MainActivity,这是因为当前的栈顶 Activity 是 OtherActivity,不是 MainActivity。在这种情况下还会创建一个新的 MainActivity 实例并将其压入返回栈中。这样的执行过程与图 3-13 所示的运行结果一致。

当按 Back 键 3 次时,在观察到显示顺序 MainActivity→OtherActivity→MainActivity 后,退出应用。

3. singleTask 方式

采用 singleTask 方式会使在同一个返回栈只有一个 Activity 实例。启动 Activity 时,分为如下情况。

如果 Activity 实例不存在,系统会创建目标 Activity 实例,并将它加入返回栈顶。

如果将要启动的目标 Activity 已经存在,则返回栈的栈顶,此时会直接复用在返回栈栈顶的 Activity。

如果要启动的目标 Activity 已经存在,但没有处于返回栈的栈顶,则系统将会把位于该 Activity 实例上面的所有其他 Activity 实例移出返回栈,以使目标 Activity 实例成为栈顶的 Activity。

若上述的两种情况采用 singleTask 加载模式,可以将系统的配置清单 AndroidManifest.xml 文件配置,代码如下:

```xml
<?xml version="1.0" encoding="utf-8"?>
<manifest xmlns: android="http://schemas.android.com/apk/res/android"
    package="chenyi.book.android.ch03_10">
    <application …><!--application 的配置省略 -->
        <!--配置 OtherActivity,加载模式 singleTask -->
        <activity
            android: name=".OtherActivity"
            android: label="@string/title_other_activity"
            android: launchMode="singleTask" />
        <!--配置 MainActivity,加载模式 singleTask -->
        <activity
            android: name=".MainActivity"
            android: label="@string/title_main_activity"
```

```xml
            android: launchMode="singleTask"
            android: exported="true">
        <intent-filter>
            <action android: name="android.intent.action.MAIN" />
            <category android: name="android.intent.category.LAUNCHER" />
        </intent-filter>
    </activity>
  </application>
</manifest>
```

在这种配置下，MainActivity 和 OtherActivity 均以 singleTask 方式启动。把 MainActivity 实例压入栈中，显示 MainActivity 界面；单击 MainActivity 按钮，进入 OtherActivity，此时是将 OtherActivity 实例推入返回栈中，然后通过 OtherActivity 的 FloatingActionButton 按钮启动 MainActivity。在这种情况下，会将第一个 MainActivity 之前所有的 Activity 推出返回栈，再将 OtherActivity 推出返回栈。这一过程中，返回栈的情况如图 3-14 所示。

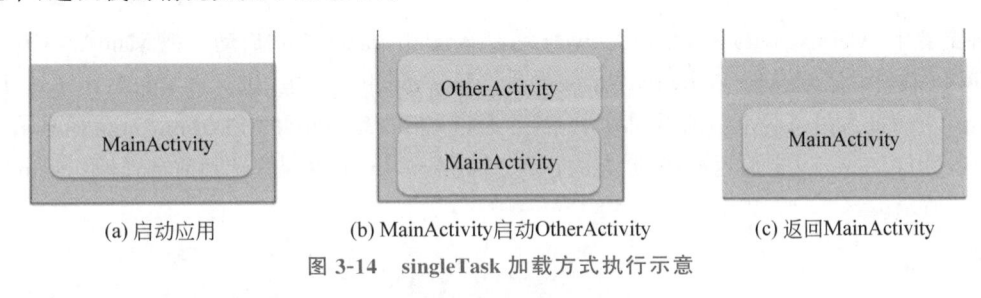

(a) 启动应用　　(b) MainActivity启动OtherActivity　　(c) 返回MainActivity

图 3-14　singleTask 加载方式执行示意

当再次显示 MainActivity 时，按 Back 键，会发现直接退出了移动应用。因为按 Back 键之前，返回栈中只有 MainActivity 实例。退出栈顶，此时返回栈为空。

4. singleInstance 方式

singleInstance 方式非常特殊。应用过程中每个 Activity 只能有一个唯一的实例，只要应用的执行过程中创建了 Activity 的实例，就不会再创建新的 Activity 实例对象。造成这样的情况是因为每次创建一个 Activity 实例，都会使用一个全新的返回栈来加载具有 singleInstance 模式的 Activity 实例。具体处理情况如下所示。

如果将要启动的目标 Activity 不存在，系统会先创建一个全新的返回栈，再创建目标 Activity 的实例，并将它加入新的返回栈的栈顶；采用 singleInstance 模式加载的 Activity 所在的返回栈将只包括该 Activity。

如果将要启动的目标 Activity 已经存在，无论它位于哪个应用程序中，无论它位于哪个返回栈中，系统都会把该 Activity 所在的返回栈转到前台，从而使用该 Activity 显示出来。

加载的目标 Activity 一直都位于返回栈的栈顶。

上述两种情况均采用 singleInstance 加载模式，可以将系统的配置清单 AndroidManifest.xml 文件配置，代码如下：

```xml
<?xml version="1.0" encoding="utf-8"?>
<manifest xmlns: android="http://schemas.android.com/apk/res/android"
    package="chenyi.book.android.ch03_10">
    <application…>　<!--application 的配置省略 -->
        <!--配置 OtherActivity,加载模式为 singleInstance -->
```

```xml
<activity
    android:name=".OtherActivity"
    android:label="@string/title_other_activity"
    android:launchMode="singleInstance" />
<!--配置MainActivity,加载模式为singleInstance-->
<activity
    android:name=".MainActivity"
    android:label="@string/title_main_activity"
    android:launchMode="singleInstance"
    android:exported="true">
    <intent-filter>
        <action android:name="android.intent.action.MAIN" />
        <category android:name="android.intent.category.LAUNCHER" />
    </intent-filter>
</activity>
</application>
</manifest>
```

在这种配置下,MainActivity 和 OtherActivity 均以 singleInstance 方式启动。把 MainActivity 实例压入一个新的返回栈中,显示 MainActivity 界面;单击 MainActivity 按钮,进入 OtherActivity。此时是将 OtherActivity 实例推入另外一个新的返回栈中,然后通过 OtherActivity 的 FloatingActionButton 按钮启动 MainActivity,OtherActivity 退出返回栈,原有的包含 MainActivity 的返回栈转到前台,显示 MainActivity 界面,如图 3-15 所示。

图 3-15　singleInstance 加载方式的执行示意

当再次显示 MainActivity 时,按 Back 键,会发现直接退出了移动应用。因为返回栈中只有 MainActivity 实例。退出栈顶,则返回栈为空。

Android 12 新增了 singleInstancePerTask 加载方式。设置 singleInstancePerTask 加载方式的 Activity 在第一次启动时会创建任务栈,并在该任务栈中创建该 Activity 的对象实例。该 Activity 的对象实例成为该任务栈的根 Activity。当再次启动这个 Activity 实例时,它不会重复启动,而是回调转向前台。加载方式为 singleInstancePerTask 的 Activity 可以在不同的任务栈的多个实例中启动。

3.4.3　Activity 的生命周期

每个 Activity 都有生命周期,生命周期有 4 种状态。Activity 所处的状态与运行过程中 Activity 的执行情况相对应。

(1) 运行状态。在启动一个新 Activity 时,会创建这个 Activity 的实例,并进入返回栈的栈顶。这时,Activity 界面会在屏幕显示,以可见状态和用户交互。

(2) 暂停状态。当 Activity 被另一个透明或者 Dialog 样式的 Activity 覆盖时，该 Activity 仍在返回栈中，但已经不再返回栈顶。它依然与窗口管理器保持连接，系统继续维护其内部状态，虽然它仍然可见，但已经失去了焦点，故不可与用户交互。

(3) 停止状态。当 Activity 被另一个 Activity 覆盖、失去焦点、不可见时，该 Activity 处于停止状态。

(4) 销毁状态。当 Activity 被系统杀死或者没有被启动时，该 Activity 处于销毁状态。这时，Activity 已经从返回栈中退出。如果某个 Activity 在返回栈的栈底，长期没有激活，则在移动终端内存不足时，Android 系统也会杀死这个 Activity 并释放空间。这时 Activity 也会从返回栈中退出。

在 Activity 的生命周期中定义了生命周期的方法，这些方法与生命周期的状态是相关的，如图 3-16 所示。

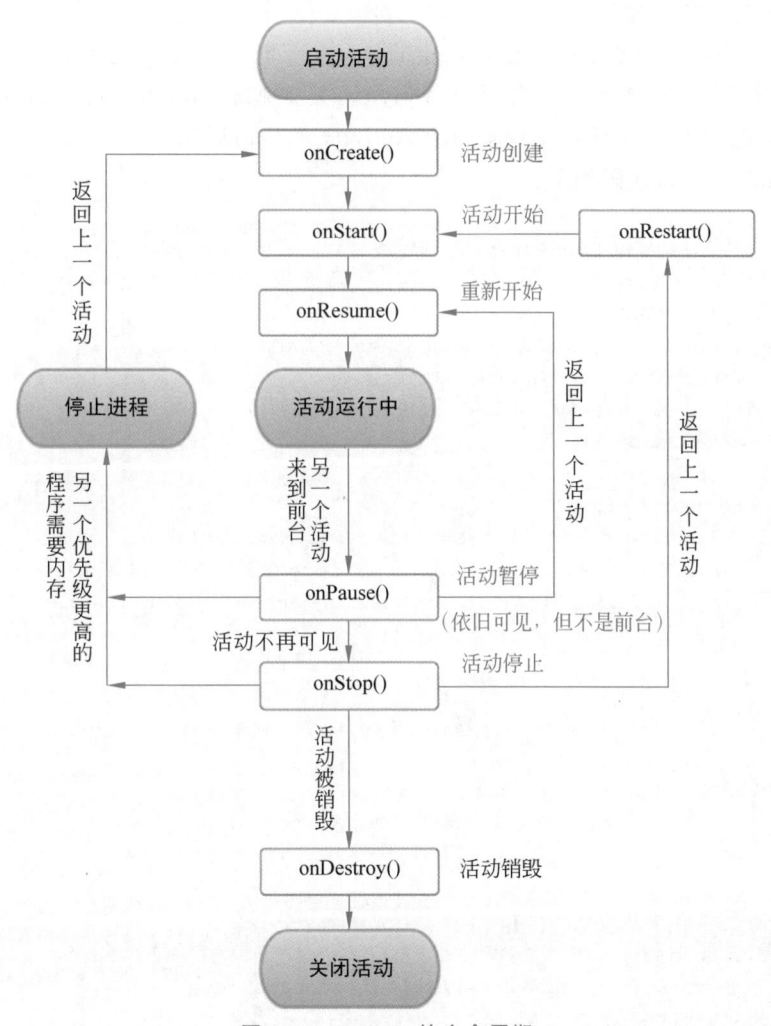

图 3-16　Activity 的生命周期

(1) onCreate(Bundle)。它是启动 Activity 第一次创建调用的函数，用来执行许多初始化工作。Bundle 作为参数传递，可能包括上一个 Activity 的动态状态信息，例如与用户界面外观关联的状态等。

(2) onStart。它在 onCreate 函数或 onRestart 函数后被立刻调用。这个函数的调用确保 Activity 对用户可见。一旦该函数被调用，如果该 Activity 移动到返回栈的栈顶，onResume 函数就会被调用，或由于事件发生 onStop 函数调用,导致其他 Activity 进入返回栈。

(3) onResume。它用于确保 Activity 在返回栈的栈顶并处于运行状态,且可以与用户交互。

(4) onPause。它表明 Activity 处于暂停状态。该函数会在 onResume 函数恢复一个 Activity 或 onStop 函数停止一个 Activity 时调用,使得 Activity 可以重返前台可见或对用户不可见。在执行该函数时可能会发生存储持久数据。为了避免 Activity 转换的延迟,耗时的操作如存储数据或执行网络连接在这个函数应该避免。

(5) onStop。当 Activity 对用户不可见时,处于停止状态。之后可执行 onRestart 函数或 onDestroy 函数。

(6) onDestroy。Activity 处于销毁状态。Activity 会执行 finish 函数或已经完成任务,运行时终止 Activity 并释放内存,或者设备配置发生变化(例如设备的方向发生变化),都会导致 onDestroy 函数的调用,使得 Activity 进入销毁状态。

(7) onRestart。当运行时重启一个已经停止的 Activity 时,该函数会被调用。

例 3-7　生命周期的应用实例。定义两个 Activity,其中 MainActivity 可以启动 OtherActivity,并将 OtherActivity 设置为 Dialog 样式。观察两个 Activity 的生命周期。

(1) 定义 MainActivity,代码如下:

```kotlin
//模块 Ch03_09 主活动的定义 MainActivity.kt
class MainActivity : ComponentActivity() {
    private val TAG="TAG"
    override fun onCreate(savedInstanceState: Bundle?) {
        super.onCreate(savedInstanceState)
        Log.d(TAG,"MainActivity:onCreate()")
        enableEdgeToEdge()
        setContent {
            Chapter03Theme {
                Scaffold(modifier=Modifier.fillMaxSize()) {
                    innerPadding ->Box(modifier=Modifier.padding(innerPadding).
                    fillMaxSize(),contentAlignment=Alignment.Center){
                        Button(onClick={
                            turnTo()
                        }){
                            Text("MainActivity跳转 OtherActivity",fontSize=24.sp)
                        }
                    }
                }
            }
        }
    }
    private fun turnTo(){
        val intent=Intent(this,OtherActivity::class.java)
        startActivity(intent)
    }
    override fun onStart() {
        super.onStart()
        Log.d(TAG,"MainActivity:onStart()")
    }
```

```kotlin
    override fun onResume() {
        super.onResume()
        Log.d(TAG,"MainActivity:onResume()")
    }
    override fun onPause() {
        super.onPause()
        Log.d(TAG,"MainActivity:onPause()")
    }
    override fun onStop() {
        super.onStop()
        Log.d(TAG,"MainActivity:onStop()")
    }
    override fun onDestroy() {
        super.onDestroy()
        Log.d(TAG,"MainActivity:onDestroy()")
    }
    override fun onRestart() {
        super.onRestart()
        Log.d(TAG,"MainActivity:onRestart()")
    }
}
```

(2) 定义 OtherActivity,代码如下:

```kotlin
//模块 Ch03_09 其他活动的定义 OtherActivity.kt
class OtherActivity : ComponentActivity() {
    private val TAG="TAG"
    override fun onCreate(savedInstanceState: Bundle?) {
        super.onCreate(savedInstanceState)
        Log.d(TAG,"OtherActivity:onCreate()")
        enableEdgeToEdge()
        setContent {
            Chapter03Theme {
                Scaffold(modifier=Modifier.fillMaxSize()) {
                    innerPadding ->Box(
                        modifier=Modifier.padding(innerPadding).fillMaxSize(),
                        contentAlignment=Alignment.Center
                    ) {
                        Text("OtherActivity活动", fontSize=36.sp)
                    }
                }
            }
        }
    }
    override fun onStart() {
        super.onStart()
        Log.d(TAG,"OtherActivity:onStart()")
    }
    override fun onResume() {
```

```kotlin
        super.onResume()
        Log.d(TAG,"OtherActivity:onResume()")
    }
    override fun onPause() {
        super.onPause()
        Log.d(TAG,"OtherActivity:onPause()")
    }
    override fun onStop() {
        super.onStop()
        Log.d(TAG,"OtherActivity:onStop()")
    }
    override fun onDestroy() {
        super.onDestroy()
        Log.d(TAG,"OtherActivity:onDestroy()")
    }
    override fun onRestart() {
        super.onRestart()
        Log.d(TAG,"OtherActivity:onRestart()")
    }
}
```

在 AndroidManifest.xml 中指定 OtherActivity 的主题 theme 为 "@android:style/Theme.DeviceDefault.Dialog",表示对话框的显示样式。AndroidManifest.xml 的部分代码如下:

```xml
<!--模块 Ch03_11 配置清单 AndroidManifest.xml -->
<?xml version="1.0" encoding="utf-8"?>
<manifest xmlns:android="http://schemas.android.com/apk/res/android"
    package="chenyi.book.android.ch03_11">
    <application……>   …   <!--省略 -->
        <activity
            android:name=".OtherActivity"
            android:label="OtherActivity界面"
            android:theme="@android:style/Theme.DeviceDefault.Dialog" />
    </application>
</manifest>
```

当启动应用时,初始运行 MainActivity,使 MainActivity 处于运行状态,可以执行单击按钮的操作。此时运行结果和日志记录如图 3-17 所示。在运行状态中,MainActivity 在屏幕前台可见。

单击 MainActivity 的按钮,启动 OtherActivity。因为 OtherActivity 配置成对话框样式,因此屏幕将 OtherActivity 以对话框的样式显示界面和日志记录,如图 3-18 所示。MainActivity 调用了 onPause 函数,在屏幕的对话框背景中可以看到 MainActivity 此时处于暂停状态。至于屏幕显示 OtherActivity 对话框,这是因为 OtherActivity 依次调用 onCreate 函数→onStart 函数→onResume 函数,进入运行状态,成为返回栈的栈顶活动。

如果按 Back 键退出 OtherActivity,则一方面,MainActivity 执行 onResume 函数恢复运行状态,成为屏幕的前台界面;另一方面,OtherActivity 依次执行 onPause 函数→onStop 函数→onDestroy,退出返回栈,执行活动销毁。日志记录如图 3-19 所示。

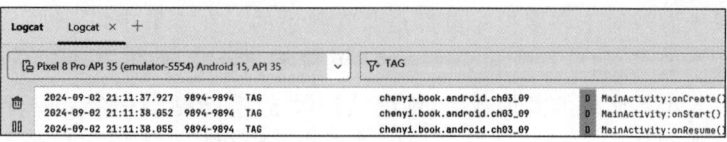

图 3-17 运行 MainActivity 的界面和日志

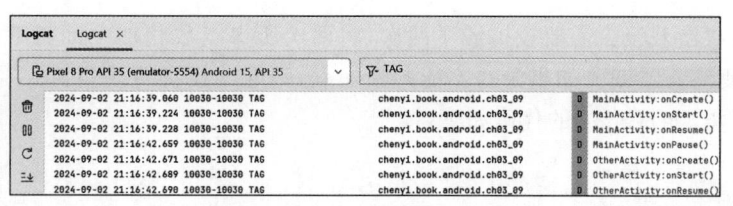

图 3-18 运行 OtherActivity 的界面和日志记录

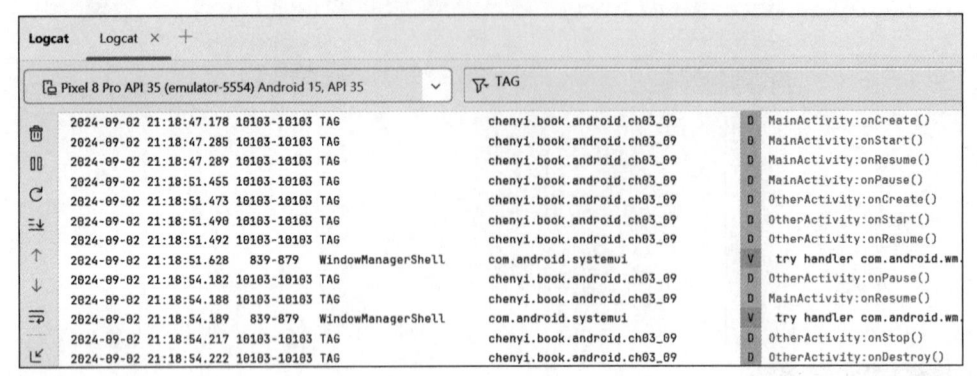

图 3-19　日志记录

3.5　掷骰子游戏

掷骰子游戏对于许多程序员都是非常熟悉的。在本节中设计如下：每个骰子有 6 个面，用 1～6 表示。掷两个骰子，如果点数和为 2、3 或 12，玩家就输了；如果点数和是 7 或者 11，玩家就赢了；如果点数和是上述 5 个数之外的其他数字，则继续掷骰子，直至抛出一个 7 或者掷出和上一次相同的点数。如果掷出的点数和是 7，玩家就输了。如果掷出的点数和与前一次掷出的点数和相同，玩家就赢了。这是一个非常简单的应用。

1. 定义游戏业务

首先，根据游戏的描述，可以发现投掷骰子游戏的结果存在 4 种状态：开始游戏、赢得游戏、输掉游戏和继续游戏。因此，根据这 4 种情况，定义一个游戏的状态的枚举类 GameStatus，代码如下：

```kotlin
//模块 dicegame 枚举类 GameStatus.kt
enum class GameStatus(val content:String,var point:Int=0){
    BEGIN("开始游戏",0),
    WIN("祝贺你,赢得了比赛",0),
    LOSE("别难过,这次输了,下次再努力",0),
    GOON("继续投掷骰子",0);
    fun changePoint(p:Int){
        this.point=p
    }
}
```

每次游戏至少需要投掷骰子两次，将两次所得点数相加判断游戏结果。如果游戏能一次获得输赢结果，则终止游戏。如果游戏不能判断输赢，则继续投掷直至得到结果。根据游戏的业务要求，定义游戏逻辑类 DiceGame 的代码如下：

```kotlin
//模块 dicegame 游戏主逻辑 DiceGame.kt
object DiceGame {
    var rand=Random()
    fun roll():Int=rand.nextInt(6)+1                              //模拟扔骰子
    fun judgeGame(first:Int,second:Int):GameStatus{               //判断游戏胜负
        var total=first+second
```

```kotlin
            if (total==7||total==11)
                return GameStatus.WIN
            else if (total==2||total==3||total==12){
                return GameStatus.LOSE
            } else {
                return GameStatus.GOON
            }
        }
        fun goOn(first:Int,second:Int,status:GameStatus):GameStatus{      //继续游戏处理
            var point=status.point
            var total=first+second
            when (total) {
                point->return GameStatus.WIN
                7->return GameStatus.LOSE
                 else ->{
                    status.changePoint(total)
                    return status
                }
            }
        }
    }
```

因为掷骰子游戏的整个过程只需要一个游戏对象对游戏进行控制和判定,因此将 DiceGame 定义为 object 对象类,即 DiceGame 类维护唯一一个自己对象,就是 DiceGame 本身。单例模式的实现。

定义一个 GameViewModel 组件,实现界面和数据模型之间的交互。DiceGameViewModel 类实现配置数据,并更新界面的作用,代码如下:

```kotlin
//DiceGameApp 的 GameViewModel.kt
class GameViewModel: ViewModel() {
    private val _first_image_output=MutableStateFlow(R.mipmap.empty)   //第一张图片
    val first_image_output=_first_image_output.asStateFlow()
    private val _second_image_output=MutableStateFlow(R.mipmap.empty)  //第二张图片
    val second_image_output=_second_image_output.asStateFlow()
    private val _gameStatus_ouput=MutableStateFlow(GameStatus.BEGIN)   //游戏状态
    val gameStatus_output=_gameStatus_ouput.asStateFlow()
    private fun getBy(dice:Int):Int=when(dice) {                       //根据点数转换成图片资源
        1->R.mipmap.one
        2->R.mipmap.two
        3->R.mipmap.three
        4->R.mipmap.four
        5->R.mipmap.five
        6->R.mipmap.six
        else ->R.mipmap.empty
    }
    fun play(){
        var first=DiceGame.roll()                                      //掷骰子
        var second=DiceGame.roll()
        _first_image_output.value=getBy(first)                         //转换成图片资源
        _second_image_output.value=getBy(second)
```

```
when(_gameStatus_ouput.value){                              //根据游戏状态处理
    GameStatus.BEGIN ->{
        _gameStatus_ouput.value=DiceGame.judgeGame(first,second)
                                                            //判断游戏
    }
    GameStatus.GOON ->{
        _gameStatus_ouput.value=DiceGame.goOn(first,second,_gameStatus_ouput
            .value)                                         //继续游戏
    }
    else ->{
        _gameStatus_ouput.value=_gameStatus_ouput.value     //记录游戏状态
    }
}
```

2. 交互控制

为了控制游戏,本次游戏设计了 3 个 Compose 可组合函数定义的界面:LogoScreen 表示启动界面时初始界面,GameScreen 表示玩游戏的界面,ResultScreen 表示显示游戏结果的界面。这 3 个 Compose 可组合函数的定义如下。

(1) 定义开始界面的 LogoScreen.kt,代码如下:

```
//模块 DiceGameApp 定义应用开始的界面 LogoScreen.kt
@Composable
fun LogoScreen(modifier:Modifier) {
    val context=LocalContext.current                        //当前的上下文
    Column(modifier=modifier.fillMaxSize(),
        horizontalAlignment=Alignment.CenterHorizontally,
        verticalArrangement=Arrangement.Center) {
        Text("骰子游戏", fontSize=54.sp)
        Spacer(modifier=Modifier.padding(bottom=50.dp))
        Button(onClick={
            val intent=Intent(context,GameActivity::class.java)
            context.startActivity(intent)
        }) {
            Text("开始游戏", fontSize=32.sp)
        }
    }
}
```

在 LogoScreen 展示了游戏名称,提供了按钮,通过单击该按钮进入游戏界面。将 LogoScreen 嵌入 MainActivity 活动,代码如下:

```
//模块 DiceGameApp 定义 MainActivity.kt
class MainActivity : ComponentActivity() {
    override fun onCreate(savedInstanceState: Bundle?) {
        super.onCreate(savedInstanceState)
```

```
        enableEdgeToEdge()
        setContent {
            Chapter03Theme {
                Scaffold(modifier=Modifier.fillMaxSize()) {
                    innerPadding ->LogoScreen(modifier=Modifier.padding(innerPadding))
                }
            }
        }
    }
}
```

运行结果如图 3-20 所示。

图 3-20 启动的初始界面

（2）定义游戏业务处理的界面 GameScreen，代码如下：

```
//模块 DiceGameApp 定义的游戏处理界面 GameScreen
@Composable
fun GameScreen(modifier: Modifier,gameViewModel: GameViewModel){
    val context=LocalContext.current
    var firstImage by remember { mutableStateOf(R.mipmap.empty) }
    var secondImage by remember { mutableStateOf(R.mipmap.empty) }
    var gameStatus by remember { mutableStateOf(GameStatus.BEGIN) }
    Box(contentAlignment=Alignment.Center,
        modifier=modifier.fillMaxSize().background(color=Color(0xf5, 0xb8, 0x44))){
        Column(modifier=Modifier.padding(10.dp)){
```

```
        Row(modifier=Modifier.padding(40.dp).fillMaxWidth(),
            verticalAlignment=Alignment.CenterVertically,
            horizontalArrangement=Arrangement.Center){
                Image(painter=painterResource(id=firstImage),
                    contentDescription="骰子 1",
                    modifier=Modifier.width(150.dp).height(150.dp))
                Spacer(modifier=Modifier.padding(end=10.dp))
                Image(painter=painterResource(id=secondImage),
                    contentDescription="骰子 2",
                    modifier=Modifier.width(150.dp).height(150.dp))
        }
        Spacer(modifier=Modifier.padding(top=100.dp))
        Button(onClick={
            gameViewModel.play()
            firstImage=gameViewModel.first_image_output.value
            secondImage=gameViewModel.second_image_output.value
            gameStatus=gameViewModel.gameStatus_output.value
            if (gameStatus==GameStatus.WIN||gameStatus==GameStatus.LOSE){
                val intent=Intent(context,ResultActivity::class.java)
                intent.putExtra("result","${gameStatus.content}")
                context.startActivity(intent)
            }
        },modifier=Modifier.fillMaxWidth()){
            Icon(imageVector=Icons.Filled.PlayArrow, contentDescription="图标")
            Text("掷骰子",fontSize=30.sp,textAlign=TextAlign.Center,color=Color
                .Green)
        }
    }
  }
}
```

运行结果如图 3-21 所示。

GameScreen 是核心可组合界面。通过单击按钮,模拟两次掷骰子,将两个骰子的点数在界面中显示出来。同时,它还承担了执行游戏逻辑处理交互的工作。如果游戏能获得输赢,则跳转到 ResultScreen,并传递输赢的结果。如果单击按钮不能一次判断成功,表示继续游戏,需要单击多次直至获得输赢结果。

(3) 定义显示游戏结果的界面 ResultScreen,代码如下:

```
//模块 DiceGameApp 定义的显示游戏结果的界面 ResultScreen
@Composable
fun ResultScreen(resultTxt:String,modifier:Modifier){
    Box(contentAlignment=Alignment.Center,modifier=modifier.fillMaxSize()
        .background(color=colorResource(id=android.R.color.holo_blue_bright))){
        Column(
            horizontalAlignment=Alignment.CenterHorizontally,
```

```
            verticalArrangement=Arrangement.Center
        ){
            Text(text="${resultTxt}",fontSize=30.sp)
        }
    }
}
```

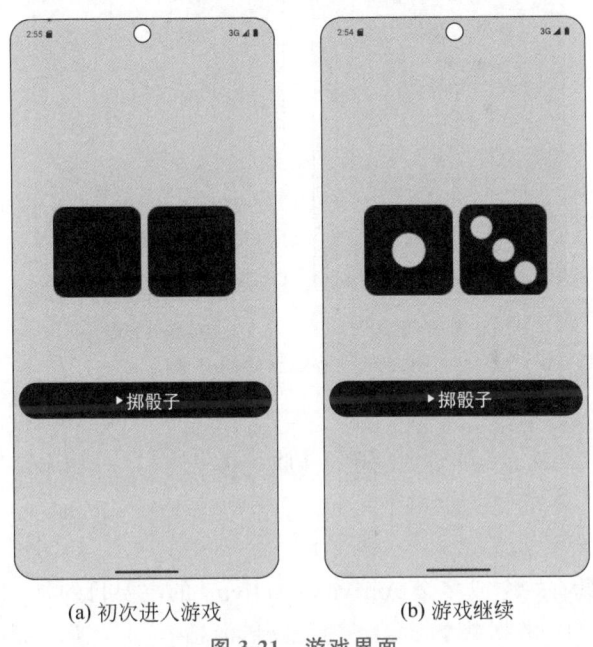

(a) 初次进入游戏　　　　(b) 游戏继续

图 3-21　游戏界面

在 ResultActivity 中嵌入 ResultScreen 界面,根据 GameActivity 活动的数据分别为 ResultScreen 加载不同的运行结果。ResultActivity 的代码如下:

```
//DiceGameApp模块的ResultActivity.kt定义:
class ResultActivity : ComponentActivity() {
    override fun onCreate(savedInstanceState: Bundle?) {
        super.onCreate(savedInstanceState)
        enableEdgeToEdge()
        val gameResult=intent.getStringExtra("result")
        setContent {
            Chapter03Theme {
                Scaffold(modifier=Modifier.fillMaxSize()) {
                    innerPadding ->ResultScreen(resultTxt=gameResult!!,
                        modifier=Modifier.padding(innerPadding) )
                }
            }
        }
    }
}
```

运行结果如图 3-22 所示。可根据传递的游戏结果,设置 ResultScreen 文本框的内容。

(a) 赢的界面　　　　(b) 输的界面

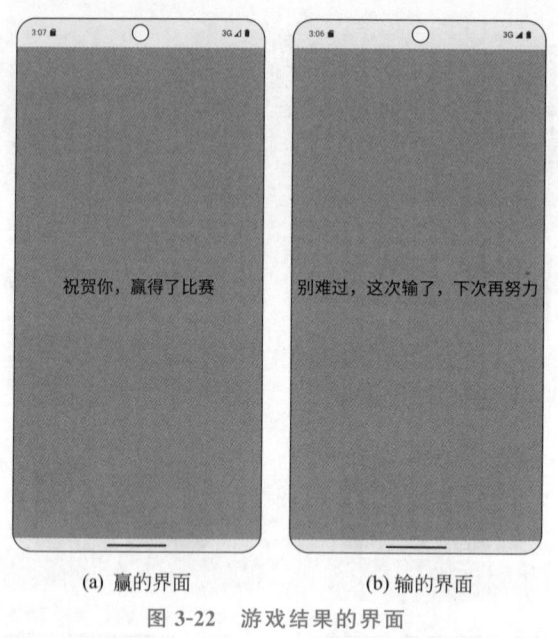

图 3-22　游戏结果的界面

习　题　3

一、选择题

1. 一个应用程序可以具有一个或多个 Activity，Activity 的主要目的是_____。
　　A. 提供后台服务　　B. 数据共享　　　C. 发送消息　　　D. 与用户互动

2. 显式 Intent 需要定义 Intent 对象以实现启动不同的组件。假设在 Activity 定义的 MainActivity.kt 中包括了如下选项代码，实现从 MainActivity 跳转到 FirstActivity。选项正确的是_____。

　　A.

　　val intent＝Intent(this,FirstActivity::class)
　　startActivity(intent)

　　B.

　　val intent＝Intent(MainActivity::class,FirstActivity::class)
　　startActivity(intent)

　　C.

　　val intent＝Intent(MainActivity@this,FirstActivity::class)
　　startActivity(intent)

　　D.

　　val intent＝Intent(MainActivity@this,FirstActivity::class.java)
　　startActivity（intent）

3. Android 的 Activity 有 4 种状态：Running(运行)状态、Paused(暂停)状态、Stopped(停止)状态和 Killed(销毁)状态。Activity 调用_____函数处于 Running(运行)状态。

A. onStart　　　　B. onResume　　　　C. onPause　　　　D. onStop

4. 为 MainActivity 设置加载模式为_____或_____,都可以创建唯一的 MainActivity 实例对象。

A. stardand　　　　　　　　　　B. singleTop
C. singleInstance　　　　　　　　D. singleTask

5. 假设启动 Android 应用的一个 MainActivity,然后用户按下移动终端的 Home 键退出该 Activity。在这个过程中该 Activity 的_____函数没有执行。

A. onCreate　　B. onPause　　　　C. onStop　　　　D. onDestroy

6. 已知创建了一个 Android 项目,定义一个 MainActivity FirstActivity,要求从 MainActivity 跳转到 FirstActivity 时,实现将字符串"hello from MainActivity"从 MainActivity 传递给 FirstActivity。假设在 MainActivity 中定义如下代码实现 Activity 的跳转:

```
val intent=Intent(this,FirstActivity::class.java)
intent.putExtra("frmMain","hello from MainActivity")
startActivity(intent)
```

选择下列选项中的_____,可以让 FirstActivity 通过 Toast 组件显示接收的数据。

A.
```
class FirstActivity:Activity() {
    override fun onCreate(savedInstanceState:Bundle?) {
        super.onCreate(savedInstanceState)
        val data=intent.getStringExtra("frmMain")
        Toast.makeText(this,data,Toast.LENGTH_LONG).show()
    }
}
```

B.
```
class FirstActivity:Activity() {
    override fun onCreate(savedInstanceState:Bundle?) {
        super.onCreate(savedInstanceState)
        var intent=Intent(this,MainActivity::class.java)
        val data=intent.getStringExtra("frmMain")
        Toast.makeText(this,data,Toast.LENGTH_LONG).show()
    }
}
```

C.
```
class FirstActivity:Activity() {
    override fun onCreate(savedInstanceState:Bundle?) {
        super.onCreate(savedInstanceState)
        val intent=Intent(this,MainActivity::class.java)
        val data=intent.getStringExtra("frmMain")
        Toast.makeText(this,data,Toast.LENGTH_LONG).show()
    }
}
```

D.
```
class FirstActivity:Activity() {
    override fun onCreate(savedInstanceState：Bundle?) {
        super.onCreate(savedInstanceState)
        val data=savedInstanceState.getStringExtra("frmMain")
        Toast.makeText(this,data,Toast.LENGTH_LONG).show()
    }
}
```

7. 已知定义数据类 Person，代码如下：

```
data class Person(val name: String,val gender: String): Serializable
```

设在 MainActivity 中发送 Person 对象给 OtherActivity，发送数据代码片段：

```
val person=Person("张三 ","男")
val intent=Intent(this,OtherActivity::class.java)
intent.putExtra("data",person)
startActivity(intent)
```

则 OtherActivity 接收一个 Person 对象的正确的代码片段应该是_____。

  A. val data=intent.getSerializableExtra("data")

  B. val data=intent.getSerializableExtra("data") as Person

  C. val data=intent.getExtra("data")

  D. val data=intent.getExtra("data") as Person

8. 已知从 MainActivity 可以根据用户的选择，分别跳转到 FirstActivity 或 SecondActivity。从 FirstActivity 或 SecondActivity 分别返回 MainActivity，那么 MainActivity 需要调用_____方法来创建 ActivityResultLauncher 对象，并调用该对象的_____方法，以实现加载意图对象。

  A. registerForResult onActivityResult

  B. registerForActivity contract

  C. registerForActivityResult launch

  D. 以上答案均不正确

二、填空题

1. 可以在配置清单文件_____中配置 Activity 的加载模式。Android 的 Activity 加载模式有_____、_____、_____ 和 _____。

2. 调用_____函数可以结束 Activity，调用_____函数可以终止移动应用。

3. Activity 的_____实现对 Activity 的管理。

三、上机实践题

1. 编程实现数据的传递。已知定义 MainActivity、FirstActivity 和 SecondActivity。要求：如果在 MainActivity 界面中单击按钮"跳转第一个页面"后，再从"第一个页面"指向的 FristActivity 返回 MainActivity，则弹出对话框，显示"从第一个页面返回"。如果在 MainActivity 界面中单击"跳转第二个页面"按钮后，再从"第二个页面"指向的 SecondActivity 返回 MainActivity，则弹出对话框，显示"从第二个页面返回"。如果定义一个 Student 类，属性有 name：String 和 birthday：LocalDate，要求将学生对象(张三,1999-02-23)从 MainActivity 传递到 FirstActivity,并在 FirstActivity 中显示该对象的信息。请分别采用 Java 的 Serializable 序列化方式和 Android 的 Parcelable 序列化方式传递对象。

2. 实现一个简易的掷骰子游戏。要求：这个游戏需要玩家同时掷 4 个骰子,每个骰子都是一个印有数字 1~6 的正方体。玩家同时掷出这 4 个骰子,如果这 4 个骰子向上面的数字之和大于或等于 10,玩家就会获得 10 分游戏积分的奖励,否则没有任何积分奖励。

3. 实现一个猜成语游戏。要求：提供 10 张猜谜的图片,要求根据图片猜出表示的成语。成功猜出成语,可以继续下一张猜谜图片进行猜谜。如果成功猜出 8 张以上的图片,则显示"非常棒";如果猜出的图片在 5 张以下,则表示"要加油";在其他情况下,显示"还不错！"。

第 4 章　Android 的界面开发

良好的移动应用开发 UI 可以提高用户体验，让移动应用的使用更加方便。在 Android 系统中定义了一系列 GUI 编程的 API 支持移动界面。特别是 Android 5.0 以后推出 Material Design（材质设计），让同款移动应用在不同的移动终端具有整体风格的移动界面。

早期的 Android 利用 XML 文件定义 UI 界面，UI 界面的更新也是由程序员使用代码主动刷新，UI 与数据并无必然的映射关系，这样的 UI 称为命令式 UI。开发命令式 UI 易错，成本高。Jim Sproch 开创性地提出了声明式界面开发框架——JetPack Compose，并于 2021 年正式推出。JetPack Compose 组件让人们开始意识到一种新的简化界面开发方式——声明式 UI。声明式 UI 的更新并非由程序员使用代码来主动刷新，而是由隐藏机制负责维护 UI 的刷新，数据的变化会导致 UI 界面发生相应的变化，使得 UI 与数据存在映射关系。JetPack Componse 库需要在 Android Studio Flamingo 及其以上版本中创建 Empty Activity 模板应用进行实现。

4.1　JetPack Compose 组件

JetPack Compose 组件是声明性 UI 开发包，将响应式编程模式与 Kotlin 语言结合，通过 Compose 组件大大简化了与构建和更新界面关联的工程任务。该技术的工作原理是从头开始重新生成整个屏幕界面，然后仅根据数据状态的变化执行必要的更改。此方法可避免手动更新有状态视图层次结构的复杂性。通俗地讲，利用 JetPack Compose 框架创建的 UI 界面就是一个视图模板，这个视图模板伴随传入数据或者内容状态的变化，使得界面进行重组，导致呈现不同的显示效果。开发基于 JetPack Compose 组件的移动 UI 界面代码量更少，开发更方便。

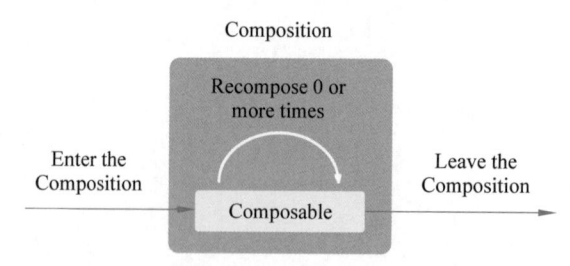

图 4-1　可组合函数的组合构成视图模板

4.1.1　可组合函数和预览函数

Compose 开发的界面是由可组合函数构成的。可组合函数是在函数前增加@Composable 标记来定义。在可组合函数中可以嵌套多个其他可组合函数定义界面层次。可组合函数的定义形式如下：

```
@Composable
fun 函数名([参数列表]) {
    …              //函数体
}
```

可组合函数带有 @Composable 标记。这个标记可告知 Compose 编译器,这个可组合函数旨在将数据转换为界面。可组合函数可以接收一些参数,这些参数可让应用逻辑描述界面。可组合函数的函数体可以调用其他可组合函数,从而构成了可组合 UI 树,因此可组合函数有时也称为"可组合项"。同时,可组合函数没有返回值。发出界面的 Compose 函数不需要返回任何内容,因为它们是描述所需的屏幕状态,而不是构造界面 Widget(UI 小部件)。

如果可组合函数没有任何参数,则可以在这种组合函数前加上再增加@Perview 标记,这种无参的可组合函数被声明成一个预览函数。可以在 Android Studio 的 Design 视图显示预览函数代码组成的界面。可单击预览函数左侧的 ▶ 按钮,将预览函数定义界面部署到模拟器显示界面。当有多个预览函数时,会生成多个按钮,根据需要选择实现模拟器预览界面。

例 4-1 可组合函数和预览函数的应用实例,代码如下:

```kotlin
//Ch04_01模块的 MainActivity.kt
class MainActivity: ComponentActivity() {
    override fun onCreate(savedInstanceState: Bundle?) {
        super.onCreate(savedInstanceState)
        enableEdgeToEdge()
        setContent {
            Chapter04Theme {
                Scaffold(modifier=Modifier.fillMaxSize()) {
                    innerPadding->DisplayScreen(modifier=Modifier.padding
                    (innerPadding))                        //调用可组合函数
                }
            }
        }
    }
}

@Composable
fun DisplayScreen(modifier: Modifier=Modifier) {
    val displayState:MutableState<Boolean>=remember{ mutableStateOf(false) }
                                                        //状态值控制显示类型切换
    Column(modifier=modifier.fillMaxSize(),
        verticalArrangement=Arrangement.Center,          //垂直居中
        horizontalAlignment=Alignment.CenterHorizontally){  //水平居中
        Text(text="可组合函数和预览函数示例",fontSize=36.sp)
        Button(onClick={                                 //点击按钮动作处理
            displayState.value=!displayState.value       //修改状态值
        }){
            Text(if (displayState.value) "显示图片" else "显示文本",fontSize=24.sp)
        }
        if (displayState.value)
            DisplayImage()                               //显示图片
        else
            DisplayText()                                //显示文本
    }
}
```

```
@Preview(showBackground=true)
@Composable
fun DisplayImage() {                                              //定义预览函数
    Chapter04Theme {
        Box(modifier=Modifier.fillMaxWidth(),contentAlignment=Alignment.Center){
            Image(painterResource(id=R.mipmap.scene),
                contentDescription="display image")
        }
    }
}

@Preview(showBackground=true)
@Composable
fun DisplayText() {                                               //定义预览函数
    Chapter04Theme {
        Box(modifier=Modifier.fillMaxWidth(),contentAlignment=Alignment.Center){
            Text("显示文字的示例", fontSize=32.sp)
        }
    }
}
```

在上述的代码中，定义了 3 个组合函数，其中最后两个可组合函数是预览函数。其中，预览函数可以直接在 Android Studio 中预览定义的效果，如图 4-2 所示。

图 4-2 预览函数的运行示例

本例中定义的示例是通过按钮实现切换显示内容。为了实现内容的切换，定义了一个状态值 displayState，这个状态值通过 remember 将值保存在内存中，只有当状态值 displayState 的值发生变换时，运行的界面才会在显示文本和显示图片之间进行切换。

运行结果如图 4-3 所示。

4.1.2 Modifier 修饰符

可以发现，在上述的例 4-1 中可组合函数中出现了 Modifier 类型。Modifier 修饰符是用于配置组

· 88 ·

图 4-3 可组合函数和预览函数示例的运行结果

件属性的工具,是标准的 Kotlin 对象,常用的 Modifier 修饰符如表 4-1 所示。Modifier 调用属性配置的函数,并返回 Modifier 对象。因此,通过 Modifier 对象的链式调用可为组件设置多种样式。修饰符可以修饰或扩充可组合项,通过 Modifier 修饰符可以实现以下功能。

表 4-1 常用的 Modifier 修饰符

修饰符名	说 明	备 注
Modifier.border	设置边框	
Modifier.padding	设置内边距和外边距	出现在 Modifier.border 前为设置外边距,在 Modifier.border 后为设置内边距
Modifer.size	设置组件大小	单位为 dp
Modifier.fillMaxSize Modifier.fillMaxWidth Modifier.fillMaxHeight	设置最大显示尺寸 设置最大的宽度 设置最大的高度	
Modifier.wrapContentSize Modifier.wrapContentWidth Modifier.wrapContentHeight	设置组件自身大小 设置组件自身的宽度 设置组件自身的高度	
Modifier.background	设置背景	可以设置背景的颜色、图片等
Modifier.offset	设置组件的偏移量	在从左到右的上下文中,正 offset 会将元素向右移,而在从右到左的上下文中,它会将元素向左移
Modifier.clickable	设置组件的点击动作	

- 更改可组合项的大小、布局、行为和外观。
- 添加无障碍标签等信息。
- 处理用户输入。
- 添加元素的点击、滚动、拖动、缩放等高级互动。

Modifier 修饰符限制在特定的作用域中使用，即 Compose 作用域限定在实现 Modifier 的安全调用，只能在特定作用域中调用修饰符。常见的特定作用域有 3 种：

```
@LayoutScopeMarker @Immutable Interface ColumnScope    //列范围,同一列
@LayoutScopeMarker @Immutable Interface RowScope       //列范围,同一行
@LayoutScopeMarker @Immutable Interface BoxScope       //帧范围,显示最后的组件
```

在上述 3 种范围中，注解 @LayoutScopeMarker 表示限定范围，不允许跨级访问。注解 @Immutable 说明是稳定类型，是不可变的。

修饰符函数调用的顺序非常重要。调用顺序不同构成不同的 Modifier 链。实质上，Modifier 对象是 Modifier 接口的内部伴随对象 Modifier，具体的结构如下所示：

```
interface Modifier {
    fun <R>foldIn(initial: R, operation: (R, Element)->R): R      //正向遍历 Modifier 链
    fun <R>foldOut(initial: R, operation: (Element, R)->R): R     //逆向遍历 Modifier 链
    fun any(predicate: (Element)->Boolean): Boolean
    fun all(predicate: (Element)->Boolean): Boolean
    infix fun then(other: Modifier): Modifier=
        if (other===Modifier) this else CombinedModifier(this, other)    //生成 Modifier 链
    @JvmDefaultWithCompatibility
    interface Element : Modifier {
        ...                                                              //略
    }
    companion object:Modifier{                                           //伴随对象
        ...                                                              //略
    }
}
```

Modifier 修饰符的连接通过 then 的调用来实现。如果不是同一个 Modifier 对象，就调用 CombinedModifier 将不同的 Modifier 对象组合在一起。例如：

```
Column(modifier=Modifier.fillMaxSize()
    .padding(30.dp).background(Color.Green){
    ...                                                                  //略
})
```

上述代码中，伴随对象 Modifier 创建 Modifier 链。依次与 FillElement、PaddingElement 和 BackgroundElement 进行链接。对应的数据结构如图 4-4 所示。

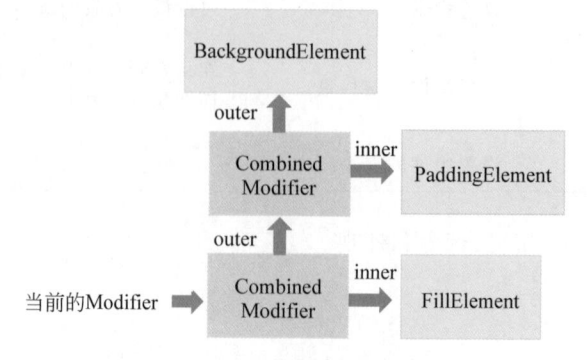

图 4-4 Modifier 对象的数据结构

由于每个函数都会对上一个函数返回的 Modifier 对象进行更改，导致顺序不同会影响最终结果。Modifier 链在渲染界面是依次遍历的，可以通过 foldIn 和 foldOut 函数遍历 Modifier 链。其中，foldIn 函数用于正向遍历 Modifier 链，上述代码定义 Modifier 链的访问顺序是 FillElement→PaddingElement→BackgroundElement。foldout() 函数用于反向遍历 Modifier 链，则上述代码定义 Modifier 链的反向顺序是 BackgroundElement→PaddinngElement→FillElement。

例 4-2 Modifier 的应用示例，代码如下：

```kotlin
//Ch04_02模块的 MainActivity.kt
class MainActivity : ComponentActivity() {
    override fun onCreate(savedInstanceState: Bundle?) {
        super.onCreate(savedInstanceState)
        enableEdgeToEdge()
        setContent {
            Chapter04Theme {
                Scaffold(modifier=Modifier.fillMaxSize()) {
                    innerPadding ->Column(modifier=Modifier.padding(innerPadding),
                    verticalArrangement=Arrangement.Center,
                    horizontalAlignment=Alignment.CenterHorizontally){
                        Row(modifier=Modifier.fillMaxWidth()
                            .background(Color.Yellow)                        //背景颜色
                            .padding(5.dp)                                   //外边距
                            .border(1.dp, Color.Blue, RoundedCornerShape(1.dp))//定义边框
                            .clickable {                                     //点击动作处理
                                displayInfo("已经点击第一行!")
                            }) { Text("HELLO JETPACK COMPOSE",fontSize=24.sp) }
                        Row(modifier=Modifier.fillMaxWidth()
                            .background(Color.Green)                         //背景颜色
                            .border(1.dp, Color.Blue, RoundedCornerShape(1.dp))//边框
                            .padding(5.dp)                                   //内边距
                        ) { Text("HELLO JETPACK COMPOSE",fontSize=24.sp) }
                        val modifier=Modifier.padding(10.dp).background(Color.Blue).size(100.
                            dp,100.dp)
                        modifier.foldIn(Unit){
                            _, element->Log.d("TAG"," Forward: ${element::class.simpleName}")
                                                                             //Modifierr链正向遍历
                        }
                        modifier.foldOut(Unit){
                            element, _ -> Log.d("TAG"," Backward: ${element::class.
                            simpleName}")
                                                                             //Modifier链反向遍历
                        }
                    }
                }
            }
        }
    }
    private fun displayInfo(info:String){
        Toast.makeText(this,info,Toast.LENGTH_LONG).show()
    }
}
```

运行结果如图 4-5 所示。

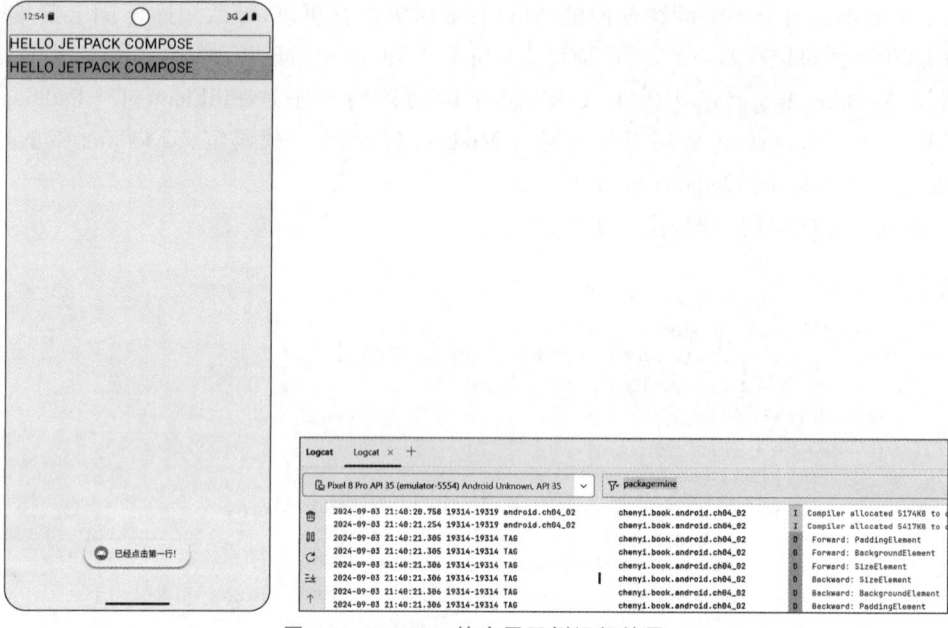

图 4-5 Modifier 的应用示例运行结果

4.1.3 Compose 常见的 UI 组件

Compose 定义的 UI 组件实质上是可组合函数。在 Compose 组件中已经定义了一些可组合函数用于实现 UI 组件。通过多次调用组合可以将这些可组合函数生成自定义可组合函数，生成用户自定义的 UI 组件。

1. 文本 Text

Text 是一个可组合函数，表示定义要显示的文本。函数规格定义如下：

```
@Composable
fun Text(
    text: String,                                               //要显示的文本
    modifier: Modifier=Modifier,                                //修饰文本的修饰符
    color: Color=Color.Unspecified,                             //文本的颜色
    fontSize: TextUnit=TextUnit.Unspecified,                    //字体尺寸
    fontStyle: FontStyle?=null,                                 //字体样式
    fontWeight: FontWeight?=null,                               //字体的粗细
    fontFamily: FontFamily?=null,                               //字体
    letterSpacing: TextUnit=TextUnit.Unspecified,               //字母之间的间距
    textDecoration: TextDecoration?=null,                       //文本绘制的装饰,例如下划线
    textAlign: TextAlign?=null,                                 //段落行内文本的对齐方式
    lineHeight: TextUnit=TextUnit.Unspecified,                  //段落中行高间距
    overflow: TextOverflow=TextOverflow.Clip,                   //文本溢出处理,默认将多余裁剪
    softWrap: Boolean=true,                                     //软换行,默认为 true
    maxLines: Int=Int.MAX_VALUE,                                //可显示的最大行数
    minLines: Int=1,                                            //可显示的最小行数
    onTextLayout: ((TextLayoutResult)->Unit)?=null,             //文本布局
    style: TextStyle=LocalTextStyle.current                     //文本样式
)
```

例 4-3 Text 的应用实例。代码如下：

```kotlin
//Ch04_03模块 MainActivity.kt
class MainActivity : ComponentActivity() {
    override fun onCreate(savedInstanceState: Bundle?) {
        super.onCreate(savedInstanceState)
        enableEdgeToEdge()
        setContent {
            Chapter04Theme {
                Scaffold(modifier=Modifier.fillMaxSize()) {
                    innerPadding ->MainScreen(
                        name="Android",
                        modifier=Modifier.padding(innerPadding)
                    )
                }
            }
        }
    }
}

@Composable
fun MainScreen(name: String, modifier: Modifier=Modifier) {
    Text(
        text="Hello $name!\nWelcome $name!",              //显示的文本
        modifier=modifier.fillMaxWidth(),                  //修饰符
        fontFamily=FontFamily.SansSerif,                   //设置字体
        letterSpacing=7.sp,                                //字母间距
        lineHeight=40.sp,                                  //行高
        textAlign=TextAlign.Center                         //文本居中
    )
}
```

运行结果如图 4-6 所示。

图 4-6 Text 的运行效果

2. 文本输入 TextField

文本输入常见的是 TextField 可组合函数,它的规格形式定义为如下的形式:

```
@Composable
fun TextField(
    value: String,                                              //显示在输入框的文本
    onValueChange: (String)->Unit,                              //文本输入变换的动作处理
    modifier: Modifier=Modifier,                                //修饰符
    enabled: Boolean=true,                                      //控制文本是否可输入的状态值
    readOnly: Boolean=false,                                    //设置文本只读
    textStyle: TextStyle=LocalTextStyle.current,                //文本的样式
    label: @Composable(()->Unit)?=null,                         //输入框提示的标签
    placeholder: @Composable(()->Unit)?=null,                   //可选占位符定义
    leadingIcon: @Composable(()->Unit)?=null,                   //设置前置的图标
    trailingIcon: @Composable(()->Unit)?=null,                  //设置后置的图标
    prefix: @Composable(()->Unit)?=null,                        //可选前缀
    suffix: @Composable(()->Unit)?=null,                        //可选后缀
    supportingText: @Composable(()->Unit)?=null,                //输入框下方显示的可选支持文本
    isError: Boolean=false,                 //输入框的错误状态值,如果false,显示错误提示
    visualTransformation: VisualTransformation=VisualTransformation.None,
                                                                //转换输入值的表示
    keyboardOptions: KeyboardOptions=KeyboardOptions.Default,   //软键盘选项
    keyboardActions: KeyboardActions=KeyboardActions.Default,   //软键盘动作处理
    singleLine: Boolean=false,              //输入单行状态值,true 则为当行滚动显示
    maxLines: Int=if (singleLine) 1 else Int.MAX_VALUE,         //输入框的最大行数
    minLines: Int=1,                                            //输入框的最小行数
    interactionSource: MutableInteractionSource=remember {
    MutableInteractionSource() },                               //输入交互流
    shape: Shape=TextFieldDefaults.shape,                       //文本框的形状
    colors: TextFieldColors=TextFieldDefaults.colors()          //输入框的颜色设置
)
```

TextField 是常见的输入框,在 JetPack Compose 组件中还定义了 BasicTextField、OutlinedTextField 等文本输入框。BasicTextField 往往是定义无太多修饰的文本输入框,而 OutlinedTextField 是定义边框的输入框。这些输入框的函数参数规格类似上述的定义。这里就不再一一介绍了。

例 4-4 TextField 输入框的应用实例,代码如下:

```
//模块 Ch04_04 的 MainActivity.kt
class MainActivity : ComponentActivity() {
    override fun onCreate(savedInstanceState: Bundle?) {
        super.onCreate(savedInstanceState)
        enableEdgeToEdge()
        setContent {
            Chapter04Theme {
                Scaffold(modifier=Modifier.fillMaxSize()) {
                    innerPadding->MainScreen(modifier=Modifier.padding(innerPadding))
                }
            }
        }
    }
```

```kotlin
    }
}

@Composable
fun MainScreen(modifier: Modifier=Modifier) {
    val passwordState:MutableState<String>=remember { mutableStateOf("") }
    val passwordShowSate:MutableState<Boolean>=remember{ mutableStateOf(true) }
    Column(modifier=modifier.fillMaxWidth(),
        horizontalAlignment=Alignment.CenterHorizontally,
        verticalArrangement=Arrangement.Center) {                    //自定义密码框
        TextField(
            modifier=Modifier.fillMaxWidth(),
            colors=TextFieldDefaults.colors(                         //设置颜色
                focusedTextColor=Color.Black,                        //获得焦点文本颜色
                focusedContainerColor=Color.White,                   //获得焦点容器颜色
                unfocusedContainerColor=Color.White,                 //未获焦点容器颜色
                unfocusedTextColor=Color.Gray                        //未获焦点文本颜色
            ),
            placeholder={                                            //占位符
                Text("输入密码",fontSize=32.sp)
            },
            label={                                                  //提示文本
                Text("密码框: ")
            },
            leadingIcon={                                            //前置图标
                Icon(Icons.Default.Edit,contentDescription="email")
            },
            trailingIcon={                                           //后置图标
                IconButton(onClick={                                 //图标按钮
                    passwordShowSate.value=!passwordShowSate.value   //切换密码可视状态
                }){
                    Icon(painterResource(id=R.mipmap.eye),contentDescription="show")
                }
            },
            value=passwordState.value,                               //输入框显示的文本
            onValueChange={                                          //输入文本变化处理
                it:String->passwordState.value=it
            },
            visualTransformation=                                    //可视转换
                if (passwordShowSate.value) PasswordVisualTransformation() else
                VisualTransformation.None
        )
        if (passwordShowSate.value)
            Text("显示输入的密码: ${passwordState.value}\n",fontSize=32.sp)
    }
}
```

运行效果如图 4-7 所示。

图 4-7 输入框的运行效果

3. 按钮

可组合函数 Button 定义按钮的效果，Button 可组合函数的规格定义如下所示：

```
@Composable
fun Button(
    onClick: ()->Unit,                                              //点击动作处理
    modifier: Modifier=Modifier,                                    //修饰符
    enabled: Boolean=true,                                          //可用状态
    shape: Shape=ButtonDefaults.shape,                              //按钮形状
    colors: ButtonColors=ButtonDefaults.buttonColors(),             //按钮颜色设置
    elevation: ButtonElevation?=ButtonDefaults.buttonElevation(),   //按钮不同状态下的标高
    border: BorderStroke?=null,                                     //按钮的边框
    contentPadding: PaddingValues=ButtonDefaults.ContentPadding,    //按钮的内容的边距
    interactionSource: MutableInteractionSource=remember{ MutableInteractionSource() },
                                                                    //处理状态的属性
    content: @Composable RowScope.()->Unit                          //按钮显示的内容
)
```

TextButton 和 IconButton 是两种常见定义按钮的可组合函数，它们分别用来定义文本按钮和图标按钮。二者的函数参数的规格与 Button 函数参数类似。在具体调用中，调用 TextButton 可组合函数处理为 content 内容调用 Text 可组合函数，使之显示文本内容。而 IconButton 是在 content 内容中调用 Icon 可组合函数，使之显示图标。

4. 选择性组件

单选按钮、复选框、三相状态复选框、单选开关和滑块都是常见的选择性组件。它们的共性就是在多种状态下切换。

可组合函数 RadioButton 用来定义单选按钮。函数参数的规格如下：

```
@Composable
fun RadioButton(
```

```
    selected: Boolean,                                                      //选择状态
    onClick: (()->Unit)?,                                                   //点击动作处理
    modifier: Modifier=Modifier,                                            //修饰符
    enabled: Boolean=true,                                                  //可用状态
    colors: RadioButtonColors=RadioButtonDefaults.colors(),                 //颜色配置
    interactionSource: MutableInteractionSource=remember { MutableInteractionSource() }
                                                                            //处理状态的属性
)
```

因为单选按钮不会单个出现,往往是多个同时使用,只有一个会被选中。因此会定义一个 Boolean 状态值来设置 RadioButton 在是否选择中进行切换。

CheckBox 可组合函数用于定义复选框。函数参数说明如下:

```
@Composable
fun Checkbox(
    checked: Boolean,                                                       //选择状态
    onCheckedChange: ((Boolean)->Unit)?,                                    //选择的动作处理
    modifier: Modifier=Modifier,                                            //修饰符
    enabled: Boolean=true,                                                  //可用状态
    colors: CheckboxColors=CheckboxDefaults.colors(),                       //颜色
    interactionSource: MutableInteractionSource=remember { MutableInteractionSource() }
                                                                            //处理状态的属性
)
```

复选框是可以多个选择,也是通过 Boolean 值来设置复选框是否选中或没有选中。

三相状态复选框是通过 TriStateCheckbox 可组合函数定义,函数参数说明如下:

```
@Composable
fun TriStateCheckbox(
    state: ToggleableState,                                                 //选择、未选或中间状态
    onClick: (()->Unit)?,                                                   //点击动作处理
    modifier: Modifier=Modifier,                                            //修饰符
    enabled: Boolean=true,                                                  //可用状态
    colors: CheckboxColors=CheckboxDefaults.colors(),                       //颜色配置
    interactionSource: MutableInteractionSource=remember { MutableInteractionSource() }
                                                                            //处理状态的属性
)
```

三相状态复选框的选择状态包括 3 种状态:ToggleableState.On 表示选中,ToggleableState.Off 表示不选中,ToggleableState.Indeterminate 表现为横杠,表示部分选中。

Switch 可组合函数用于定义单选开关,在打开开关和关闭开关之间进行切换。函数参数说明如下:

```
@Composable
@Suppress("ComposableLambdaParameterNaming", "ComposableLambdaParameterPosition")
fun Switch(
    checked: Boolean,                                                       //开关状态
    onCheckedChange: ((Boolean)->Unit)?,                                    //选择动作处理
    modifier: Modifier=Modifier,                                            //修饰符
    thumbContent: (@Composable()->Unit)?=null,                              //内部内容
    enabled: Boolean=true,                                                  //可用状态
    colors: SwitchColors=SwitchDefaults.colors(),                           //配置颜色
    interactionSource: MutableInteractionSource=remember { MutableInteractionSource() }
                                                                            //处理状态的属性
)
```

Slider 可组合函数定义了滑块，通常表示进度。函数参数说明如下：

```
@OptIn(ExperimentalMaterial3Api::class)
@Composable
fun Slider(
    value: Float,                                                       //进度取值
    onValueChange: (Float)->Unit,                                       //取值变化动作处理
    modifier: Modifier=Modifier,                                        //修饰符
    enabled: Boolean=true,                                              //可用状态
    valueRange: ClosedFloatingPointRange<Float>=0f..1f,                 //每次值变化幅度
    @IntRange(from=0)                                                   //初始从 0 开始
    steps: Int=0,                                                       //步长
    onValueChangeFinished: (()->Unit)?=null,                            //取值变化后动作处理
    colors: SliderColors=SliderDefaults.colors(),                       //配置颜色
    interactionSource: MutableInteractionSource=remember { MutableInteractionSource() }
                                                                        //处理状态的属性
)
```

例 4-5 选择性组件的应用实例，代码如下：

```
//模块 Ch04_05 的 ChoiceScreen.kt
@Preview
@Composable
fun ChoiceScreen(){
    //复选按钮状态
    var checkedState by remember { mutableStateOf(false) }
    //单选按钮状态
    var selectedState by remember { mutableStateOf(false) }
    //三相复选按钮状态
    val (triState1,onStateChange1)=remember { mutableStateOf(false) }
    val (triState2,onStateChange2)=remember { mutableStateOf(false) }
    val triState=remember(triState1,triState2){
        if (triState1 && triState2) ToggleableState.On
        else if (!triState1 && !triState2) ToggleableState.Off
        else ToggleableState.Indeterminate
    }
    //switch 开关
    var openState by remember{ mutableStateOf(false) }
    //sliderProgress 值
    var sliderProgress by remember{ mutableStateOf(0.35f) }
    Column(modifier=Modifier.fillMaxSize().fillMaxSize().background(colorResource(id=
        R.color.teal_200)),verticalArrangement=Arrangement.Center){
        Text(text="常见选择项相关组件",fontSize=40.sp,
            textAlign=TextAlign.Center,color=Color.White,
            modifier=Modifier.fillMaxWidth()
        )
        Row(verticalAlignment=Alignment.CenterVertically){
            Checkbox(checked=checkedState,
                onCheckedChange={it:Boolean->checkedState=it},
                colors=CheckboxDefaults.colors(checkedColor=Color.Green)
            )
```

```kotlin
        Text("复选框",fontSize=30.sp)
    }
    //单选按钮
    var selectedColor=Color.Yellow
    Row(verticalAlignment=Alignment.CenterVertically){
        RadioButton(selected=selectedState,
            onClick={ selectedState=!selectedState },
            colors=RadioButtonDefaults.colors(selectedColor=selectedColor)
        )
        Text("单选按钮",fontSize=30.sp)
    }
    //三相状态复选框
    Column(modifier=Modifier.fillMaxWidth()){
        Row(verticalAlignment=Alignment.CenterVertically){
            TriStateCheckbox(state=triState,
                onClick={
                    val s=triState!=ToggleableState.On
                    onStateChange1(s)
                    onStateChange2(s)
                },
                colors=CheckboxDefaults.colors(checkedColor=MaterialTheme
                    .colorScheme.primary)
            )
            Text("三相状态复选框",fontSize=30.sp)
        }
        Column(modifier=Modifier.padding(10.dp)){
            Row(verticalAlignment=Alignment.CenterVertically) {
                Checkbox(triState1, onStateChange1)
                Text("选项 1",fontSize=30.sp)
            }
            Row(verticalAlignment=Alignment.CenterVertically){
                Checkbox(triState2,onStateChange2)
                Text("选项 2",fontSize=30.sp)
            }
        }
    }
    //Switch 单选开关
    Row(verticalAlignment=Alignment.CenterVertically){
        Switch(checked=openState,
            onCheckedChange={it:Boolean->openState=it},
            colors=SwitchDefaults.colors(disabledCheckedTrackColor=Color.DarkGray,
                checkedTrackColor=Color.Red)
        )
        Text(if (openState) "打开" else "关闭",fontSize=30.sp)
    }
    //定义 Slider 组件
    Column(horizontalAlignment=Alignment.CenterHorizontally){
        Slider(value=sliderProgress,
            onValueChange={ it:Float->sliderProgress=it },
```

```
            colors=SliderDefaults.colors(activeTrackColor=Color.Yellow)
        )
        Text(text="进度：%.1f%%".format(sliderProgress*100),fontSize=30.sp)
    }
}
```

运行效果如图 4-8 所示。

图 4-8 选择性组件的运行结果

5. 图片和图标 Image、Icon 和 AsyImage

要显示图片可以调用 Image 可组合函数。Image 可组合函数有 3 种形式。

形式 1：

```
@Composable
fun Image(
    painter: Painter,                                    //图片资源编号
    contentDescription: String?,                         //图片的文本描述
    modifier: Modifier=Modifier,                         //修饰符
    alignment: Alignment=Alignment.Center,               //图片对齐方式
    contentScale: ContentScale=ContentScale.Fit,         //图片缩放
    alpha: Float=DefaultAlpha,                           //透明度
    colorFilter: ColorFilter?=null                       //颜色过滤器
)
```

形式 2：

```
@Composable
@NonRestartableComposable
fun Image(
    imageVector: ImageVector,                            //矢量图片
    contentDescription: String?,                         //图片的文本描述
    modifier: Modifier=Modifier,                         //修饰符
    alignment: Alignment=Alignment.Center,               //图片对齐方式
```

```
    contentScale: ContentScale=ContentScale.Fit,       //图片缩放
    alpha: Float=DefaultAlpha,                         //透明度
    colorFilter: ColorFilter?=null                     //颜色过滤器
)
```

形式 3：

```
@Composable
@NonRestartableComposable
fun Image(
    bitmap: ImageBitmap,                               //位图
    contentDescription: String?,                       //图片的文本描述
    modifier: Modifier=Modifier,                       //修饰符
    alignment: Alignment=Alignment.Center,             //图片对齐方式
    contentScale: ContentScale=ContentScale.Fit,       //图片缩放
    alpha: Float=DefaultAlpha,                         //透明度
    colorFilter: ColorFilter?=null                     //颜色过滤器
)
```

注解@NonRestartableComposable 可应用于可组合函数，该注解应用于函数或属性 getter 时，使得可组合函数成为一个不可重新启动。该注解以防止生成允许跳过或重新启动可组合函数执行的代码。

Icon 这个可组合函数用来显示图标。Icon 可组合函数也有 3 种形式。

形式 1：

```
@Composable
fun Icon(
    painter: Painter,                                  //图片资源编号
    contentDescription: String?,                       //图标的文本描述
    modifier: Modifier=Modifier,                       //修饰符
    tint: Color=LocalContentColor.current              //设置图标的颜色
)
```

形式 2：

```
@Composable
fun Icon(
    imageVector: ImageVector,                          //矢量图片
    contentDescription: String?,                       //图标的文本描述
    modifier: Modifier=Modifier,                       //修饰符
    tint: Color=LocalContentColor.current              //设置图标的颜色
)
```

形式 3：

```
@Composable
fun Icon(
    bitmap: ImageBitmap,                               //位图
    contentDescription: String?,                       //图标的文本描述
    modifier: Modifier=Modifier,                       //修饰符
    tint: Color=LocalContentColor.current              //设置图标的颜色
)
```

有时需要使用线上的图片,可以利用 Coil 库提供了大量的可组合函数实现在线图片的显示,本节抛砖引玉,介绍 Coil 库的 AsyncImage 来显示在线图片。其中,需要在项目模块中配置 Coil 库("io.coil-kt:coil-compose",当前版本是 2.7.0)。

在项目的 libs.versions.toml 文件中增加如下配置:

```
[versions]
coil="2.7.0"
[libraries]
coil-compose={group="io.coil-kt", name="coil-compose",version.ref="coil"}
```

在项目模块的 build.gradle.kts 文件中增加如下依赖:

```
dependencies {
    implementation(libs.coil.compose)
}
```

又因为 AsyncImage 使用在线图片。因此,需要在 AndroidManifest.xml 增加配置访问互联网权限。配置如下:

```
<uses-permission android:name="android.permission.INTERNET" />
```

配置成功后就可以使用 Coil 库的 AsyncImage 可组合函数定义加载异步图片,AsyncImage 可组合函数定义的规格如下:

```
@Composable
@NonRestartableComposable
fun AsyncImage(
    model: Any?,                                                    //在线资源的 URL
    contentDescription: String?,                                    //图片的文本描述
    modifier: Modifier=Modifier,                                    //修饰符
    transform: (State)->State=DefaultTransform,                     //变形状态
    onState: ((State)->Unit)?=null,                                 //图片变换的调用
    alignment: Alignment=Alignment.Center,                          //图片排列
    contentScale: ContentScale=ContentScale.Fit,                    //图片缩放
    alpha: Float=DefaultAlpha,                                      //透明度
    colorFilter: ColorFilter?=null,                                 //颜色过滤器
    filterQuality: FilterQuality=DefaultFilterQuality,              //过滤质量
    clipToBounds: Boolean=true,                                     //图片裁剪状态
    modelEqualityDelegate: EqualityDelegate=DefaultModelEqualityDelegate
                                                                    //model 的相等性
)
```

例 4-6 显示图片和图标的应用实例,代码如下:

```
//模块 Ch04_05 MainActivity.kt
class MainActivity : ComponentActivity() {
    override fun onCreate(savedInstanceState: Bundle?) {
        super.onCreate(savedInstanceState)
        enableEdgeToEdge()
        setContent {
```

```kotlin
            Chapter04Theme {
                Scaffold(modifier=Modifier.fillMaxSize()) {
                    innerPadding->MainScreen(modifier=Modifier.padding(innerPadding))
                }
            }
        }
    }
}
@Composable
fun MainScreen(modifier: Modifier=Modifier) {
    val imageUrl="www.example.com/image.png"              //图片 url 可以自行定义
    Column(modifier=modifier.fillMaxSize().padding(30.dp),
        horizontalAlignment=Alignment.CenterHorizontally){
        AsyncImage(model=imageUrl, contentDescription="async image")   //可以替换
        Image(painter=painterResource(R.mipmap.scene1),
            modifier=Modifier.size(100.dp,100.dp).padding(10.dp),
            contentDescription="currentImage1")
        Image(ImageBitmap.imageResource(id=R.mipmap.scene1),
            modifier=Modifier.size(100.dp,100.dp).padding(10.dp),
            contentDescription="currentImage2")
        Image(Icons.Default.Home, contentDescription="home icon",
            modifier=Modifier.size(100.dp,100.dp))            //默认为黑色的图片
        Icon(Icons.Default.Home, contentDescription="edit icon",
            modifier=Modifier.size(100.dp,100.dp),
            tint=Color.Green)                                 //设置图标为绿色
        Icon(painter=painterResource(id=android.R.mipmap.sym_def_app_icon),
            modifier=Modifier.size(100.dp,100.dp),contentDescription="system icon",
            tint=Color.Unspecified)            //设置图标颜色不指定,则使用图标自身的颜色
    }
}
```

运行结果如图 4-9 所示。

图 4-9　图片和图标的运行效果

6. 基本布局 Column、Row 和 Box

常见的基本布局有可组合函数 Column、Row 和 Box 来定义,它们分别表示列布局、行布局和帧布局。

Column 可组合函数定义列布局,使得在 ColumnScope 范围的所有的 UI 组件(可组合函数构成的 UI 组件)能按照列依次排列。Column 可组合函数的规格定义如下:

```
@Composable
inline fun Column(                                                            //内联函数
    modifier: Modifier=Modifier,                                              //修饰符
    verticalArrangement: Arrangement.Vertical=Arrangement.Top,                //垂直排列
    horizontalAlignment: Alignment.Horizontal=Alignment.Start,                //水平布局
    content: @Composable ColumnScope.()->Unit                                 //ColumnScope 设置中心区
)
```

Row 可组合函数定义行布局,使得在 RowScope 范围的所有 UI 组件在同一行中排列。Row 可组合函数的规格定义如下:

```
@Composable
inline fun Row(                                                               //内联函数
    modifier: Modifier=Modifier,                                              //修饰符
    horizontalArrangement: Arrangement.Horizontal=Arrangement.Start,          //水平排列
    verticalAlignment: Alignment.Vertical=Alignment.Top,                      //垂直布局
    content: @Composable RowScope.()->Unit                                    //RowScope 设置中心区
)
```

Box 可组合函数定义帧布局。帧布局的最典型的显示效果,就是所有的调用的可组合函数构成的 UI 组件在没有特殊的配置情况下,会在 BoxScope 范围的左上角添加 UI 组件,最后添加的组件才会被显示出来。Box 可组合函数的规格定义如下:

```
@Composable
inline fun Box(
    modifier: Modifier=Modifier,                                              //修饰符
    contentAlignment: Alignment=Alignment.TopStart,                           //内容排列
    propagateMinConstraints: Boolean=false,                                   //按测量尺寸排列,否则 true
    content: @Composable BoxScope.()->Unit                                    //BoxScope 范围设置中心区
)
```

7. 列表和网格

在移动应用中需要定义列表控件或网格控件限制 UI 组件的布局。在 JetPack Compose 组件中定义了延迟列表 LazyColumn 和 LazyRow。二者最大的不同就是滚动的方向不同,LazyColumn 是垂直方向滚动,LazyRow 是水平方向滚动。

LazyColumn 可组合函数定义的函数规格如下:

```
@Composable
fun LazyColumn(
    modifier: Modifier=Modifier,                                              //修饰符
    state: LazyListState=rememberLazyListState(),                             //控制观察列表状态
    contentPadding: PaddingValues=PaddingValues(0.dp),                        //内容的边距
```

```
    reverseLayout: Boolean=false,                                      //反向滚动和布局状态
    verticalArrangement: Arrangement.Vertical=
        if (!reverseLayout) Arrangement.Top else Arrangement.Bottom,   //垂直布局
    horizontalAlignment: Alignment.Horizontal=Alignment.Start,         //水平排列
    flingBehavior: FlingBehavior=ScrollableDefaults.flingBehavior(),   //滑动行为,默认滚动
    userScrollEnabled: Boolean=true,                                   //运行滚动
    content: LazyListScope.()->Unit                                    //LazyListScope 范围设置
)
```

LazyRow 可组合函数定义的函数规格如下:

```
@Composable
fun LazyRow(
    modifier: Modifier=Modifier,                                       //修饰符
    state: LazyListState=rememberLazyListState(),                      //控制观察列表状态
    contentPadding: PaddingValues=PaddingValues(0.dp),                 //内容边距
    reverseLayout: Boolean=false,                                      //反向滚动和布局状态
    horizontalArrangement: Arrangement.Horizontal=
        if (!reverseLayout) Arrangement.Start else Arrangement.End,    //水平布局
    verticalAlignment: Alignment.Vertical=Alignment.Top,               //垂直排列
    flingBehavior: FlingBehavior=ScrollableDefaults.flingBehavior(),   //滑动行为
    userScrollEnabled: Boolean=true,                                   //滚动状态
    content: LazyListScope.()->Unit                                    //LazyListScope 范围设置
)
```

延迟组件与 Compose 中的大多数布局组件渲染不同。延迟组件不是通过接收 @Composable 内容块参数生成可组合项,而是提供了一个 LazyListScope.() 块。在 LazyListScope 块提供一个语言标准规范,按照规范来定义和描述列表项内容。然后,延迟组件负责按照布局和滚动位置的要求添加每个列表项的内容。以 LazyColumn 为例:

```
LazyColumn{
    //添加单项
    Item {
        ...
    }
    //添加多项
    items{
        item ->
        ...
    }
}
```

在 LazyColumn 或 LazyRow 中通过 item 添加列表单项,也可以调用 LazyListScope 的内联函数 items 根据传递的列表或数组添加多个列表项。也可以通过 itemsIndexed 函数通过列表或数组的索引定义列表的单项。

例 4-7 列表的应用实例。定义一个图书列表,代码如下:

```kotlin
//模块 Ch04_06 Book.kt 定义图书实体类
data class Book(
    val imageId:Int,                                              //图书图片资源编号
    val name:String,                                              //书名
    val author:String,                                            //作者
    val publisher:String,                                         //出版社
    val description:String)                                       //图书的描述
//模块 Ch04_6 定义 BookCard.kt,定义显示单本图书布局
@Composable
fun BookCard(book:Book){
    Card(modifier=Modifier.fillMaxWidth().padding(10.dp).height(200.dp),
                                                                  //屏幕宽度高度 200dp 外边距 5dp
        elevation=CardDefaults.elevatedCardElevation(hoveredElevation=10.dp),
                                                                  //设置浮动抬高高度
        colors=CardDefaults.cardColors(
            containerColor=Color(0xffe3882f),                     //设置容器颜色
            contentColor=Color(0xff49548A))) {                    //设置显示内容颜色
        Row(verticalAlignment=Alignment.CenterVertically){
            Image(painter=painterResource(id=book.imageId),       //设置书的图片
                alignment=Alignment.Center,
                modifier=Modifier.size(200.dp,160.dp),
                contentDescription="${book.name}")
            Column{                                               //定义文本描述
                Text("${book.name}",fontSize=28.sp)
                Text("作者: ${book.author}",fontSize=20.sp)
                Text("出版社: ${book.publisher}",fontSize=20.sp)
                Text("${book.description}", fontSize=20.sp)
            }
        }
    }
}

//模块 Ch04_06 定义 BookColumnScreen 按上下滚动显示
@Composable
fun BookColumnScreen(bookList:List<Book>){
    LazyColumn {                                                  //定义延迟列列表
        items(bookList){                                          //依次遍历 bookList
            book:Book ->BookCard(book)                            //添加列表单项
        }
    }
}

//模块 Ch04_06 定义 BookRowScreen 按左右滚动显示
@Composable
fun BookRowScreen(bookList:List<Book>){
    LazyRow {                                                     //定义延迟行列表
        items(bookList){                                          //依次遍历 bookList
            book:Book ->BookCard(book)                            //添加列表单项
        }
    }
}
```

```kotlin
//模块 Ch04_06 定义 BookScreen.kt,定义图书列表
@Composable
fun BookScreen(){
    val aStr="陈轶等"
    val pStr="清华大学出版社"
    val books=listOf(
        Book(R.mipmap.book1, stringResource(id=R.string.book1_name),aStr,pStr,
            stringResource(id=R.string.book1_desc)),
        Book(R.mipmap.book2, stringResource(id=R.string.book2_name),aStr,pStr,
            stringResource(id=R.string.book2_desc)),
        Book(R.mipmap.book3, stringResource(id=R.string.book3_name),aStr,pStr,
            stringResource(id=R.string.book3_desc)),
        Book(R.mipmap.book4, stringResource(id=R.string.book4_name),aStr,pStr,
            stringResource(id=R.string.book4_desc)),
        Book(R.mipmap.book5, stringResource(id=R.string.book5_name),aStr,pStr,
            stringResource(id=R.string.book5_desc))
    )
    val convertState=remember{ mutableStateOf(true) }
    Scaffold(
        floatingActionButton={
            FloatingActionButton(                              //定义悬浮按钮
                onClick={
                    convertState.value=!convertState.value     //切换状态值
                }
            ){
                Icon(Icons.AutoMirrored.Filled.ArrowBack, contentDescription="convert")
            }
        }
    ){ innerPadding->Box(modifier=Modifier.padding(innerPadding)) {
        if (convertState.value){
            BookColumnScreen(books)                            //上下滚动显示
        } else {
            BookRowScreen(bookList=books)                      //左右滚动显示
        }
    }
    }
}

//模块 Ch04_06 定义 MainActivity.kt 在主活动调用列表并显示
class MainActivity : ComponentActivity() {
    override fun onCreate(savedInstanceState: Bundle?) {
        super.onCreate(savedInstanceState)
        enableEdgeToEdge()
        setContent {
            Chapter04Theme {
                BookScreen()
            }
        }
    }
}
```

运行结果如图 4-10 所示。

(a) LazyColumn形式　　　　(b) LazyRow形式

图 4-10　列表的显示效果

通过 LazyVerticalGrid(延迟垂直网格)、LazyHorizontalGrid(延迟水平网格)、LazyVerticalStaggeredGrid(延迟垂直交错网格)和 LazyHorizontalStaggeredGrid(延迟水平交错网格)分别定义不同形态的网格结构，如图 4-11 所示。

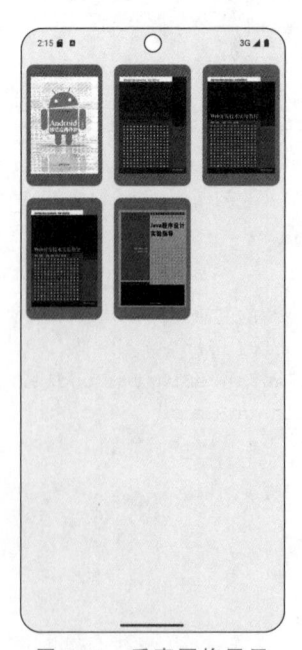

图 4-11　垂直网格显示

修改上述的 BookScreen，将列表显示替换成延迟垂直网格，代码如下：

· 108 ·

```
@Composable
fun BookScreen(){
    ...                                                         //直接量定义同上略
    ...                                                         //bookList 列表定义同上略
    Scaffold {
        innerPadding->Box(modifier=Modifier.padding(innerPadding)) {
            LazyVerticalGrid(
                columns=GridCells.Adaptive(minSize=120.dp)
            ) {
                items(books){
                    book:Book->BookCard(book)
                }
            }
        }
    }
}
```

也可以将上述的 BookScreen 代码修改成垂直交错网格形式,代码如下:

```
@Composable
fun BookScreen(){
    ...                                                         //直接量定义同上略
    ...                                                         //bookList 列表定义同上略
    Scaffold {
        innerPadding ->Box(modifier=Modifier.padding(innerPadding)){
            LazyVerticalStaggeredGrid(                          //定义延迟垂直交错网格
                columns=StaggeredGridCells.Adaptive(180.dp),    //列单元格
                contentPadding=PaddingValues(5.dp),             //内边距
                verticalItemSpacing=4.dp,                       //垂直单项的间距
                horizontalArrangement=Arrangement.spacedBy(5.dp),//水平布局间距
                content={
                    items(books) {                              //遍历数据列表
                        book:Book->BookCard(book)               //添加单项
                    }
                },
                modifier=Modifier.fillMaxSize()
            )
        }
    }
}
```

4.1.4 ConstraintLayout

ConstraintLayout 定义约束布局,它将 UI 组件按照相对的约束来放置可组合项。ConstraintLayout 布局安排 UI 可组合项的位置,比 Row 布局、Column 布局和 Box 布局更便利。

要使用 Compose 的 ConstraintLayout 需要增加依赖项,为此在项目的 libs.versions.toml 文件中增加以下代码:

```
[versions]
constraintlayoutCompose="1.0.1"
[libraries]
```

```
androidx-constraintlayout-compose={ group =
    "androidx.constraintlayout", name="constraintlayout-compose",
    version.ref="constraintlayoutCompose" }
```

然后在模块的 build.gradle.kts 中的增加依赖，代码如下：

```
dependencies {
    ...                                                              //略
    implementation(libs.androidx.constraintlayout.compose)
}
```

要理解 ConstraintLayout 布局，需要了解几个基本概念。

(1) 要配置可组合项到 ConstraintLayout，需要为可组合项定义相应的引用。通过这些引用设置相关的约束条件，可以确定可组合项的位置。通过 createRefs 函数批量化创建一组引用对象，也可以通过 createRefFor( ) 函数创建特定可组合项的引用对象。

(2) 调用修饰符的 constrainAs 函数来设置引用对象的约束条件。具体的约束条件通过 linkTo( ) 函数来构建。

(3) 引导线是设计布局的辅组工具。有两种不同引导线：垂直引导线和水平引导线。垂直引导线分别是 top 和 bottom，水平引导线分别从 start 和 end 来创建。

```
ConstraintLayout {
    //距离 start 的 0.1f 的位置创建水平引导线
    val startGuideline=createGuidelineFromStart(0.1f)
    //距离 end 的 0.1f 的位置创建水平引导线
    val endGuideline=createGuidelineFromEnd(0.1f)
    //距离 top 的 16dp 的位置创建垂直引导线
    val topGuideline=createGuidelineFromTop(16.dp)
    //距离 end 的 16dp 的位置创建垂直引导线
    val bottomGuideline=createGuidelineFromBottom(16.dp)
}
```

(4) ConstraintLayout 链，将不同的可组合项构建双向联系。可以调用函数 createVerticalChain 或 createHorizontalChain 分别创建垂直链和水平链。这种链定义在 Modifier.constrainAs 块中，用来创建约束。通过不同的 ChainStyles 链样式，设置不同可组空间处理的方式。

ChainStyle.Spread：空间在所有可组合项之间均匀分配。

ChainStyle.SpreadInside：空间在所有可组合项之间均匀分配，但不包括第一个可组合项之前和最后一个可组合项之后。

ChainStyle.Packed：可组合项紧密排列在一起，没有间隙。

例 4-8　ConstraintLayout 应用实例，代码如下：

```
//Ch04_07 模块的 ConstrainLayoutScreen
@Preview
@Composable
fun ConstrainLayoutScreen(){
    ConstraintLayout(
        modifier=Modifier.fillMaxSize()){
        val (textRef1,textRef2)=remember { createRefs() }        //创建两个引用
```

```
val startGuideline=createGuidelineFromStart(0.5f)    //创建水平引导线距离开始位置 50%
createHorizontalChain(textRef1,textRef2,chainStyle=ChainStyle.Packed)
                                                    //创建水平链,指定样式
Text(modifier=Modifier.constrainAs(textRef1){       //设置约束
    start.linkTo(parent.start)                      //文本 start 与布局 start 链接
    end.linkTo(startGuideline)                      //文本 end 与引导线链接
    top.linkTo(parent.top)                          //文本 top 与布局 top 链接
    bottom.linkTo(parent.bottom)                    //文本 bottom 与布局 bottom 链接
}.background(Color.Yellow).border(2.dp,Color.Black, RectangleShape),
    text="文本 1",fontSize=32.sp)
Text(modifier=Modifier.constrainAs(textRef2){
    start.linkTo(startGuideline)                    //文本 start 与引导线链接
    end.linkTo(parent.end)                          //文本 end 与布局 end 链接
    top.linkTo(parent.top)
    bottom.linkTo(parent.bottom)
}.background(Color.Green).border(2.dp,Color.Black, RectangleShape),
    text="文本 2",fontSize=32.sp)
    }
}
```

运行结果如图 4-12 所示。

(a) Spread样式　　　　(b) SpreadInside样式　　　　(c) Packed样式

图 4-12　链样式类型

注意：此处代码的 parent 是 ConstraintLayout 本身约束条件的引用。图 4-12 是上述代码在设置不同的链样式的运行效果。

4.2　搭建 Scaffold

Scaffold(脚手架)是 Material Design 布局基本结构,为复杂的用户界面提供标准化平台。它将 UI 的不同部分,包含顶部栏、底部栏、悬浮按钮等多种 UI 元素结合在一起,为应用提供一致的外观和感觉。Scaffold 可组合函数的函数规范如下所示：

```
@Composable
fun Scaffold(
    modifier: Modifier=Modifier,                                            //修饰符
    topBar: @Composable()->Unit={},                                         //定义头部栏
    bottomBar: @Composable()->Unit={},                                      //定义底部栏
    snackbarHost: @Composable()->Unit={},                                   //信息交互框
    floatingActionButton: @Composable()->Unit={},                           //定义浮动按钮
    floatingActionButtonPosition: FabPosition=FabPosition.End,              //浮动按钮位置
    containerColor: Color=MaterialTheme.colorScheme.background,             //容器的颜色
    contentColor: Color=contentColorFor(containerColor),                    //内容的颜色
    contentWindowInsets: WindowInsets=ScaffoldDefaults.contentWindowInsets,
                                                                            //内容窗口插值
    content: @Composable(PaddingValues)->Unit                               //定义中心区
)
```

要搭建脚手架 Scaffold,则需要定义 Scaffold 的各个部分,Scaffold 的常见结构如下：

```
Scaffold(
    topBar={
        //定义顶部栏
    },
    bottomBar={
        //定义底部栏
    },
    floatingActionButton={
        //定义悬浮按钮
    },
    snackbarHost={
        //定义信息交互栏
    },
    content={innerPadding->
        //中心区内容
    }
)
```

或

```
Scaffold(
    topBar={
        //定义顶部栏
    },
    bottomBar={
        //定义底部栏
    },
    floatingActionButton={
        //定义悬浮按钮
    },
    snackbarHost={
        //定义信息交互部件
```

```
}){innerPadding->
    //定义中心区
}
```

为了更好说明 Scaffold 应用,将结合具体实例来展示 Scaffold 搭建界面结构。图 4-13 显示了一个典型的 Scaffold 搭建的界面结构。在这个结构中,头部栏包含了顶部导航图标和文本以及在顶部栏右侧图标按钮。单击顶部导航图标显示侧滑菜单(有时也称抽屉布局)。单击右侧图标按钮可以用于展示顶部下拉菜单。底部栏包含了 3 个带有文本的图标按钮,用于不同的界面的切换。中间的主要区是中心区用来定义当前单个显示的界面。在右下方定义悬浮按钮的位置。

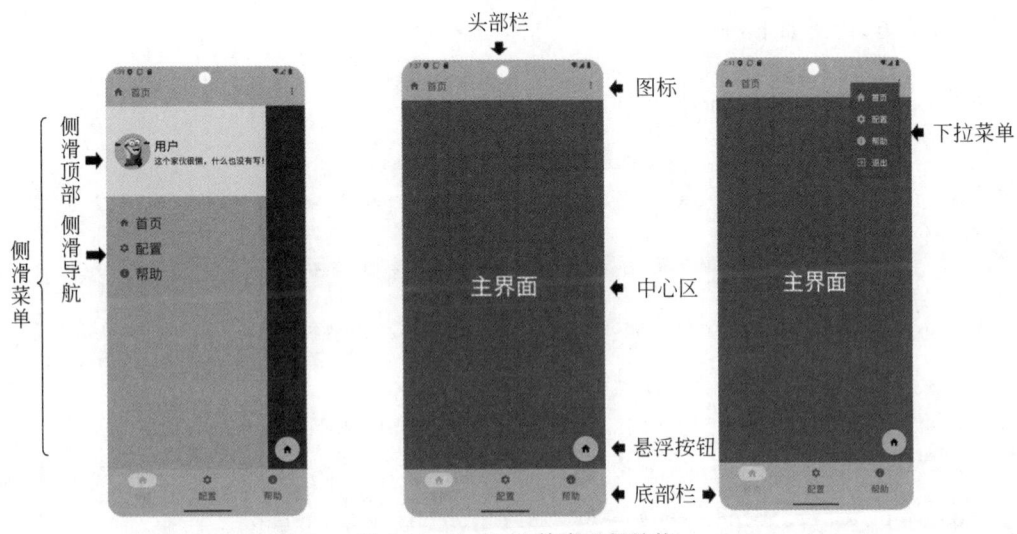

图 4-13 Scaffold 的脚手架结构

(1) 定义通用显示文本的组件 DisplayScreen,代码如下:

```
/** @param title String: 文本内容
 *  @param colorId Int: 颜色资源编号
 *  @param backgroundColorId Int: 背景颜色资源编号 */
@Composable
fun DisplayText(title:String,colorId:Int,backgroundColorId:Int){
    Box(contentAlignment=Alignment.Center, modifier=Modifier.fillMaxSize()
        .background(colorResource(id=backgroundColorId))){
        Text(title,fontSize=50.sp, textAlign=TextAlign.Center,
        color=colorResource(id=colorId), modifier=Modifier.fillMaxWidth())
    }
}
```

(2) 创建 3 个应用界面,代码如下:

```
@Composable
fun HomeScreen(){
    DisplayScreen(title="主界面",colorId=R.color.white, backgroundColorId=R.color
        .teal_700)
}

@Composable
```

· 113 ·

```
fun ConfigScreen(){
    DisplayScreen(title="配置界面",colorId=R.color.white,
        backgroundColorId=R.color.purple_500)
}

@Composable
fun HelpScreen(){
    DisplayScreen(title="帮助界面",colorId=R.color.white,
        backgroundColorId=R.color.teal_200)
}
```

(3) 定义屏幕类,代码如下:

```
/* * @property route String:路径
 *   @property title String:标题
 *   @property icon ImageVector:图标矢量图
 *   @property loadScreen [@androidx.compose.runtime.Composable] Function0<Unit>:加载屏
 *   幕 */
sealed class Screen (val route:String, val title:String, val icon: ImageVector,
    val loadScreen:@Composable()->Unit){
    data object HomePage:Screen("Home","首页", Icons.Filled.Home,{HomeScreen()
                                                                //加载 HomeScreen 界面
    })
    data object ConfigPage:Screen( "Config","配置", Icons.Filled.Settings,{
        ConfigScreen()                                          //加载 ConfigScreen 界面
    })
    data object HelpPage:Screen("Help","帮助", Icons.Filled.Info,{
        HelpScreen()                                            //加载 HelpScreen 界面
    })
}
```

该密封类指定要进行切换的 3 个不同界面对象,为了方便操作,定义全局的切换界面的列表,代码如下:

```
val screens:List<Screen>=listOf<Screen>(Screen.HomePage, Screen.ConfigPage,
    Screen.HelpPage)
```

(4) 定义底部导航栏,代码如下:

```
/* *定义应用窗口的底部组件
 * @param currentScreen MutableState<Screen>:记录当前屏幕界面的状态 */
@Composable
fun BottomViews(currentScreen: MutableState<Screen>){
    BottomAppBar(                                               //定义底部应用栏
        containerColor=Color(0xff,0xbb,0x19),                   //容器颜色
        contentColor=Color.Green) {                             //内容颜色
        screens.forEach {                                       //遍历屏幕界面列表
            screen:Screen->NavigationBarItem(                   //导航栏单项
                selected=screen.route==currentScreen.value.route, //是否是当前界面路径
                onClick={                                       //点击动作处理
```

```
                    currentScreen.value=screen                    //加载当前屏幕界面
                },
                colors=NavigationBarItemDefaults.colors(          //导航栏颜色设置
                    selectedIconColor=Color(0xa1,0xcf,0x00),      //选中图标颜色
                    unselectedIconColor=Color(0x4b,0x81,0x30),    //未选图标颜色
                    selectedTextColor=Color(0xa1,0xcf,0x00),      //选中文本颜色
                    unselectedTextColor=Color(0x4b,0x81,0x30)     //未选文本颜色
                ),
                label={                                           //导航标签文本
                    Text(text=screen.title,fontSize=20.sp)
                },
                icon={                                            //图标设置
                    Icon(imageVector=screen.icon,
                        contentDescription=null)
                }
            )
        }
    }
}
```

（5）定义 MainScreen 可组合函数。下面将在 MainScreen 中搭建 Scaffold 脚手架，并将上述定义的 BottomViews 可组合项在 bottomBar 处调用，代码如下：

```
@Preview
@Composable
fun MainScreen(){
    val currentScreen:MutableState<Screen>=remember{ mutableStateOf(Screen.HelpPage) }
                                                                  //记录当前的屏幕状态
    Scaffold(
        topBar={ },
        bottomBar={
            BottomViews(currentScreen =currentScreen )            //调用 BottomViews
        },
        floatingActionButton={ },
        snackbarHost={ }
    ) {
        innerPadding->Box(modifier=Modifier.padding(innerPadding)){
            currentScreen.value.loadScreen()                      //加载当前的屏幕
        }
    }
}
```

注意：在上面的代码中必须将 currentScreen 的类型限制为 MutableState<Screen>，因为无法根据赋值来推断 currentScreen 的类型。然后预览 MainScreen 界面，可以通过 BottomVIews 定义的底部应用栏实现不同界面的切换。运行结果如图 4-14 所示。

（6）定义顶部栏。顶部栏的情况比较复杂，它需要在顶部的左方定义导航图标和文本提示，并在右部定义一个图标按钮，通过图标按钮显示下列菜单。

图 4-14 底部栏界面切换的运行结果

首先定义顶部的下列菜单,代码如下:

```
/**定义下拉菜单
  * @param expandState MutableState<Boolean>:扩展菜单状态
  * @param currentScreen MutableState<Screen>:当前显示的界面状态 */
@Composable
fun TopMenuViews(expandState: MutableState<Boolean>,        //定义可扩展状态,控制菜单是否下拉
    currentScreen: MutableState<Screen>){                   //定义当前界面状态
    Column{
        DropdownMenu(expanded=expandState.value,            //定义下列菜单
            onDismissRequest={                              //取消菜单
                expandState.value=false                     //设置扩展状态为 false
            },
            modifier=Modifier.clickable {                   //定义点击动作
                expandState.value=!expandState.value        //点击菜单变更可扩展状态值
            }.background(color=Color(0x4b,0x81,0x30))) {

            screens.forEach {                               //遍历界面列表
                screen: Screen ->DropdownMenuItem(          //定义菜单单项
                    text={                                  //定义菜单单项的文本
                        Text(screen.title,fontSize=18.sp,color=Color(0xe6,0xee,0xca) )
                    },
                    leadingIcon={                           //定义菜单单项的前导图标
                        Icon(imageVector=screen.icon,
                            contentDescription=screen.title,
                            tint=Color(0xfb,0xbb,0x19)
                        )
                    },
                    onClick={                               //点击菜单单项的动作
                        currentScreen.value=screen          //修改当前的界面状态值
```

```
                })
        }

            DropdownMenuItem(                               //增加一个菜单单项用户处理退出应用
                text={
                    Text("退出",fontSize=18.sp,color=Color(0xe6,0xee,0xca))
                },
                leadingIcon={
                    Icon(
                        Icons.AutoMirrored.Filled.ExitToApp,
                        contentDescription="退出应用",
                        tint=Color(0xfb,0xbb,0x19)
                    )
                },
                onClick={
                    exitProcess(0)                          //调用退出应用
                }
            )
        }
    }
}
```

然后定义 TopBarViews,定义顶部栏,代码如下:

```
@OptIn(ExperimentalMaterial3Api::class)
@Composable
fun TopBarViews(currentScreen: MutableState<Screen>){
    val scope=rememberCoroutineScope()                      //定义协程的范围
    val expandState=remember{ mutableStateOf(false) }
    TopAppBar(                                              //定义顶部栏
        title={                                             //定义顶部栏标题
            Text("${currentScreen.value.title}",
                color=Color(0x4b,0x81,0x30)
            )
        },
        colors=TopAppBarColors(                             //设置顶部栏的颜色
            containerColor=Color(0xff,0xbb,0x19),           //容器颜色
            titleContentColor=Color(0xff,0xbb,0x19),        //标题颜色
            scrolledContainerColor=Color(0xff,0xbb,0x19),   //滚动容器颜色
            actionIconContentColor=Color(0xff,0xbb,0x19),   //动作交互图标颜色
            navigationIconContentColor=Color(0xff,0xbb,0x19)//导航图标颜色
        ),
        navigationIcon={                                    //定义左上方导航的图标
            IconButton(onClick={                            //导航按钮
                                                            //动作处理,暂空
            }){
                Icon(imageVector=Icons.Filled.Home,         //定义图标
                    contentDescription="头部导航图标",
                    tint=Color(0x4b,0x81,0x30)
```

```
                )
            }
        },
        actions={                                              //顶部栏右上方交互动作处理
            IconButton(
                onClick={
                    expandState.value=!expandState.value       //变更下拉菜单扩展项
                }){
                Icon(imageVector=Icons.Filled.MoreVert,
                    contentDescription=null,
                    tint=Color(0x4b,0x81,0x30))                //图标颜色为绿色
                if (expandState.value) {                       //扩展状态值为真,显示下列菜单
                    TopMenuViews(expandState=expandState,      //调用顶部的菜单
                        currentScreen=currentScreen)
                }
            }
        },
        modifier=Modifier.fillMaxWidth().wrapContentHeight())  //宽度满屏,高度与组件自身尺寸
}
```

修改 MainScreen 界面,将 TopBarViews 添加到 MainScreen 脚手架的 topBar 部分。这样 MainScreen 增加了顶部栏,顶部栏右上方可以控制下列菜单,通过下列菜单实现不同界面的切换。代码如下:

```
@Preview
@Composable
fun MainScreen(){
    val currentScreen:MutableState<Screen>=remember{ mutableStateOf(Screen.HelpPage) }
                                                                //记录当前的屏幕状态
    Scaffold(
        topBar={
            TopBarViews(currentScreen=currentScreen)
        },
        bottomBar={
            BottomViews(currentScreen=currentScreen)
        },
        floatingActionButton={

        },
        snackbarHost={

        }
    ) {
        innerPadding->Box(modifier=Modifier.padding(innerPadding)){
            currentScreen.value.loadScreen()                    //加载当前的屏幕
        }
    }
}
```

这时,预览 MainScreen 的运行效果如图 4-15 所示。

图 4-15 配置顶部栏的运行效果

(7)增加侧滑菜单。目前的 MainScreen 已经具备了头部栏的主要部件。但是,在常见的应用中,往往会单击左上方的导航图标按钮显示一个侧滑菜单。为了与常见移动应用的侧滑菜单一致,在这里,将侧滑菜单的显示部分分为两部分:侧滑菜单介绍信息的上部分内容,侧滑菜单的下部定义的菜单,并可以实现界面切换的功能。

首先定义侧滑菜单的上部,代码如下:

```
/* *侧滑的顶部内容*/
@Preview
@Composable
fun DrawerHeaderViews(){
    ConstraintLayout(modifier=Modifier.fillMaxWidth().height(200.dp)    //定义受限布局
        .background(Color(0xe6,0xee,0xca))){
        val vGuideLine=createGuidelineFromTop(0.5f)                     //垂直导航线
        val hGuideLine=createGuidelineFromStart(0.3f)                   //水平导航线
        val(imageRef,titleRef,contentRef)=remember {createRefs()}
                                                                        //创建多组受限布局引用参数
        createVerticalChain(titleRef,contentRef, chainStyle=ChainStyle.Packed)
                                                                        //创建垂直约束链
        Icon(modifier=Modifier.size(80.dp, 80.dp)                       //定义图标
            .constrainAs(imageRef) {                                    //设置约束条件
                top.linkTo(parent.top)
                start.linkTo(parent.start,10.dp)
                end.linkTo(hGuideLine)
                bottom.linkTo(parent.bottom)
            }.background(Color(0xf8,0xbb,0x19), CircleShape),
            painter=painterResource(id=R.mipmap.happy),
            tint=Color.Unspecified,
            contentDescription=null)
```

```
        Text(modifier=Modifier.constrainAs(titleRef){           //定义用户文本
            top.linkTo(parent.top)                              //设置约束条件
            start.linkTo(hGuideLine)
            bottom.linkTo(vGuideLine)
        },
        text="用户",fontSize=28.sp)
        Text(modifier=Modifier.constrainAs(contentRef){         //定义介绍文本
            start.linkTo(hGuideLine)
            top.linkTo(vGuideLine)
        },
        text="这个家伙很懒,什么也没有写!",fontSize=18.sp
        )
    }
}
```

然后定义侧滑菜单下部的导航部分,代码如下：

```
/**定义侧滑的下面菜单部分
 * @param drawerState DrawerState: 侧滑状态
 * @param currentScreen MutableState<Screen>: 当前界面状态
 */
@Composable
fun DrawerContentViews(drawerState: DrawerState,currentScreen:
MutableState<Screen>){
    val scope=rememberCoroutineScope()                          //获取协程范围
    screens.forEach {                                           //遍历界面列表
        screen:Screen->NavigationDrawerItem(                    //定义导航侧滑单项
            modifier=Modifier.padding(NavigationDrawerItemDefaults.ItemPadding),
            icon={
                Icon(screen.icon,contentDescription=null,tint=Color(0x4b,0x81,0x30))
                                                                //导航项的图标
            },
            label={                                             //导航项的标签
                Text(screen.title,fontSize=30.sp)
            },
            selected=screen==currentScreen,                     //判断当前界面是否选中
            shape=RectangleShape,                               //形状设置正方形
            colors=NavigationDrawerItemDefaults.colors(         //设置颜色
                selectedIconColor=Color(0x4b,0x81,0x30),
                unselectedIconColor=Color(0xe6,0xee,0xca),
                selectedContainerColor=Color.Unspecified,
                unselectedContainerColor=Color(0xa1,0xcf,0x00)
            ),
            onClick={                                           //点击导航单项的动作处理
                scope.launch {                                  //加载协程
                    currentScreen.value=screen                  //变更当前界面
                    drawerState.close()                         //关闭侧滑菜单
                }
            }
        )
    }
}
```

上述代码中，drawerState.close 函数关闭侧滑菜单是一个异步处理，必须在协程范围中完成。调用了代码 rememberCoroutineScope 函数，获得协程范围，并在协程范围中执行了侧滑菜单关闭。

最后将侧滑菜单的内容组合起来构成一个完整的侧滑菜单内容，代码如下：

```kotlin
/* * 定义侧滑菜单
 * @param drawerState DrawerState: 控制侧滑菜单显示
 * @param currentScreen MutableState<Screen>: 当前显示界面状态 */
@Composable
fun DrawerViews(drawerState: DrawerState,currentScreen: MutableState<Screen>){
    ModalNavigationDrawer(                                          //定义侧滑菜单
        drawerState=drawerState,                                    //设置侧滑状态
        drawerContent={                                             //定义并组装侧滑菜单
            Column(modifier=Modifier.fillMaxHeight().background(Color(0xa1,0xcf,0x00))
                .width(360.dp)){
                DrawerHeaderViews()                                 //侧滑菜单的头部
                Spacer(modifier=Modifier.padding(top=30.dp))
                DrawerContentViews(                                 //侧滑菜单的导航内容
                    currentScreen=currentScreen,
                    drawerState=drawerState
                )
            }
        }
    ){
        currentScreen.value.loadScreen()                            //主要显示的内容
    }
}
```

在上述的基础上，修改 TopBarViews，增加 drawerState 用于控制侧滑菜单的显示和关闭。代码如下：

```kotlin
/**
 * 定义头部内容
 * @param drawerState DrawerState: 侧滑状态
 * @param currentScreen MutableState<Screen>: 当前界面状态
 */
@OptIn(ExperimentalMaterial3Api::class)
@Composable
fun TopBarViews(drawerState: DrawerState,currentScreen: MutableState<Screen>){
    val scope=rememberCoroutineScope()                              //定义协程的范围
    val expandState=remember{ mutableStateOf(false) }

    TopAppBar(                                                      //头部的标题
        ...                                                         //略
        navigationIcon={                                            //定义导航的图标
            IconButton(onClick={                                    //导航按钮
                scope.launch {                                      //设置侧滑菜单状态打开,使之显示侧滑菜单
                    if (drawerState.isClosed)
                        drawerState.open()                          //打开侧滑菜单
                    else
                        drawerState.close()                         //关闭侧滑菜单
                }
            }){
                Icon(imageVector=Icons.Filled.Home,
```

```
                    contentDescription="头部导航图标",
                    tint=Color(0x4b,0x81,0x30)
                )
            }
        },
        actions={                                                    //定义交互动作处理
            ...                                                      //略
        },
        modifier=Modifier.fillMaxWidth()
            .wrapContentHeight()
    )
}
```

修改 MainScreen,代码如下:

```
@Preview
@Composable
fun MainScreen(){
    ...                                                              //略
    val drawerState=rememberDrawerState(initialValue =DrawerValue.Closed)   //侧滑状态
    Scaffold(
        ...                                                          //略
    ) {
        innerPadding->Box(modifier=Modifier.padding(innerPadding)){
            DrawerViews(drawerState,currentScreen)                   //加载侧滑菜单的内容
        }
    }
}
```

上述代码通过调用 rememberDrawerState()获得侧滑状态,通过侧滑状态的 open 和 close 的控制,实现侧滑菜单的显示和关闭,并在脚手架的中心区调用 DrawerViews()函数。这样,当侧滑状态为打开状态,则显示侧滑菜单,否则关闭侧滑菜单,直接显示当前界面。这样修改后,预览并运行 MainScreen,运行结果如图 4-16 所示。

图 4-16 增加侧滑菜单

(8)增加悬浮按钮和信息交互部件。通过悬浮按钮控制信息提示栏的显示。直接修改 MainScreen 的脚手架,代码如下:

```
@Preview
@Composable
fun MainScreen(){
    ...                                                          //略
    val snackbarState=remember{ mutableStateOf(false) }          //提示信息栏状态
    Scaffold(
        ...                                                      //略
        floatingActionButton={
            FloatingActionButton(
                onClick={
                    snackbarState.value=!snackbarState.value     //控制信息提示栏显示
                },
                modifier=Modifier.shadow(2.dp,shape=CircleShape),
                containerColor=Color(0xe6,0xee,0xca)
            ){
                Icon(imageVector=Icons.Filled.Home,
                    contentDescription="返回",
                    tint=Color(0x4b,0x81,0x30)
                )
            }
        },
        snackbarHost={
            if (snackbarState.value){
                Snackbar(
                    modifier=Modifier.fillMaxWidth(),
                    action={ },
                    dismissAction={ }
                ) {
                    Row(horizontalArrangement=Arrangement.Center){
                        Icon(Icons.Filled.Warning, contentDescription="警告信息")
                        Text("信息提示内容")
                    }
                }
            }
        }
    ) {
        innerPadding->Box(modifier=Modifier.padding(innerPadding)){
            DrawerViews(drawerState,currentScreen)                //加载当前的屏幕
        }
    }
}
```

这样,完整的脚手架搭建完成,最终的运行效果如图 4-17 所示。

图 4-17　脚手架的运行效果

4.3　Compose 组件的状态管理和重组

在使用 JetPack Compose 定义 UI 界面时,可以发现界面的变换往往与 Compose 组件内部的状态相关,当状态值发生变化时,Compose 构成的可组合的界面也会刷新发生相应的变化。非状态值是不能使可组合项内部自身发生变化的。

4.3.1　可组合项的状态

JetPack Compose 采用了单向数据流设计思想。定义界面的可组合函数本身没有任何返回值,也没有像类一样封装内部的私有状态。因此通过可组合函数的状态,使得可组合函数关联的界面可以观察是否发生了变化。Kotlin 语言中定义了一个接口 MutableState,代码如下:

```
interface MutableState : State {
    override var value: T
}
```

实现 MutableState 接口的任何类型的对象就是一个状态,状态是可变的,每个状态中保存一个 value 值。在执行可组合函数期间读取 value 属性,如果 value 属性值发生了变化,则可组合函数会发生重构,如果 value 属性值没有变化,则不会产生可组合函数的重构。Compose 组件可以通过 mutableStateOf 函数来获得一个这样的状态对象。例如:

```
val someState=mutableStateOf(true)
```

例如在上述的定义中,someState 就会被解析为一个可以存储 Boolean 布尔真值的可变状态值。

Android 结合 remember API 可以将状态值保存到内存中,当在内存中记住这个状态值。这样的好处就是,系统会在初始组合期间将由 remember 计算的值存储在组合中,并在重组期间返回存储的值。

当 remember 和状态值结合,会非常容易对可组合函数的重构产生作用,因为 remember 记住的状态值在内存中。当然,remember 不仅仅与可变的状态值组合,也可以与非可变值组合。在可组合项中声明 MutableState 对象的方法有 3 种。

方式 1:

```
val mutableState=remember { mutableStateOf(默认值) }
```

这种方式是直接通过状态的引用来获取或设置 value 属性值,需要导入以下代码:

```
import androidx.compose.runtime.remember;
```

方式 2:

```
var value by remember { mutableStateOf(默认值) }
```

在这种方式中,采用了代理的方式来直接获取或设置状态内部包含的 value 属性值。在这种方式中必须导入以下代码:

```
import androidx.compose.runtime.getValue
import androidx.compose.runtime.setValue
import androidx.compose.runtime.remember
```

方式 3:

```
val (value, setValue)=remember{ mutableStateOf(默认值) }
```

这种方式是非传统表示形式,value 对应的是状态的 value 属性的值,而设置状态的 value 属性是通过指定的 setValue 来实现的。

例 4-9　可组合项状态的应用实例,代码如下:

```
@Composable
fun DisplayScreen(){
    val contentState=remember{ mutableStateOf("") }              //方式 1
    var showedIcon by remember{ mutableStateOf(false) }          //方式 2
    val (showedImage,setValue)=remember { mutableStateOf(false) } //方式 3
    Column(modifier=Modifier.fillMaxSize().padding(30.dp),
        horizontalAlignment=Alignment.CenterHorizontally,
        verticalArrangement=Arrangement.Center){
            Card(modifier=Modifier.fillMaxWidth().wrapContentHeight()){
                Text("方式 1: 状态值${contentState.value}",fontSize=36.sp)
            }
            Card(modifier=Modifier.fillMaxWidth()){
                if (showedIcon){
                    Text("方式 2 状态切换", fontSize=24.sp)
                    Icon(Icons.Filled.Home,contentDescription="icon")
                }
                Button(onClick={
                    showedIcon=!showedIcon                       //修改 showedIcon 状态值
                    contentState.value="图标显示: ${showedIcon}-图片显示: ${showedImage}"
                                                                 //修改 contentState 状态值
                }){
```

```
                Text("方式 2 状态切换",fontSize=18.sp)
            }
        }
        Card(modifier=Modifier.fillMaxWidth()){
            if (showedImage){
                Text("方式 3 状态切换",fontSize=24.sp)
                Image(painterResource(id=R.mipmap.scene),contentDescription="image")
            }
            Button(onClick={
                setValue(!showedImage)                              //修改状态值
                contentState.value="图标显示：${showedIcon}-图片显示：${showedImage}"
                                                                    //修改 contentState 状态值
            }){
                Text("方式 3 状态切换",fontSize=18.sp)
            }
        }
    }
}
```

运行结果如图 4-18 所示。

图 4-18　可组合项状态的应用实例

4.3.2　无状态的可组合函数和有状态的可组合函数

可组合函数有两种形式：Stateless Composable 无状态可组合和 State Composable 有状态可组合。

无状态的可组合形式就是函数定义形参，通过调用时依赖传递的实参，实现界面的重构。这种可组合形式称为无状态的可组合。代码如下：

```
@Composable
fun CountScreen(counter:Int){
    Box(contentAlignment=Alignment.Center,modifier=Modifier.size(300.dp,200.dp)){
        Text(text="点击的次数：${counter}",fontSize=20.sp)
    }
}
```

有状态的可组合形式就是函数没有定义形参,通过定义内部的状态值,如果状态值发生变化,会导致界面进行重构。代码如下:

```
@Preview
@Composable
fun DisplayScreen(){
    var counter by remember{mutableStateOf(0)}        //定义状态值
    Column(modifier=Modifier.fillMaxSize(),horizontalAlignment=Alignment.
        CenterHorizontally){
        CountScreen(counter)                          //调用无状态的可组合函数
        CountScreen
        Button(onClick={
            counter+=1
        }){
            Text("点击按钮")
        }
    }
}
```

上述定义的 DisplayScreen 就是一个有状态的可组合函数。点击按钮会增加 counter 的取值,使得状态值发生变化,导致界面的重构。在该函数中实现对上述无状态可组合函数 CountScreen 的调用。

4.3.3 状态提升

Stateful(有状态)的可组合函数内部包含某些状态,函数内部状态值的变化,导致可组合进行界面的重组。函数调用方无须控制状态值,但是具有内部状态的可组合项难以复用,也很难测试。

如果其他可组合项共用界面元素状态,并在不同位置将界面逻辑应用到状态,则这时需要在界面层次结构中提升状态所在的层次。这样做会使可组合项的可重用性更高,并且更易于测试。具体表现形式是:将有状态的可组合函数中的状态移至可组合项的调用方,使得原来的有状态的可组合函数变成无状态的形式。状态提升是对相应的状态需要考虑替换成可组合函数的两个参数:

```
value:T:需要修改的状态的值
action(T)->Unit:请求修改值的事件
```

下列代码是有状态的可组合函数:

```
@Composable
fun DisplayInput(){
    val emailText=remember{mutableStateOf("")}
    Box(modifier=Modifier.fillMaxSize(),
        contentAlignment=Alignment.Center){
        TextField(
            value=emailText.value,
            label={
                Text("电子邮件")
            },
            leadingIcon={
                Icon(Icons.Filled.Email,contentDescription=null)
            },
            onValueChange={
                it:String->emailText.value=it
            }
        )
```

```
        }
    }
```

将上述代码进行状态提升,修改为无状态的可组合函数,将状态的值转换成函数的按时,同时增加一个修改值的动作处理,DisplayInput 修改如下:

```
@Composable
fun DisplayInput(input:String,inputChanged:(String)->Unit){
    Box(modifier=Modifier.fillMaxSize(),contentAlignment=Alignment.Center){
        TextField(
            label={Text("输入框测试: ")},
            value=input,
            onValueChange={
                it:String->inputChanged(it)
            }
        )
    }
}
```

通过这样的定义,DisplayInput 这个可组合函数就很容易被其他可组合项共享和调用。下列 TestScreen 可组合函数调用了上述的 DisplayInput 可组合项,代码如下:

```
@Composable
fun TestScreen(){
    var input by remember{mutableStateOf("请输入")}
    DisplayInput(
        input=input,
        inputChanged={
            it:String->input=it
        }
    )
}
```

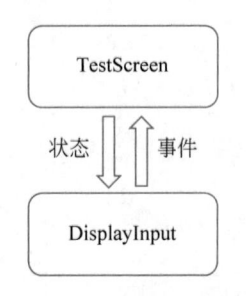

图 4-19 状态提升后可组合项的调用示意

如图 4-19 所示,在状态提升后,可组合项 DisplayInput 处于无状态的可组合项。其他可组合项如 TestScreen 要调用 DisplayInput 无状态可组合项,则需要将状态的值和状态变更的动作作为参数传递给无状态的可组合项。当无状态的可组合项的状态变更的事件发生后,又会将调用的界面包含的状态值进行变更,导致整个 UI 界面的刷新和重构。观察上面实现的状态提升,可以发现状态提升存在如下特点。

单一可信来源:通过传递状态,而不是复制状态,可确保只有一个可信来源。这有助于避免 bug。

封装:有状态可组合项只在内部修改其状态。

可共享:可与多个可组合项共享提升的状态。

可拦截:无状态可组合项的调用方可以在更改状态之前决定忽略或修改事件。

解耦:无状态的状态可以存储在任何位置。

4.3.4 状态丢失和状态保留

任何 Android 应用都可能因为活动 Activity 重新创建或者进程，导致丢失界面的状态。最常见的一种界面状态丢失的情况就是旋转移动屏幕，导致重新创建当前移动应用的活动，导致状态丢失。要解决重新创建活动或进程导致状态的丢失问题，则可以通过 rememberSaveable 来保留状态，使得重新创建活动或进程依然可以使用原有的状态。

rememberSaveable 通过保存的实例状态机制将界面元素状态存储在 Bundle 中。有以下几种情况可以实现状态保留。

- 自动将基元类型存储到 Bundle 中。如果是自定义的类实现 Parcelable，实现序列化，可以通过 Bundle 来传递数据。
- 使用 listSaver 和 mapSaver 等 ComposeAPI。
- 实现会扩展 Compose 运行时 Saver 类的自定义 Saver 类。

方式1：自动将基元类型和实现 Parcelable 接口的类型的数据存储到 Bundle 中。任何基元类型如 String、Int、Double、Float、Boolean、Short、Long 等以及实现 parcelable 接口自定义类型的对象，可以通过 rememberSaveable 中的状态会随着 onSaveInstanceState 以 Bundle 的键值对的形式进行存储。这里的关键字就是 Composable 函数在编译期确定的唯一标识。通过这个唯一标识，可以将数据按照键值对保存在 Bundle，并通过这个关键字进行数据恢复。

首先，自定义一个数据类 Employee，代码如下：

```
@Parcelize
data class Employee(val name:String,val gender:String,var salary:Double): Parcelable
```

定义保留状态的可组合项，代码如下：

```
@Preview
@Composable
fun EmployeeScreen(){
    val userState=rememberSaveable {mutableStateOf(Employee("张三","男",5000.0)) }
                                                                        //可保存的状态
    var salary by rememberSaveable{mutableStateOf(0.0) }                //可保存的状态
    Box(contentAlignment=Alignment.Center,modifier=Modifier.fillMaxSize()){
        Column{
            Text(userState.value.toString())
            TextField(
                value="${salary}",
                label={
                    Text("修改工资：")
                },
                leadingIcon={
                    Icon(imageVector=Icons.Filled.Info,contentDescription="工资")
                },
                onValueChange={
```

```
                    salary=it.toDouble()
                })
            Button(onClick={
                userState.value.salary=salary
            }){
                Text("修改工资")
            }
        }
    }
}
```

通过输入文本到文本框,然后旋转屏幕,可以发现第一行显示可保存状态没有丢失。

方式 2:实现会扩展 Compose 运行时 Saver 类的自定义 Saver 类。

自定义 Saver 类,自定义保存状态值的逻辑。通过自定义的 Saver 类定制数据保存的方式和数据恢复的方式。下面定义一个对应上例 Employee 数据类的 EmployeeSaver 类定制保存和恢复 Employee 数据的逻辑,代码如下:

```
object EmployeeSaver: Saver<Employee, Bundle>{
    override fun restore(value: Bundle) : Employee? {              //恢复成 Employee 对象
        return value.getString("name")?.let{
            name:String->value.getString("gender")?.let{
                gender:String->value.getDouble("salary")?.let{
                    salary:Double->Employee(name,gender,salary)
                }
            }
        }
    }
    override fun SaverScope.save(value: Employee) : Bundle? {      //保存到 Bundle 中
        return Bundle().apply{
            putString("name",value.name)
            putString("gender",value.gender)
            putDouble("salary",value.salary)
        }
    }
}
```

然后修改 EmployeeScreen 可组合函数,将 Employee 对象的存储和恢复按照 EmployeeSaver 指定的逻辑进行,代码如下:

```
@OptIn(ExperimentalMaterial3Api::class)
@Preview
@Composable
fun EmployeeScreen(){
    val userState=rememberSaveable(stateSaver=EmployeeSaver) {
        mutableStateOf(Employee("张三","男",5000.0))
    }
    var salary by rememberSaveable{mutableStateOf(0.0)}
    Box(contentAlignment=Alignment.Center,modifier=Modifier.fillMaxSize()){
        Column{
            Text("${userState.value}")
```

```
                TextField(
                    value="${salary}",
                    label={
                        Text("修改工资：")
                    },
                    leadingIcon={
                        Icon(imageVector=Icons.Filled.Info,contentDescription="工资")
                    },
                    onValueChange={
                        salary=it.toDouble()
                    }
                )
                Button(onClick={
                    userState.value.salary=salary
                }){
                    Text("修改工资")
                }
            }
        }
    }
}
```

方式3：使用 listSaver 和 mapSaver 等 ComposeAPI 进行数据保存和恢复。

通过 listSaver 和 mapSaver 等 Compose API 定制保存和恢复数据的逻辑，修改上述的 EmployeeScreen 函数，代码如下：

```
@Composable
fun EmployeeScreen(){
    val employeeSaver=run{                  //定义保存和恢复数据的逻辑
        mapSaver(save={                     //定义映射的方式指定键值对进行数据保存逻辑
            mapOf("name" to it.name,"gender" to it.gender,"salary" to it.salary)
        },
        restore={                           //定义数据根据映射恢复数据的逻辑
            Employee(it["name"] as String,it["gender"] as String,it["salary"] as Double)
        })
    }
    val userState=rememberSaveable(stateSaver=employeeSaver) {
        mutableStateOf(Employee("张三","男",5000.0))
    }
    var salary by rememberSaveable{mutableStateOf(0.0)}

    Box(contentAlignment=Alignment.Center,modifier=Modifier.fillMaxSize()){
        Column{
            Text("${userState.value}")
            TextField(
                value="${salary}",
                label={
                    Text("修改工资：")
                },
                leadingIcon={
                    Icon(imageVector=Icons.Filled.Info,contentDescription="工资")
                },
```

```
            onValueChange={
                salary=it.toDouble()
            }
        )
        Button(onClick={
            userState.value.salary=salary
        }){
            Text("修改工资")
        }
    }
}
```

4.3.5 状态容器

在 4.2 节搭建脚手架 Scaffold 中为了控制侧滑菜单、顶部下拉菜单切换以及当前界面的设置,都与状态有关。这些状态值一致通过单独的调用传参操作繁杂。为了方便状态的管理,可以定义一个状态容器。通过状态容器直接对不同的状态进行设置和获取,操作方便,代码简洁可读。下面定义一个状态容器类,代码如下:

```
/**定义状态容器类*/
class StateHolder (
    val currentScreen: MutableState<Screen>,     //当前界面状体
    val drawerState: DrawerState,                 //侧滑菜单
    val expandState: MutableState<Boolean>,       //下列菜单可扩展状态
    val snarkbarState : MutableState<Boolean>     //信息交互状态
)
```

为了方便调用各个状态,并确保每个状态是唯一的,可以定义保存所有状态到状态容器对象中,代码如下:

```
@Composable
fun rememberState(
    currentScreen: MutableState<Screen>=remember {mutableStateOf(Screen.HomePage) },
                                                //设定默认值为首页
    drawerState:DrawerState=rememberDrawerState(initialValue=DrawerValue.Closed),
                                                //设定关闭抽屉
    expandState: MutableState<Boolean>=remember{mutableStateOf(false) },
                                                //设置扩展关闭
    snarkbarState: MutableState<Boolean>=remember{mutableStateOf(false) }
                                                //设置信息栏关闭
):StateHolder=remember(currentScreen,drawerState,expandState,snarkbarState){
    StateHolder(currentScreen,drawerState,expandState,snarkbarState) //创建状态容器对象
}
```

经过这样处理后,可以将界面进行相应的修改。4.2 节的配置脚手架的 MainScreen 代码可以修改为

```
@Composable
```

```kotlin
fun MainScreen(){
    val currentState=rememberState()                                  //创建容器对象
    Scaffold (
        topBar={                                                      //定义头部
            TopBarViews(drawerState=currentState.drawerState,
                currentScreen=currentState.currentScreen)             //引用状态容器管理状态
        },
        bottomBar={                                                   //定义底部导航栏
            BottomViews(currentScreen=currentState.currentScreen)     //引用状态容器管理状态
        },
        snackbarHost={                                                //定义 Snackbar
            if (currentState.snarkbarState.value){                    //引用状态容器管理状态
                Snackbar(
                    modifier=Modifier.fillMaxWidth(),
                    action={ },
                    dismissAction={}
                ) {
                    Row(horizontalArrangement=Arrangement.Center){
                        Icon(Icons.Filled.Warning, contentDescription ="警告信息")
                        Text("信息提示内容")
                    }
                }
            }
        },
        content={                                                     //定义中心区
            innerPadding->Box(modifier=Modifier.padding(innerPadding)){
                DrawerViews(currentState.drawerState,currentState.currentScreen)
                                //定义侧滑菜单,利用 drawerState 来控制显示
            }
        },
        floatingActionButton={                                        //定义悬浮按钮
            FloatingActionButton(
                onClick={
                    currentState.snarkbarState.value=!currentState.snarkbarState.value
                },                                                    //控制信息提示栏的显示
                modifier=Modifier.shadow(2.dp,shape=CircleShape),
                containerColor=Color(0xe6,0xee,0xca)
            ){
                Icon(imageVector=Icons.Filled.Home, contentDescription="返回",
                    tint=Color(0x4b,0x81,0x30))
            }
        }
    )
}
```

4.4 ViewModel 组件

用户界面的控制器往往由 Activity 活动和 Compose 可组合项构成。在这些 UI Controller 中,往往存在丢失临时性数据的问题。例如,在横纵屏幕切换时,往往会导致屏幕临时性数据的丢失。其次,在 Activity 或 UI 片段发生异步处理的情况下,一些异步调用的结果需要一段时间才能返回,为了避免内存泄漏,需要大量的管理性的工作,导致 UI Controller 的任务负担过重。基于这些原因,通过

ViewModel 组件保存视图中需要的数据。ViewModel 组件将与用户界面相关的数据模型和应用程序的逻辑与负责实际显示和管理用户界面以及与操作系统交互的代码分离开，为 UI 界面管理数据。ViewModel 组件在整个 Activity 生命周期都会存在，在这些 UI Controller 的生命周期中，不会发生数据的丢失，如图 4-20 所示。

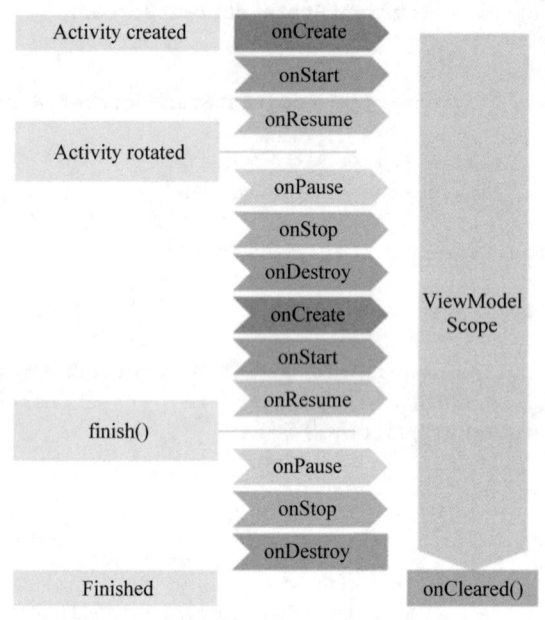

图 4-20　Activity 中 ViewModel 的范围

下面，介绍 ViewModel 组件的用法。

要实现 ViewModel，需要定义一个 ViewModel 类的子类，形式如下：

```
class MyViewModel: ViewModel() {
    ...                                                                            //定义状态
}
```

要获取一个 ViewModel 的对象，则需要执行以下代码：

```
val viewModel:MyViewModel=ViewModelProvider(上下文对象).get(MyViewModel::class.java)
```

或

```
val viewModel:MyViewModel=ViewModelProvider(上下文对象)[MyViewModel::class.java]
```

通过 ViewModel 组件用于保存视图中需要的数据。常见的管理方式是 StateFlow 方式。StateFlow 是一个状态容器可观测的数据流，可发出当前状态和新状态，更新其收集器也可以通过设置 value 属性。StateFlow 作为数据模型，可以表示任何状态。MutableStateFlow 是一种可变的 StateFlow，可更新状态并将其发送到数据流。

具体做法是在 ViewModel 组件中对每个单向输出的状态值定义的 MutableStateFlow 对象，然后通过调用 MutableStateFlow 对象的 asStateFlow 函数获得单向的状态流。应用的视图通过 ViewModel 对象获取这个单向状态流，调用 collectAsState()获得具体的状态。只要从状态流获取的状态发生变更，界面就会刷新重构。

要了解 ViewModel 组件的应用,就不得不谈到 MVVM 模式和 MVI 模式。下面将详细介绍。

4.4.1 MVVM 模式

MVVM 模式是 Model-View-ViewModel(模型-视图-视图模型)的简称,MVVM 模式将 View 的状态和行为抽象化,将视图 UI 和业务逻辑分开。图 4-21 展示了 MVVM 模式的结构。

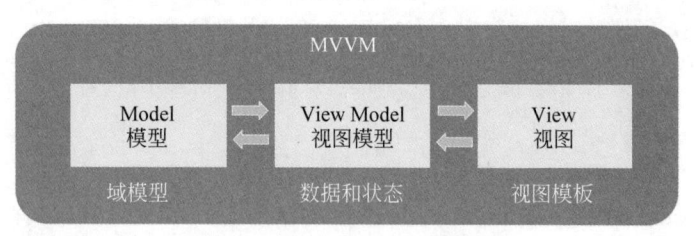

图 4-21 MVVM 模式的结构

Model(模型):表示的是域模型,保存了应用中的数据。

View(视图):表示视图模板,当接收的数据发生变化时,视图模板渲染的界面也随之变化。

ViewModel(视图模型):表示的是 Model 和 View 之间的中间,是实际业务的处理者。一方面,它接收模型的数据,修改模型,从而使得界面的状态也随之变化。

例 4-10 ViewModel 在 MVVM 模式的应用实例。定义一个彩票生成的简单应用,通过屏幕输入彩票每个数字开始和结束的范围(1~30),以及输入彩票由多少个数字构成。这些参数也可以通过输入设置。根据提供的相应参数,随机生成彩票。

(1) 定义彩票模型,代码如下:

```
/* * 彩票实体类
 * @property number Int: 彩票数字的个数
 * @property startRange Int: 彩票每个数字的开始范围
 * @property endRange Int: 彩票每个数字的结束范围 */
data class LotteryTicket(val number:Int=6,val startRange:Int=10, val endRange:Int=30){
    fun generate():ArrayList<Int>{                          //生成保存彩票数字的列表
        if (number<1||startRange>=endRange)
            throw RuntimeException("参数传递错误,创建彩票失败!")   //参数不正确抛出异常
        val result=ArrayList<Int>()
        for (i in 1..number){
            result.add(Random.nextInt(startRange,endRange))  //随机生成的数字加入列表
        }
        return result
    }
}
```

(2) 定义 LotteryViewModel,代码如下:

```
class LotteryViewModel: ViewModel() {
    private val _output=MutableStateFlow(ArrayList<Int>())         //内部修改的状态值,私有
    val output: StateFlow<ArrayList<Int>>=_output.asStateFlow()    //获取状态流,只能为只读
    var numberInput by mutableStateOf("6")                         //界面彩票数字状态
    var startInput by mutableStateOf("1")                          //界面彩票开始数字状态
    var endInput by mutableStateOf("30")                           //界面彩票结束数字状态
```

```kotlin
    fun changeUI(numberStr:String,startStr:String,endStr:String) {        //用于修改界面
        numberInput=numberStr
        startInput=startStr
        endInput=endStr
    }
    private fun generateLottery(number: Int, startRange: Int=1, endRange: Int=30):
        LotteryTicket=LotteryTicket(number,startRange,endRange)            //生成彩票对象
    fun get(number: Int, startRange: Int, endRange: Int) {
        _output.value=generateLottery(number,startRange,endRange).generate()  //生成状态
    }
}
```

在 LotteryViewModel 类中_output 是一个私有 MutableStateFlow，包含保存彩票所有数字的列表。这个私有值可以在 LotteryViewModel 中修改变更，但外界是无法调用它的。为此，在这个 ViewModel 组件中，调用_output.asStateFlow()创建了对应的 output 的 StateFlow 对象。output 这个唯一作用就是为外界提供单向流，提供_output 变化的数据。

(3) 定义界面，代码如下：

```kotlin
@Composable
fun HomeScreen(modifier: Modifier, viewModel: LotteryViewModel) {
    val output: State<ArrayList<Int>>=viewModel.output.collectAsState()
                                                                       //采集状态流的状态
                                                                       //定义界面的状态
    var number by remember { mutableStateOf("6") }
    var startRange by remember { mutableStateOf("1") }
    var endRange by remember { mutableStateOf("30") }
    var isGenerated by remember { mutableStateOf(false) }
    Box(
        modifier=modifier
            .fillMaxSize()
            .padding(30.dp),
        contentAlignment=Alignment.Center
    ) {
        Column(
            horizontalAlignment=Alignment.CenterHorizontally,
            modifier=Modifier.align(Alignment.Center)
        ) {
            Text("彩票模拟生成器", fontSize=30.sp, color=colorResource(id=R.color.teal_
                200))
            Row(verticalAlignment=Alignment.CenterVertically) {
                Text("彩票的个数", color=colorResource(id=R.color.teal_200))
                OutlinedTextField(value=number,                         //输入数字文本框
                    onValueChange={
                        number=it
                        viewModel.changeUI(number, startRange, endRange)  //修改 UI
                    }
                )
            }
            Row(verticalAlignment=Alignment.CenterVertically) {         //开始范围文本框
```

```kotlin
            Text("彩票数字的范围：", color=colorResource(id=R.color.teal_200))
            OutlinedTextField(
                value=startRange,
                onValueChange={
                    startRange=it
                    viewModel.changeUI(number, startRange, endRange)    //修改UI
                },modifier =Modifier.width(100.dp)
            )
            Text("--")
            OutlinedTextField(                                          //结束范围文本框
                value=endRange,
                onValueChange={
                    endRange=it
                    viewModel.changeUI(number, startRange, endRange)    //修改UI
                }, modifier=Modifier.width(100.dp)
            )
        }
        Button(
            onClick={
                var n=viewModel.numberInput.toInt()
                val start=viewModel.startInput.toInt()
                val end=viewModel.endInput.toInt()
                viewModel.get(n, start, end)                            //生成彩票
                isGenerated=!isGenerated                                //修改生成状态值
            }, colors=ButtonColors(
                containerColor=colorResource(id=R.color.teal_700),
                contentColor=Color.White,
                disabledContentColor=Color.Blue,
                disabledContainerColor=Color.DarkGray
            )
        ) {Text("生成彩票") }
        if (isGenerated) {
           Row(modifier=Modifier.fillMaxWidth(),
               verticalAlignment=Alignment.CenterVertically) {
               for (n in output.value) {                                //遍历状态保存的彩票数字列表
                   Text("$n", textAlign=TextAlign.Center, fontSize=20.sp,
                       color=Color.White,
                       modifier=Modifier.background(colorResource(R.color.teal_
                       700), shape=CircleShape).width(30.dp).height(30.dp)
                   )
               }
           }
        }
    }
}
```

在可组合项 HomeScreen 内调用 viewModel.output.collectAsState，获得 StateFlow 的状态，从而得到了 LotteryViewModel 的单向流，达到更新界面的目的。

（4）MainActivity 生成 ViewModel 对象并调用界面，用 ViewModel 对象处理业务逻辑。代码如下：

```
class MainActivity : ComponentActivity() {
    override fun onCreate(savedInstanceState: Bundle?) {
        super.onCreate(savedInstanceState)
        enableEdgeToEdge()
        val viewModel=ViewModelProvider(this)[LotteryViewModel::class.java]
                                                        //获取 ViewModel 对象
        setContent {
            Chapter04Theme {
                Scaffold(modifier=Modifier.fillMaxSize()) {
                    innerPadding ->HomeScreen(modifier=Modifier.padding(innerPadding),
                        viewModel=viewModel)
                }
            }
        }
    }
}
```

运行结果如图 4-22 所示。

图 4-22 彩票生成运行结果

4.4.2 MVI 模式

MVI(Model-View-Intent)模式是一种单向数据流处理方式。

虽然 Model 在 MVI 模式中,翻译为模型,但表示的并不是传统所理解的数据模型,而是状态 (State),即变更 UI 界面的状态(State)。一般定义成类:

```
data classAppState(var 状态值:数据类型,val 状态: MutableState<数据类型>,…)
```

View 表示视图。这个视图可以是 Activity,也可以是 Fragment,也可以是加载在 Activity/ Fragment 中的基于 Compose 定义 UI 界面的可组合项。

Intent 表示意图。这个意图并不是 Activity 之间跳转的意图,它所表示的是在移动应用中需要做出的操作。用户的任何希望执行的操作,都可以定义为 Intent 意图。执行用户操作的 Intent 意图可以封装需要传递的数据。一般定义成如下形式：

```
sealed class Intent {                                    //定义意图
    data class UserIntent1(状态:数据类型):TestIntent()    //定义带参的用户意图
    data object UserIntent2:TestIntent()                 //定义不带参的用户意图
    …
}
```

在定义意图的前提下，又因为本书采用 Compose 组件构建 UI 界面，在交互动作处理时，封装 Intent 对象，将传递相应的状态给 Intent。请求 viewModel 处理用户意图。可以将视图的处理成类似以下形式：

```
@Composable
fun SomeScreen(viewModel:SomeViewModel){
    val output=viewModel.output.collectAsState()
    …
    Button(onClick={
        viewModel.processIntent(UserIntent1(状态))        //请求处理意图
    }){
        …
    }
}
```

如图 4-23 所示，用户交互操作中，将封装状态数据的 Intent 通知给 Model。Model 从接受的 Intent 中获取更新的状态。View 订阅了 Model 变化的状态，接受到变化的状态，UI 界面从而刷新重构。视图模型中承担请求处理意图的动作。具体处理意图的动作由 ViewModel 来承担。在 ViewModel 根据不同的用户意图，依次做出处理，自定义 ViewModel 类似如下处理方式：

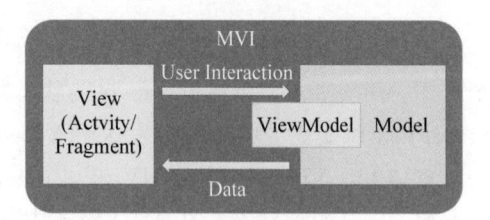

图 4-23　MVI 模式结构

```
class SomeViewModel:ViewModel(){
    private var _output:MutableStateFlow=…
    val output:StateFlow=_output.asStateFlow()

    fun processIntent(intent:UserIntent){                //实际处理意图
        when(intent){                                    //根据意图类别依次处理
            is UserIntent1->业务处理 1                    //在处理的过程中会修改_output 保存状态
            is UserIntent2->业务处理 2
            …
        }
    }
}
```

这样的方式使得数据状态在一个环形的单向数据流中流动，如图 4-24 所示。

在 MVI 模式中，也从上面的 ViewModel 定义形式，可知 ViewModel 起到非常重要的作用。在图 4-25 的 MVI 总体架构中展示了 ViewModel 和其他成员的相互关系。

· 139 ·

Model主要指UI状态（State）

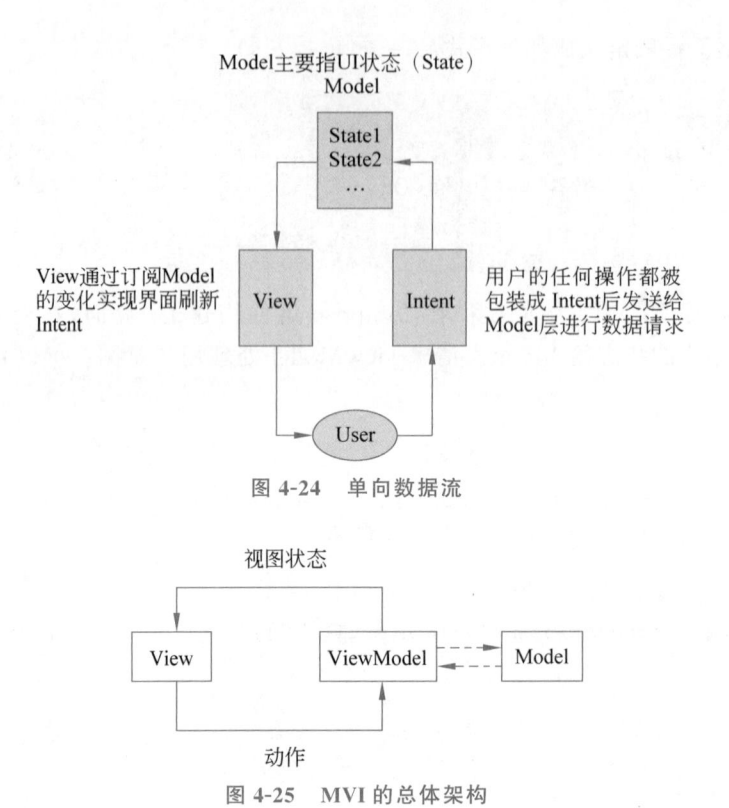

图 4-24　单向数据流

图 4-25　MVI 的总体架构

ViewModel 仍起到 View 和 Model 的交互的中间媒介。当用户通过 View 发起一个动作，执行一个意图，ViewModel 会将 Intent 封装状态数据，并发送给 Model。Model 变更后，再通过 ViewModel 的 StateFlow 将变更的状态数据发送给 View 视图，实现 View 界面更新。

例 4-11　ViewModel 在 MVI 模式的应用实例。定义一个彩票生成的简单应用，其他要求同例 4-10。

（1）定义彩票状态。在 MVI 模式中，M 字母虽然代表模式，但是它表示的是状态模型。在彩票状态模型中定义了输入的彩票数字个数和范围，以及获取最后的彩票的结果，代码如下：

```kotlin
//LotteryTicket.kt
data class LotteryTicket(
var number:Int=10,var start:Int=1,
    var end:Int=13,var result:List<String>=listOf()
)
```

（2）定义意图。这里的意图是彩票生成中需要完成的业务，不是 Activity 中定义的 Intent。此处的业务包括修改界面的意图和展示生成彩票的意图，代码如下：

```kotlin
//LotteryIntent.kt
sealed class LotteryIntent {
    data class UpdateIntent(val number:Int,val start:Int,var end:Int):LotteryIntent()
                                                                    //修改输入界面意图
    data object DisplayResultIntent:LotteryIntent()                 //显示结果意图
}
```

（3）定义视图模式。在视图中需要实现定义输入彩票的状态来更新界面。在视图中需要通过

ViewModel(视图模型)实现与界面与状态的交互。这里定义 LotteryViewModel,它处理不同意图。处理意图的过程中会修改对应的彩票状态值,为后续界面的更新提供保障,代码如下:

```kotlin
//LotteryViewModel.kt
class LotteryViewModel: ViewModel() {
    private var _output:MutableStateFlow<LotteryTicket>=MutableStateFlow(LotteryTicket())
                                                                                //内部修改
    val output=_output.asStateFlow()                                            //提供外部
    fun handleIntent(intent:LotteryIntent){
        val currentState=_output.value
        when(intent){
            is LotteryIntent.UpdateIntent->{                                    //处理修改业务
                val newState=currentState.copy(intent.number,intent.start,intent.end)
                                                                                //从意图获取数据
                _output.value=newState                                          //变更状态
            }
            is LotteryIntent.DisplayResultIntent->{                             //展示结果业务
                val newState=currentState.copy()
                newState.result=generateTicket()                                //生成彩票字符串
                _output.value=newState                                          //变更状态
            }
        }
    }

    private fun generateTicket():String{                                        //生成彩票数字
        var result=mutableListOf<String>()
        val rand=Random()
        val currentState=_output.value
        val diff=currentState.end-currentState.start
        for (i in currentState.start until currentState.end+1)
            result.add("${rand.nextInt(diff)+currentState.start} ")
        return result
    }
}
```

在视图中定义了 LotteryView 可组合函数,它专门用于处理输入彩票的状态,代码如下:

```kotlin
//LottertyView 可组合函数
@Composable
fun LotteryView(ticket:LotteryTicket,onReceivedIntent:(LotteryIntent)->Unit) {
    Column(modifier=Modifier.fillMaxWidth().wrapContentSize(),
        horizontalAlignment=Alignment.CenterHorizontally,
        verticalArrangement=Arrangement.Center ) {
        Row (verticalAlignment=Alignment.CenterVertically){
            Text("彩票个数: ")
            OutlinedTextField (
                modifier=Modifier.padding(end=10.dp),
                value="${ticket.number}",
                onValueChange={                                                 //接受修改界面意图
                    onReceivedIntent.invoke (LotteryIntent.UpdateIntent (it.toInt (),
                        ticket.start,ticket.end))
                }
            )
        }
        Row(verticalAlignment=Alignment.CenterVertically){
```

```
            Text("范围: ")
            OutlinedTextField (
                modifier=Modifier.size(100.dp,50.dp),
                value="${ticket.start}",
                onValueChange={                                          //接受修改界面意图
                    onReceivedIntent(LotteryIntent.UpdateIntent(ticket.number,
                        it.toInt(),ticket.end))
                }
            )
            OutlinedTextField (
                modifier=Modifier.size(100.dp,50.dp).padding(start=10.dp),
                value="${ticket.end}",
                onValueChange={                                          //接受修改界面意图
                    onReceivedIntent(LotteryIntent.UpdateIntent(ticket.number,
                        ticket.start,it.toInt()))
                }
            )
        }
        Button(onClick={
            onReceivedIntent(LotteryIntent.DisplayResultIntent)          //接受显示结果意图
        },colors=ButtonColors(
            containerColor=colorResource(id=R.color.teal_700),
            contentColor=Color.White,
            disabledContentColor=Color.Blue,
            disabledContainerColor=Color.DarkGray)
        ){
            Text("生成彩票")
        }
    }
}
```

另外定义 LotteryScreen 可组合函数, 它调用 LotteryView 可组合函数处理输入界面, 同时也展示最后的彩票生成的结果, 代码如下:

```
//LotteryScreen.kt
@Composable
fun LotteryScreen(modifier: Modifier,viewModel:LotteryViewModel) {
    val output=viewModel.output.collectAsState()
    Column(
        modifier=modifier.fillMaxSize(),
        horizontalAlignment=Alignment.CenterHorizontally,
        verticalArrangement=Arrangement.Center
    ) {
        Text("彩票模拟生成应用", fontSize=32.sp)
        LotteryView(output.value) {
            viewModel.handleIntent(it)
        }
        Row(verticalAlignment=Alignment.CenterVertically) {
            for (n in output.value.result) {                             //遍历彩票数字
                Text ("$n", textAlign=TextAlign.Center, fontSize=20.sp,
                    color=Color.White,
                    modifier=Modifier.background(colorResource(R.color.teal_700),
                        shape=CircleShape).size(30.dp,30.dp)
```

```
            )
        }
    }
}
```

（4）MainActivity 渲染最后的显示界面。在 MainActivity 中创建视图模型 LotteryViewModel 的对象实例,并传参给要调用的 LotteryScreen,用于处理核心业务,更新界面,代码如下:

```
//MainActivity.kt
    class MainActivity : ComponentActivity() {
    override fun onCreate(savedInstanceState: Bundle?) {
        super.onCreate(savedInstanceState)
        enableEdgeToEdge()
        val viewModel=ViewModelProvider(this).get(LotteryViewModel::class.java)
                                                                    //创建视图模型对象
        setContent {
            Chapter04_Theme {
                Scaffold(modifier=Modifier.fillMaxSize()) {    //调用视图
                    innerPadding ->LotteryScreen(modifier=Modifier.padding(innerPadding),
                        viewModel=viewModel)
                }
            }
        }
    }
}
```

4.5　Navigation 组件

一个移动应用 App 由多个内容界面构成,它们是基于 Compose 组件定义的可组合项,它们都可视为一个导航的目的地。这些界面之间的切换让移动应用呈现出多种功能。Navigation 组件实现可组合项界面之间的切换。通过"返回堆栈"来实现管理界面的切换。如果希望某个界面成为当前界面,需要将这个界面推进到"返回堆栈"中,并在导航堆栈中处于栈顶,这时它为当前的界面,当某个界面从"返回堆栈"中出栈后,处于栈顶的其他界面将成为新的当前界面。图 4-26 展示了这样的"返回堆栈"管理界面切换的情况。

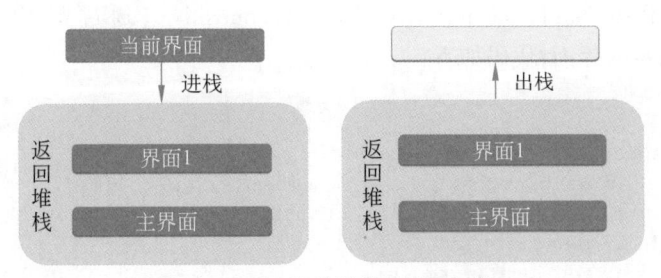

图 4-26　导航堆栈管理界面切换

Navigation 组件来处理导航业务。Navigation 组件中执行导航需要设置。
（1）NavHost(导航宿主),通过该可组合项定义导航图,明确移动应用的路线对应的显示的屏幕界面。

（2）NavController（导航控制器），它管理"返回堆栈"。NavController 可跟踪返回堆栈可组合项、使堆栈向前移动、支持对返回堆栈执行操作，以及执行在不同屏幕界面的导航。NavController 还有一个子类 NavHostController，为 NavHost（导航宿主）中使用。通过调用 rememberNavController() 可以获得一个 NavController 对象。

通过它们就能用最少量的代码实现导航。

要应用 Navigation 组件，对项目的 libs.versions.toml 配置如下内容：

```
[versions]
navigationRuntimeKtx="2.7.7"
navigationCompose="2.7.7"
[libraries]
androidx-navigation-runtime-ktx={
    group="androidx.navigation", name="navigation-runtime-ktx",
        version.ref="navigationRuntimeKtx"
}
androidx-navigation-compose ={
    group="androidx.navigation", name="navigation-compose",version.ref="navigationCompose"
}
```

模块的构建配置文件 build.gradle 增加如下依赖：

```
implementation(libs.androidx.navigation.runtime.ktx)
```

4.5.1 页面导航的实现

为了更好地理解 Navigation 组件实现不同界面的导航切换。本节通过一个展示机器人相关信息的简单应用来说明导航组件的实现。在这个应用中定义两个界面，一个界面显示机器人列表，通过单击机器人列表的单项，导航到特定机器人单项的具体显示。

首先，定义数据实体类 Robot，表示机器人，代码如下：

```
/** Robot 实体类
  * @property iconId Int：图标资源编号
  * @property title String：标题
  * @property content String：内容描述 */
data class Robot(val iconId:Int,val title:String,val content:String)
```

定义导航图界面，使之支持在不同界面的切换。为了实现不同界面的切换，需要调用 rememberNavController 获得 NavHostController 对象 navController。通过调用 navController.navigation("路径")，跳转到路径指定的界面，代码如下：

```
@Composable
fun NavigationGraphScreen(){
    val navController:NavHostController=rememberNavController()        //创建一个宿主导航器
    NavHost(navController=navController,startDestination="robots"){    //定义宿主
        composable(route="robots"){
            RobotsScreen(navController)
        }
        composable(route="robotDetail") {
            it: NavBackStackEntry ->
                val robot=Robot(android.R.mipmap.sym_def_app_icon,"测试机器人",
```

```
            "测试 Android 机器人介绍")
            RobotDetailScreen(robot,navController)
        }
    }
}
```

NavigationGraphScreen 中定义的导航图中,用 NavHost 定义了宿主,在宿主中指定了开始的路径为"robots"。在 NavigationGraphScreen 设置了路径"robots"对应可组合项 RobotsScreen 和路径"robotDetail"对应可组合项 RobotDetailScreen。为了在不同界面导航,将 navController 对象传入不同的界面中,方便这些界面实现导航。

定义 RotbotsScreen,展示机器人列表的可组合项构成的界面,代码如下:

```
@Composable
fun RobotsScreen(navController: NavController){
    val robots:MutableList<Robot>=mutableListOf()
    for (i in 1..20){
        robots.add(Robot(android.R.mipmap.sym_def_app_icon,"机器人$i",
            "第${i}个 Android 机器人介绍"))
    }
    Box(modifier=Modifier.fillMaxSize().background(colorResource(id=R.color.teal_700))
        .padding(20.dp),contentAlignment=Alignment.Center){
        Column{
            Text("首页界面",fontSize=36.sp,color=Color.White)
            LazyColumn{
                items(robots){
                    robot:Robot->RobotCard(robot,navController)
                                                    //定义机器人单项,也接受 navController
                }
            }
        }
    }
}
```

定义展示一个机器人单项的可组合项 RobotCard,代码如下:

```
@Composable
fun RobotCard(robot: Robot,navController: NavController){
    ConstraintLayout(modifier=Modifier.fillMaxWidth().padding(10.dp)
        .background(Color(0xf3,0xa2,0x13), RoundedCornerShape(20.dp))
        .border(BorderStroke(2.dp, Color(0xff,0xec,0xcb)), RoundedCornerShape(20.dp))){
        val vGuideLine=createGuidelineFromTop(0.5f)
        val hGuideLine=createGuidelineFromStart(0.3f)
        val (imageRef,titleRef,contentRef) = remember{createRefs()}
        createVerticalChain(titleRef,contentRef, chainStyle=ChainStyle.Packed)
        Image(
            painter=painterResource(id=robot.iconId),
            contentDescription=robot.title,
            modifier=Modifier.constrainAs(imageRef){
                top.linkTo(parent.top)
                start.linkTo(parent.start,20.dp)
                bottom.linkTo(parent.bottom)
                end.linkTo(hGuideLine)
```

```
            }.size(100.dp,100.dp).clickable{
                navController.navigate("robotDetail") //导航到"robotDetail"路径指向可组合项
            }
        )
        Text(
            text="${robot.title}",fontSize=24.sp,
            modifier=Modifier.constrainAs(titleRef){
                top.linkTo(parent.top)
                start.linkTo(hGuideLine,5.dp)
                bottom.linkTo(vGuideLine)
                end.linkTo(parent.end)
            }
        )
        Text(
            text="${robot.content}",fontSize=18.sp, minLines=5,
            modifier=Modifier.constrainAs(contentRef){
                top.linkTo(vGuideLine)
                start.linkTo(hGuideLine)
                end.linkTo(parent.end)
                bottom.linkTo(parent.bottom)
            }
        )
    }
}
```

RobotCard 中为 Image 可组合项增加了 clickable 的点击动作,在动作内部定义了调用 navController.navigate("robotDetail"),导航到路径"robotDetail"对应的界面 RobotDetailScreen 中。

定义 RobotDetailScreen,展示某个给机器人单项的具体信息界面,代码如下:

```
@Composable
fun RobotDetailScreen(robot: Robot,navController: NavController){
    Scaffold(
        floatingActionButton={
            FloatingActionButton(onClick={
                navController.navigate("robots")           //导航到路径"robots"指定的界面
            }) {
                Icon(Icons.Filled.Home,"返回")
            }
        }
    ){
      innerPadding->Column(modifier=Modifier
        .fillMaxSize()
        .background(colorResource(id=R.color.teal_700))
        .padding(innerPadding),
        horizontalAlignment=Alignment.CenterHorizontally,
        verticalArrangement=Arrangement.Center){
            Text("${robot.title}介绍",fontSize=36.sp, textAlign=TextAlign.Center,color
                =Color.White)
            Image(painterResource(id=robot.iconId),contentDescription="${robot.
                title}")
            Text("${robot.content}",fontSize=24.sp,color=Color.White)
        }
    }
}
```

RobotDetailScreen 界面的悬浮按钮定义了动作 navController.navigate("robots"),实现从当前界

面导航到路径名"robots"对应的可组合项 RobotsScreen。

然后定义 MainActivity，调用 NavigationGraphScreen 可组合项，代码如下：

```kotlin
class MainActivity: ComponentActivity() {
    override fun onCreate(savedInstanceState: Bundle?) {
        super.onCreate(savedInstanceState)
        enableEdgeToEdge()
        setContent {
            Chapter04Theme {
                NavigationGraphScreen()
            }
        }
    }
}
```

运行结果如图 4-27 所示。

图 4-27　导航运行结果

4.5.2　在目的地之间安全传递数据

观察运行结果，发现交互存在严重问题。点击机器人列表的单项只能转换到固定的机器人单项的内容，这与实际情况不符。因此，需要将具体的机器人对象作为参数传递给 RobotDetailsScreen 界面，使之能实现根据参数传递的情况，展示具体机器人的详细界面。

要实现不同界面在导航时，传递相应的参数。这时参数需要转换为 json 类型数据，这样传递数据时不会丢失，可以很容易地将数据恢复。需要做 3 方面的处理。

（1）为了使用 json 数据，增加 Gson 依赖库。

在项目 libs.versions.toml 中增加 Gson 版本配置，代码如下：

```
[versions]
gson="2.10"
[libraries]
gson={module="com.google.code.gson:gson",version.ref="gson"}
```

· 147 ·

在项目的模块对应的 build.gradle.kts 中增加依赖库,代码如下:

```
implementation(libs.gson)
```

(2) 数据的发送方:

```
val jsonStr=Gson().toJson(obj)                    //数据对象转换成 json 字符串
navController.navigate("导航路径/${jsonStr}")      //导航路径后面增加传递参数的 json 字符串
```

(3) 导航图中配置数据接收方接受数据的说明:

```
NavHost{
    composable(route="导航路径"+ "/{参数名}",                    //路径增加参数名
        arguments=listOf(navArgument("参数名 1"){              //配置参数
        type=NavType.StringType                              //指定参数的类型
    })) {
        it: NavBackStackEntry->val jsonStr=it.arguments?.getString("参数名 1") ?: "Error"
                                                             //按照参数接收 json 字符串
        val object=Gson().fromJson(jsonStr, 数据类::class.java)
                                                             //将 json 字符串转换成对象
        SomeScreen(object,…)                                 //将对象传递到屏幕界面
    }
    …                                                        //略
}
```

根据上述处理规则,将 4.5.1 节中机器人简单应用的导航做两处修改。

修改导航图 NavigationGraphScreen 可组合项,为 RobotDetailScreen 配置接收数据参数,代码如下:

```
@Composable
fun NavigationGraphScreen(){
    val navController:NavHostController=rememberNavController()           //创建一个宿主导航器
    NavHost(navController=navController,startDestination="robots"){       //定义宿主
        composable(route="robots"){
            RobotsScreen(navController)
        }
        composable(route="robotDetail/{robot}",
            arguments=listOf(navArgument("robot"){                         //导航参数
            type=NavType.StringType                                        //参数类型
        })) {
            it: NavBackStackEntry ->                                       //接收数据的处理
            val robotStr=it.arguments?.getString("robot") ?: "Error"       //接收 json 字符串
            val robot=Gson().fromJson(robotStr, Robot::class.java)
                                                                           //将 json 字符串转换为对象
            RobotDetailScreen(robot,navController =navController)          //将 robot 参数传递给可组合项
        }
    }
}
```

修改 RobotsScreen 调用的 RobotCard 显示机器人单项,在 RobotCard 中增加数据发送的处理,代码如下:

```
@Composable
fun RobotCard(robot: Robot,navController: NavController){
    ConstraintLayout(modifier=Modifier.fillMaxWidth()
        .padding(10.dp).background(Color(0xf3,0xa2,0x13), RoundedCornerShape(20.dp))
```

```
        .border(BorderStroke(2.dp, Color(0xff,0xec,0xcb)), RoundedCornerShape(20.dp))){
    val vGuideLine=createGuidelineFromTop(0.5f)
    val hGuideLine=createGuidelineFromStart(0.3f)
    val (imageRef,titleRef,contentRef)=remember{createRefs()}
    createVerticalChain(titleRef,contentRef, chainStyle=ChainStyle.Packed)
    Image(painter=painterResource(id=robot.iconId),
        contentDescription=robot.title, modifier=Modifier.constrainAs(imageRef){
            top.linkTo(parent.top)
            start.linkTo(parent.start,20.dp)
            bottom.linkTo(parent.bottom)
            end.linkTo(hGuideLine)
        }.size(100.dp,100.dp).clickable {
            val jsonStr=Gson().toJson(robot)          //将robot对象转换成json字符串
            navController.navigate("robotDetail/${jsonStr}")
                                                //带参数导航到robotDetail路径对应的界面
        }
    )
    Text(text="${robot.title}",fontSize=24.sp,
        modifier=Modifier.constrainAs(titleRef){
            top.linkTo(parent.top)
            start.linkTo(hGuideLine,5.dp)
            bottom.linkTo(vGuideLine)
            end.linkTo(parent.end)
        }
    )
    Text(text="${robot.content}",fontSize=18.sp, minLines=5,
        modifier=Modifier.constrainAs(contentRef){
            top.linkTo(vGuideLine)
            start.linkTo(hGuideLine)
            end.linkTo(parent.end)
            bottom.linkTo(parent.bottom)
        }
    )
  }
}
```

经过修改后,重新运行应用,可以发现在两个界面之间的切换符合实际要求。运行结果如图4-28所示。

图4-28 参数传递的导航实现

4.6 心理测试移动应用实例

4.6.1 项目说明

设计一个用于心理测试的移动应用。要求用户根据心理测试的题目进行选择,每题的答案只能选择一个,单选的选择是"A.很符合自己的情况""B.比较符合自己的情况""C.介于符合与不符合之间""D.不大符合自己的情况"和"E.很不符合自己的情况"。回答题号为奇数的题目时,选中选项 A 得 5 分,选中选项 B 得 4 分,选中选项 C 得 3 分,选中选项 D 得 2 分,选中选项 E 1 分;回答题号为偶数的题目时,题号选中选项 A 得 1 分,选中选项 B 得 2 分,选中选项 C 得 3 分,选中选项 D 得 4 分,选中选项 E 得 5 分。评测完毕提交结果,统计评测的总分,可以判断并显示心理测试的等级。移动测试的结果可以分享。此外,要求移动应用能对心理测试的相关情况进行介绍,以及对心理测试的移动应用的开发情况进行介绍。根据需求描述,绘制如图 4-29 所示的心理测试的用例图。

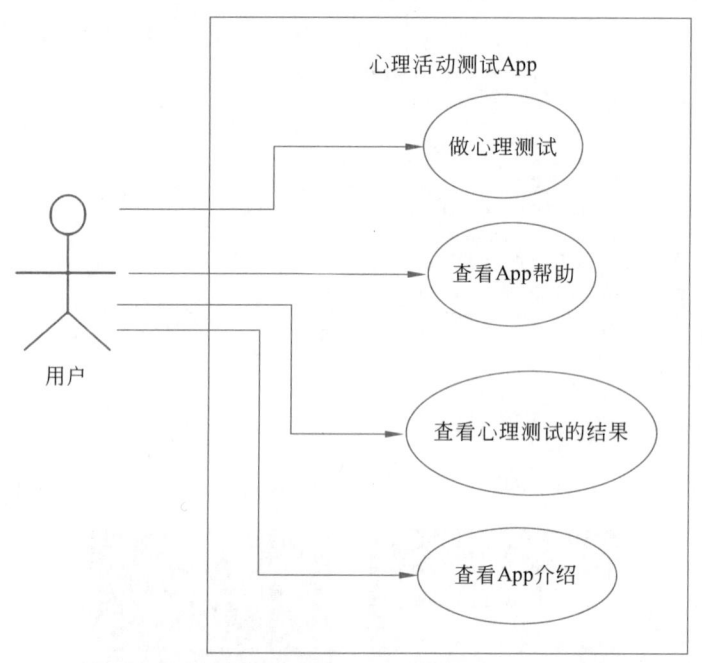

图 4-29 心理测试的用例图

4.6.2 心理测试移动应用的功能实现

为了心理测试能顺利进行,将所有的测试题目编辑到 res/values/arrays.xml 资源中,形式如下:

```
<!--模块 TestApp 测试题目定义 arrays.xml 局部 -->
<resources>
    <string-array name="questions">
        <item>1.我很喜欢…</item>
        <item>2.我给自己订的计划…</item>
        …
        <!--其他选项略-->
```

```xml
    </string-array>
</resources>
```

1. 状态模型的定义

在心理测试应用中，需要所有心理测试问题、记录用户回答的情况、统计用户心理测试的分数、根据统计的分数判断心理测试的分析结果、退出应用的控制状态。根据这些内容，定义状态模型类 TestState 类，代码如下：

```kotlin
data class TestState(
    val questions:MutableList<String>=mutableListOf(),      //加载问题
    val answers:MutableList<String>=mutableListOf(),        //记录答案
    var score:Int=0,                                        //总分
    var result:String="",                                   //分数对应的结论
    var exitState:Boolean=false)                            //退出应用
```

为了方便处理，res/values/strings.xml 定义了备选答案形如：

```xml
<string name="choice_a">A.很符合自己的情况</string>
<string name="choice_b">B.比较符合自己的情况</string>
<string name="choice_c">C.介于符合与不符合之间</string>
<string name="choice_d">D.不大符合自己的情况</string>
<string name="choice_e">E.很不符合自己的情况</string>
```

定义如下函数获取备选答案的所有数据：

```kotlin
@Composable
fun getPossibleAnswers()=listOf(
    stringResource(id=R.string.choice_a),
    stringResource(id=R.string.choice_b),
    stringResource(id=R.string.choice_c),
    stringResource(id=R.string.choice_d),
    stringResource(id=R.string.choice_e))
```

为了方便管理屏幕的各种显示的各种状态的设置，这里定义 StateHolder 类，以及获得唯一的 StateHolder 的对象的 rememberState 函数，代码如下：

```kotlin
data class StateHolder(
    val navController: NavHostController,                   //导航控件
    val drawerState:DrawerState,                            //控制侧滑的菜单
    val exitState: MutableState<Boolean>,                   //退出控制
    val currentRouteState: MutableState<String>,            //当前导航的路径
    val openState:MutableState<Boolean>,                    //对话框启动
    val returnState:MutableState<Boolean>                   //悬浮按钮的返回
)

@Composable
fun rememberState(navController: NavHostController=rememberNavController(),
    drawerState: DrawerState=rememberDrawerState(initialValue=DrawerValue.Closed),
    exitState:MutableState<Boolean>=remember {mutableStateOf(false)},
```

```
            currentRouteState: MutableState<String>=remember{mutableStateOf("")},
            openState:MutableState<Boolean>=remember {mutableStateOf(false)},
            returnState:MutableState<Boolean>=remember{mutableStateOf(false)})
        =remember(navController,drawerState,exitState,currentRouteState,openState,returnState){
            StateHolder(navController,drawerState,exitState,currentRouteState,openState,
                returnState)
        }
```

2. 定义用户意图

用户在心理测试中,需要解决的问题包括初始化心理测试的意图、回答测试的意图、查看测试解答情况的意图、显示测试分析结果的意图、将心理测试结果分享的意图以及退出应用的意图。据此定义 TestIntent 类来表示这些意图:

```
sealed class TestIntent {
    //初始化心理测试
    data class InitialTestIntent(val questions:MutableList<String>,
        val answers:MutableList<String>):TestIntent()
    //回答测试
    data class AnswerTestIntent(val answers:MutableList<String>):TestIntent()
    //显示分数
    data class DisplayAnswerIntent(val answers:MutableList<String>):TestIntent()
    //显示测试结果
    data class DisplayResultIntent(val answers:MutableList<String>):TestIntent()
    //分享测试结果
    data class ShareResultIntent(val result:String):TestIntent()
    //退出应用
    data object ExitAppIntent:TestIntent()
}
```

3. 定义视图模型

ViewModel 是模型和视图交互的媒介。定义 TestViewModel,在 TestViewModel 中定义保存 TestState 状态的私有 MutableStateFlow,在 TestViewModel 内部根据不同的用户操作意图,对包含的 TestState 状态做出修改。同时,将 MutableStateFlow 的 StateFlow 提取出来,为外界的 View 视图提供数据模型。TestViewModel 的代码如下:

```
class TestViewModel: ViewModel() {
    private val _state=MutableStateFlow(TestState())     //包含 TestState 可变状态流,内部修改
    val output: StateFlow<TestState>=_state.asStateFlow()           //为视图提供单向数据流
    fun processIntents(context:Context, stateHolder: StateHolder, intent: TestIntent){
                                                                  //处理意图
        val currentState:TestState=_state.value
        when (intent) {                                           //根据不同意图分别处理
            is TestIntent.InitialTestIntent ->{                   //初始化心理测试
                val newState=currentState.copy(questions=intent.questions, answers=
                    intent.answers)                               //复制变更的状态值
                initTest(context,newState)
                changeState(newState)
                stateHolder.navController.navigate(Screen.TestPage.route)
```

```kotlin
                }
                is TestIntent.AnswerTestIntent ->{                    //回答测试意图处理
                    val newState=currentState.copy(answers=intent.answers)
                    changeState(newState)
                    Log.d("TAG t","answer:${_state.value.answers}")
                }
                is TestIntent.DisplayAnswerIntent ->{                 //显示分数意图处理
                    val newState=currentState.copy(answers=intent.answers)
                    newState.score=countScore(newState.answers)
                    changeState(newState)
                }
                is TestIntent.DisplayResultIntent ->{                 //显示测试结果意图处理
                    val newState=currentState.copy(answers=intent.answers)
                    val score=countScore(newState.answers)
                    newState.score=score
                    newState.result=getResult(score)
                    changeState(newState)
                    stateHolder.navController.navigate(Screen.ResultPage.route)
                }
                is TestIntent.ShareResultIntent ->{                   //分享测试结果意图处理
                    val newState=currentState.copy(result=intent.result)
                    changeState(newState)
                    shareResult(context,_state.value.result)
                }
                is TestIntent.ExitAppIntent ->{                       //退出应用意图处理
                    val newState=currentState.copy(exitState=true)
                    changeState(newState)
                    if (_state.value.exitState) {
                        exitApp()
                    }
                }
            }
        }
    }
    private fun initTest(context:Context,newState:TestState){          //加载题库
        newState.questions.clear()
        val questions=context.resources.getStringArray(R.array.questions)
        for (question in questions){
            newState.questions.add(question)
            newState.answers.add("")
        }
        newState.result=""
        newState.score=0
        newState.exitState=false
    }
    private fun evaluateAnswer(questionId:Int,scoreStr:String):Int{    //评判答案的分数
        var score=0
        if (questionId%2==0) {
            score=when (scoreStr) {
```

```kotlin
                    "A"->5
                    "B"->4
                    "C"->3
                    "D"->2
                    else ->1
                }
            }else{
                score=when(scoreStr){
                    "A" ->1
                    "B" ->2
                    "C" ->3
                    "D" ->4
                    else ->5
                }
            }
            return score
        }
        private fun countScore(scoreArr:List<String>):Int{            //统计分值
            var result=0
            for(i in scoreArr.indices){
                result+=evaluateAnswer(i,scoreArr[i])
            }
            return result
        }
        private fun getResult(score:Int):String=when(score){          //根据分值获取测试的结论
            in 90..130 ->"得分：${score}分,自制力很强"
            in 60..89->"得分：${score}分,自制力尚可"
            else ->"得分：${score}分,自制力不足"
        }
        private fun shareResult(context: Context, message:String){    //分享最后的结果
            val intent=Intent(Intent.ACTION_SEND)
            intent.apply {
                type="text/plain"
                putExtra(Intent.EXTRA_SUBJECT, "主题")
                putExtra(Intent.EXTRA_TEXT,"分享：$message")
                setFlags(Intent.FLAG_ACTIVITY_NEW_TASK)
            }
            context.startActivity(intent)
        }
        private fun changeState(newState:TestState){                  //修改状态
            _state.value=newState
        }
        private fun exitApp(){                                        //退出应用
            exitProcess(0)
        }
    }
```

4. 视图的定义

视图中定义了多个可组合项，构成不同的屏幕界面。这些屏幕界面包括：应用开始界面的界面 LogoScreen、关于应用介绍的界面 AboutAppScreen、包括测试介绍的界面 AboutTestScreen、心理测试

界面 TestScreen、显示答题情况的界面 AnswersScreen、显示心理测试结果的界面 Result.Screen、导航图界面 NavigationGraphScreen 以及搭建脚手架的主界面 HomeScreen。为了构建这样的界面，还需要定义其他可组合项。在后续一一介绍。

（1）定义 Screen 类表示界面的类。将在底部栏需要导航处理的界面定义在一个界面的列表中，代码如下：

```
/** 定义屏幕的描述
 * @property route String: 屏幕路径
 * @property title String: 屏幕标题
 * @property icon Int: 图标资源编号 */
sealed class Screen(val route:String, val title:String, val icon: Int){
    data object LogoPage:Screen(route="logo",title = "Logo 显示",icon=R.mipmap.logo)
    data object TestPage:Screen(route="test",title="心理测试", icon=R.mipmap.test)
    data object ResultPage:Screen(route="result",title="测试结果",icon=R.mipmap.result)
    data object AnswerPage:Screen(route="answer",title="已选择答案",icon =R.mipmap.logo_
        test2)
    data object AboutAppPage:Screen(route="aboutApp",title="心理测试 APP 介绍",
        icon=R.mipmap.about)
    data object AboutTestPage:Screen(route="aboutTest",title="心理测试说明",icon=
        R.mipmap.desc)
}

val screens=listOf(Screen.TestPage,Screen.ResultPage,Screen.AboutAppPage)
```

（2）定义 HomeScreen。HomeScreen 定义的脚手架 Scaffold，表示了整个移动应用的 UI 结构，包括了顶部栏、底部栏、侧滑菜单、悬浮按钮和中心区，代码如下：

```
@Composable
fun HomeScreen(viewModel: TestViewModel){
    val context=LocalContext.current                              //当前的上下文
    val stateHolder: StateHolder=rememberState()                  //状态容器
    val outputState=viewModel.output.collectAsState()             //从 viewModel 获取状态
    val scope=rememberCoroutineScope()                            //获取协程范围
    Scaffold(                                                     //定义脚手架
        topBar={                                                  //定义顶部栏
            TopViews(stateHolder,outputState.value){
                viewModel.processIntents(context,stateHolder,it)  //处理意图
            }
        },
        bottomBar={                                               //定义底部栏
            BottomView(stateHolder){
                viewModel.processIntents(context,stateHolder,it)  //处理意图
            }
        },
        content={                                                 //定义中心区
            it: PaddingValues->Box(modifier=Modifier.padding(it)){

                if (stateHolder.openState.value)
```

```
                ExitDialog(stateHolder){                           //定义退出对话框
                    viewModel.processIntents(context,stateHolder,it)
                }
            DrawerViews(stateHolder,viewModel)                     //中心区的核心内容
        }
    },
    floatingActionButton={                                         //定义悬浮按钮
        FloatingActionButton(
            onClick={
                stateHolder.returnState.value=!stateHolder.returnState.value
                scope.launch {
                    if (stateHolder.returnState.value)             //导航到显示答题情况界面
                        stateHolder.navController.navigate(Screen.AnswerPage.route)
                    else                                           //导航到测试界面
                        stateHolder.navController.navigate(Screen.TestPage.route)
                }
            },
            containerColor=Color(0xFF408D58),
            contentColor=Color.White,
            modifier=Modifier.shadow(2.dp,shape=CircleShape)
        ) {
            if (!stateHolder.returnState.value)
                Icon(Icons.Filled.Info,contentDescription="show")
            else
                Icon(Icons.AutoMirrored.Filled.ArrowBack,contentDescription="return")
        }
    }
)
}
```

上述定义的 Scaffold 中包含了脚手架的所有的元素,显示的效果如图 4-30 所示。

图 4-30　脚手架结构的运行效果

TopViews 定义了顶部栏，代码如下：

```
/* * TestApp 模块的 TopViews */
@OptIn(ExperimentalMaterial3Api::class)
@Composable
fun TopViews(stateHolder: StateHolder, testState: TestState, onReceivedIntent:
    (TestIntent)->Unit){
    val scope=rememberCoroutineScope()                              //协程范围
    val context=LocalContext.current                                //上下文
    TopAppBar(modifier=Modifier.fillMaxWidth().wrapContentHeight(),
        title={
            Text(stringResource(id=R.string.app_name), fontSize=20.sp)
        },
        navigationIcon={                                            //定义顶部导航按钮
            IconButton(onClick={
                scope.launch {
                    stateHolder.drawerState.run{                    //控制侧滑菜单打开关闭
                        if (isClosed)
                            open()
                        else
                            close()
                    }
                }
            }){
                Icon(painterResource(id=R.mipmap.logo_test2), contentDescription=
                    "logo", tint=Color.Unspecified)
            }
        },
        actions={                                                   //顶部栏右侧动作处理
            IconButton(onClick={
                scope.launch {                                      //请求执行分享意图
                    onReceivedIntent(TestIntent.ShareResultIntent(testState.result))
                }
            }){
                Icon(Icons.Filled.Share, contentDescription="Share")
            }
        },
        colors=TopAppBarDefaults.centerAlignedTopAppBarColors(containerColor=
            Color(0xFF94D37D))
    )
}
```

TopViews 可组合项中需要传递函数对象 onReceivedIntent：(TestIntent)—>Unit 来发送 TestIntent.ShareResultIntent(testState.result)，要求处理分享意图，并封装了相关的状态数据。在这里是测试的结果。

BottomViews 定义了底部栏，代码如下：

```
@Composable
fun BottomView(stateHolder: StateHolder, onReceivedIntent: (TestIntent)->Unit){
    val scope=rememberCoroutineScope()                              //定义协程范围
    BottomAppBar(containerColor=Color(0xFF94D370) ) {                //定义底部应用栏
        screens.forEach{                                            //定义底部的导航项
            it: Screen->
```

```
                NavigationBarItem(                                  //定义导航栏单项
                    selected=it.route==stateHolder.currentRouteState.value,
                    onClick={
                        scope.launch {
                            stateHolder.currentRouteState.value=it.route
                            if (it.route==Screen.TestPage.route) {   //初始化心理测试意图
                                onReceivedIntent(TestIntent.InitialTestIntent(mutableListOf(),
                                    mutableListOf()))
                            } else {
                                stateHolder.navController.navigate(it.route)    //导航当前路径
                            }
                            stateHolder.drawerState.close()          //关闭侧滑菜单
                        }
                    },
                    colors=NavigationBarItemDefaults.colors(         //设置导航栏单项颜色
                        indicatorColor=Color(0xFF94D370),
                        selectedIconColor=Color.White,
                        selectedTextColor=Color.White),
                    label={                                          //设置文本
                        Text(text="${it.title}")
                    },
                    icon={                                           //设置图标
                        Icon(painter=painterResource(id=it.icon),contentDescription=null)
                    }
                )
            }
        }
    }
```

BottomViews 定义的底部导航中做了一个判断,如果导航心理测试 TestScreen 界面的路径,操作设置 onReceivedIntent(TestIntent.InitialTestIntent(mutableListOf(),mutableListOf()))意图。在 TestIntent.InitialTestIntent 初始化测试意图中封装的两个动态列表分别保存了所有题目数据和答题情况的数据。

DrawerViews 定义侧滑菜单已经相关的导航设置,代码如下:

```
//DrawerHeaderView 侧滑菜单的上部内容
@Composable
fun DrawerHeaderView(){
    ConstraintLayout(modifier=Modifier.size(300.dp,200.dp).background(Color
        (0xFF9FFE39))){
        val hguildline=createGuidelineFromStart(0.2f)                //水平导航线
        val (imageRef,titleRef)=remember {createRefs()}              //引用参数
        createHorizontalChain(imageRef,titleRef, chainStyle=ChainStyle.Packed)    //水平链
        Icon(modifier=Modifier.size(50.dp, 50.dp).constrainAs(imageRef) {
            top.linkTo(parent.top)
            start.linkTo(parent.start, 20.dp)
            end.linkTo(hguildline)
            bottom.linkTo(parent.bottom)
        },
        painter=painterResource(id=R.mipmap.logo_test2), tint=Color.Unspecified,
            contentDescription=null)
        Text(modifier=Modifier.constrainAs(titleRef){
```

```kotlin
            top.linkTo(parent.top)
            bottom.linkTo(parent.bottom)
            end.linkTo(parent.end)
            start.linkTo(hguildline)
        },
        text=stringResource(id=R.string.title_app_name),fontSize=30.sp)
    }
}
//侧滑菜单的中心内容,定义导航部分
@Composable
fun DrawerContentView(appStates: StateHolder){
    val scope=rememberCoroutineScope()                          //定义协程范围
    Column(modifier=Modifier.size(300.dp,1500.dp)
        .background(Color.White, shape=RectangleShape)){
        NavigationDrawerItem(                                   //定义导航侧滑单项
            modifier=Modifier.padding(NavigationDrawerItemDefaults.ItemPadding),
            label={                                             //定义提示标签
                Text("关于心理测试应用",fontSize=20.sp)
            },
            icon={                                              //定义图标
                Icon(painterResource(id=Screen.AboutTestPage.icon),
                    contentDescription=null,tint=Color.Unspecified)
            },
            selected=false,
            onClick={
                scope.launch {
                    appStates.navController.navigate("aboutApp")//导航到关于应用界面
                    appStates.drawerState.close()               //关闭侧滑菜单
                }
            },
            colors=NavigationDrawerItemDefaults.colors(         //配置颜色
                selectedContainerColor=Color.White,
                unselectedContainerColor=Color.White
            )
        )
        NavigationDrawerItem(                                   //导航侧滑单项
            label={                                             //提示标签
                Text("退出应用",fontSize=20.sp)
            },
            icon={                                              //图标
                Icon(painter=painterResource(id=R.mipmap.exit),contentDescription=null)
            },
            selected=false,
            onClick={                                           //动作处理
                appStates.openState.value=true                  //侧滑状态为打开
            },
            colors=NavigationDrawerItemDefaults.colors(         //设置颜色
                selectedContainerColor=Color.White,
                unselectedContainerColor=Color.White
            )
        )
```

```
            }
        }
        /* * 组装侧滑菜单DrawerViews*/
        @Composable
        fun DrawerViews(stateHolder: StateHolder, viewModel: TestViewModel){
            ModalNavigationDrawer(modifier=Modifier.fillMaxHeight().wrapContentWidth(),
                scrimColor=Color.Unspecified,
                drawerContent={                                                      //侧滑内容
                    Column{
                        DrawerHeaderView()
                        DrawerContentView(stateHolder)
                    }
                },
                drawerState=stateHolder.drawerState){                                //中心区内容
                NavigationGraphScreen(stateHolder,viewModel)                         //导航处理
            }
        }
```

在DrawerViews中的核心内容包括的导航图界面NavigationGraphScreen,通过它实现在脚手架的中心区切换不同界面可组合项。导航图界面的NavigationGraphScreen是应用的核心内容,它定义了导航的路径对应各自界面,并将整个导航路径展示出来,代码如下:

```
        /* *面向界面中心部分的导航图
         * @param stateHolder StateHolder: 状态容器
         * @param viewModel TestViewModel: 视图模型 */
        @Composable
        fun NavigationGraphScreen(stateHolder: StateHolder, viewModel: TestViewModel){
            val context=LocalContext.current                                         //当前上下文
            //定义导航图
            NavHost(modifier=Modifier.fillMaxSize(),                                 //修饰符
                navController=stateHolder.navController,                             //导航控制器
                startDestination=Screen.LogoPage.route){                             //开始的路径
                composable(route=Screen.LogoPage.route){                             //设置启动路径
                    LogoScreen(stateHolder,viewModel){
                        viewModel.processIntents(context,stateHolder,it)             //请求处理意图
                    }
                }
                composable(route=Screen.TestPage.route){                             //设置心理测试路径
                    TestScreen(stateHolder,viewModel){
                        viewModel.processIntents(context,stateHolder,it)             //请求处理意图
                    }
                }
                composable(route=Screen.AnswerPage.route){                           //设置答题情况显示路径
                    AnswersScreen(viewModel)
                }
                composable(route=Screen.ResultPage.route){                           //设置心理测试结果路径
                    ResultScreen(viewModel)
                }
                composable(route=Screen.AboutAppPage.route){                         //设置关于应用路径
```

```
            AboutAppScreen()
        }
        composable(Screen.AboutTestPage.route){                    //设置关于测试的路径
            AboutTestScreen()
        }
    }
}
```

ExitDialog 定义了退出的对话框,代码如下:

```
/**退出应用的对话框*/
@Composable
fun ExitDialog(appStates: StateHolder, onReceivedIntent:(TestIntent)->Unit) {
    val scope=rememberCoroutineScope()                             //协程范围
    AlertDialog(                                                   //定义警告对话框
        containerColor=Color.White,
        titleContentColor=Color(0xFF408D58),
        textContentColor=Color(0xFF408D58),
        icon={Icon(painter=painterResource(id=R.mipmap.logo_test),
            tint=Color.Unspecified,contentDescription="心理测试")},
        onDismissRequest={appStates.openState.value=!appStates.openState.value},
        title={
            Text("心理测试对话框",fontSize=20.sp)
        },
        text={
            Text("是否确定退出应用",fontSize=18.sp)
        },
        confirmButton={                                            //对话框确定按钮定义
            TextButton(onClick={
                scope.launch {
                    appStates.openState.value=true
                    appStates.drawerState.close()                  //侧滑菜单关闭
                    onReceivedIntent(TestIntent.ExitAppIntent)     //请求退出意图
                }
            }){
                Row{
                    Icon(imageVector=Icons.Filled.Check,contentDescription="OK")
                    Text("确定")
                }
            }
        },
        dismissButton={
            TextButton(onClick={
                scope.launch {
                    appStates.drawerState.close()
                    appStates.openState.value=false
                }
            }){
                Row{
                    Icon(imageVector=Icons.Filled.Clear,contentDescription="CANCEL")
```

```
                    Text("取消")
                }
            }
        }
    )
}
```

运行效果如图 4-31 所示。

图 4-31　退出对话框的运行效果

（3）TestScreen 定义的是心理测试的界面。在这个界面的最大特点就是使用延迟列表经所有的心理问题以水平滚动的方式进行展示。每个列表单项定义 QuestionCard 可组合项。题目的列表显示为了符合操作的习惯,通过按钮的控制,也可以实现题目的切换。代码如下：

```
/**定义单个问题卡片*/
@Composable
fun QuestionCard(questionId:Int, question:String, viewModel: TestViewModel,
    onReceivedIntent:(TestIntent)->Unit){
val output=viewModel.output.collectAsState()
val possibleAnswers=getPossibleAnswers()
val selectedChoice=remember{ mutableStateOf("null") }
Card(modifier=Modifier.fillMaxWidth().height(380.dp)
    .shadow(elevation=3.dp).background(Color.White),
    colors=CardDefaults.cardColors(containerColor=Color.White)、){
    Column(
        modifier=Modifier.width(440.dp).wrapContentHeight(),
        verticalArrangement=Arrangement.Center){
        Text(text="${question}",
            fontSize=24.sp,
            modifier=Modifier.fillMaxWidth().wrapContentHeight(),
            overflow=TextOverflow.Ellipsis,
            maxLines=10)
        possibleAnswers.forEach{
```

```kotlin
                    it:String->Row{
                        RadioButton(
                            selected=selectedChoice.value==it||output.value.answers
                                [questionId]==it.substring(0,1),
                            enabled=true,
                            onClick={
                            output.value.answers[questionId]=it.substring(0,1)
                            selectedChoice.value=it
                            onReceivedIntent(TestIntent.AnswerTestIntent(output.value
                                .answers))
                            }
                        )
                        Text(text=it,fontSize=24.sp,color=Color(0xFF408D58))
                    }
                }
            }
        }
}

/**定义测试界面*/
@Composable
fun TestScreen(stateHolder: StateHolder, viewModel: TestViewModel, onReceivedIntent:
    (TestIntent)->Unit){
val questions=stringArrayResource(id=R.array.questions)    //获取所有问题保存在字符串数组中
val scope=rememberCoroutineScope()                         //协程范围
val listState=rememberLazyListState()                      //列表状态
val scrollState=rememberScrollState(initial=0)             //滚动状态
val context=LocalContext.current
val testState=viewModel.output.collectAsState()            //获取单向流
val output=viewModel.output.collectAsState()               //获取单向流

if (output.value.questions.size==0)
    for (q in questions){
        testState.value.answers.add("")                    //初始化答题列表
        testState.value.questions.add(q)                   //初始化问题列表
    }
Box( modifier=Modifier.fillMaxSize(),contentAlignment=Alignment.Center){
    Column{
        LazyRow(modifier=Modifier.fillMaxWidth().wrapContentHeight()
            .scrollable(scrollState, Orientation.Horizontal, reverseDirection=true),
            state=listState) {
            itemsIndexed(questions){
                i: Int, s: String ->QuestionCard(i,s,viewModel) {
                    viewModel.processIntents(context, stateHolder,it)   //处理意图
                }
            }
        }
        Row(modifier=Modifier.fillMaxWidth().padding(5.dp),
            horizontalArrangement=Arrangement.Center){
            TextButton(onClick={                                        //定义上一题按钮
```

```kotlin
            scope.launch {
                if (scrollState.value in 1..25){
                    listState.scrollToItem(scrollState.value-1)    //滚动当前单项
                    scrollState.scrollTo(scrollState.value-1)
                }
            }
        },
        colors=ButtonDefaults.buttonColors(containerColor=Color(0xFF408D58))) {
            Icon(Icons.AutoMirrored.Filled.KeyboardArrowLeft,contentDescription
                =null)
            Text("上一题",fontSize=16.sp)
        }
        TextButton(onClick={                                      //定义下一题按钮
            scope.launch {
                if (scrollState.value in 0..24) {
                    listState.scrollToItem(scrollState.value+1)
                    scrollState.scrollTo(scrollState.value+1)
                }
            }
        },colors=ButtonDefaults.buttonColors(containerColor=Color
            (0xFF408D58))){
            Icon(Icons.AutoMirrored.Filled.KeyboardArrowRight,contentDescription=
                null)
            Text("下一题",fontSize=16.sp)
        }
        TextButton(                                               //定义提交按钮
            onClick={
                scope.launch{
                    onReceivedIntent(TestIntent.DisplayResultIntent(testState.
                        value.answers))
                                                                  //请求显示结果意图
                }
            },
            colors=ButtonDefaults.buttonColors(containerColor=Color
                (0xFF408D58))){
            Icon(Icons.AutoMirrored.Filled.Send,contentDescription=null)
            Text("提交",fontSize=16.sp)
        }
    }
  }
 }
}
```

运行效果如图 4-32 所示。

（4）AnswersScreen 定义显示答题情况的界面，采用了延迟网格的方式，将已经答题情况显示出来。延迟网格的每个单项定义在 AnswerCard，表示每个答题项。代码如下：

图 4-32　心理测试界面的运行效果

```
/* *AnswerCard答题单项定义*/
@Composable
fun AnswerCard(answerId:Int,answer: String){
    Card(modifier=Modifier.width(50.dp).height(80.dp).background(Color.White),
        colors=CardDefaults.cardColors(containerColor =Color.White)){
        Column(modifier=Modifier.background(Color.White),
            horizontalAlignment=Alignment.CenterHorizontally){
            Box(contentAlignment=Alignment.Center){
                Image(modifier=Modifier.width(40.dp).height(40.dp),
                    painter=painterResource(id=R.mipmap.circle),contentDescription=null)
                Text(answer,color=Color.Black,fontSize=18.sp)
            }
            Text("${answerId}",color=Color.Black,fontSize=14.sp)
        }
    }
}
/* *Answers screen
 *展示所有已经选择答案的结果*/
@Composable
fun AnswersScreen(viewModel: TestViewModel){
    val output=viewModel.output.collectAsState()
    Column(modifier=Modifier.fillMaxSize().fillMaxHeight().background(Color.White),
        horizontalAlignment=Alignment.CenterHorizontally){
        LazyVerticalGrid(
            modifier=Modifier.background(Color.White).padding(5.dp),
            columns=GridCells.Fixed(5), content={
            itemsIndexed(output.value.answers){
                i: Int, s: String->AnswerCard(answerId=i+1, answer=s )    //显示单项
            }
        }
        )
    }
}
```

运行效果如图 4-33 所示。

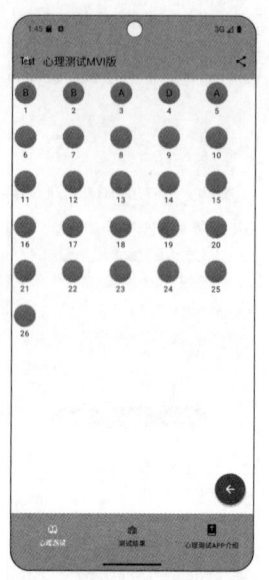

图 4-33 答题情况的运行效果

（5）ResultScreen 定义的显示测试结果的界面，代码如下：

```
@Composable
fun ResultScreen(viewModel: TestViewModel){
    val state=viewModel.output.collectAsState()              //获取测试状态
    Box(modifier=Modifier.fillMaxSize().background(Color.White),
        contentAlignment=Alignment.Center){
        Column{
            Text("测试结果",fontSize=30.sp)
            Text("${state.value.result}",fontSize=20.sp)     //显示状态中的测试结果
        }
    }
}
```

运行结果如图 4-34 所示。

图 4-34 测试结果的运行结果

(6) 定义关于应用介绍的界面 AboutAppScreen 和测试介绍界面 AboutTestScreen。这两个界面操作简单,就是定义显示文本说明。因此定义一个通用的可组合项 DisplayTextView,然后分别在 AboutAppScreen 和 AboutTestScreen 中调用它。运行效果如图 4-35 所示。

图 4-35 介绍界面的运行效果

DisplayTextView 定义文本显示,代码如下:

```
@Composable
fun DisplayTextView(textId:Int,colorId:Int){
    Box(modifier=Modifier.fillMaxWidth().padding(10.dp),
        contentAlignment=Alignment.TopStart){
        Text(text=stringResource(id=textId), fontSize=28.sp,
            color=colorResource(id=colorId), textAlign=TextAlign.Justify
        )
    }
}
```

应用介绍的界面 AboutAppScreen,代码如下:

```
@Composable
fun AboutAppScreen(){
    DisplayTextView(R.string.title_app_introduction,R.color.dark_green)
}
```

测试介绍的界面 AboutTestScreen,代码如下:

```
@Composable
fun AboutTestScreen(){
    DisplayTextView(textId=R.string.title_about_test, colorId=R.color.dark_green)
}
```

习 题 4

1. 结合 Compose 库的 LazyColumn 可组合项开发一个显示当前播放电影的简介(包括图片、电影名、电影简介)。为界面定义一个悬浮动作按钮,单击后能将手机屏幕向上滚动到第一个电影记录。定义一个菜单,菜单包括两个菜单项,其一实现随机重复增加一部电影记录;其二实现随机删减一部电影。无论执行哪个操作,要求:显示的 LazyColumn 中的电影信息能及时更新,LazyColumn 的每个单项请使用 Card 可组合项来定义。将 LazyColumn 分别换成 LazyVerticalStaggeredGrid、LazyHorizontalStaggeredGrid 来实现电影的网格显示,观察运行结果。

2. 设计一个类似 QQ 聊天的用户注册和登录的移动界面。请分别使用多种基本布局和约束布局 ConstraintLayout 来实现。

3. 设计一个类似墨迹天气的天气预报的移动界面,可以由多个界面构成。要求:
① 结合脚手架结构底部菜单导航。
② 显示当天的天气包括温度、阴晴、空气质量、湿度。
③ 结合 LazyColumn 和 Card 设计近 7 天的天气预报的界面。

4. 设计一个新闻列表显示的移动界面。要求:
① 显示新闻列表。
② 根据从新闻列表中选择新闻,将该新闻显示。

5. 设计并完成一个"音乐专辑"具有类似 QQ 音乐歌单的应用的界面,结合 Scaffold 搭建应用的 UI 界面。

第 5 章　Android 的并发处理

在移动应用中，常常需要多个任务同时处理。对于访问在线资源、访问数据库、解析数据、加载音频视频等特别耗时的操作或者在移动应用中必须同时处理多任务的情况下，可以使用 Android 系统的多线程、Handler 机制及协程来执行并发处理。Compose 组件的附带效应与协程是有关联的。此外，RxJava 库也是常见用来处理并发任务的一种方式。因此，本章将结合多线程、Handler 消息处理机制、协程、RxJava 来解决多任务。并在本章介绍 Compose 组件的附带效应。

5.1　多　线　程

线程是操作系统调度的最小单位，是依附进程而存在的。Kotlin 对 Java 的线程进行了封装，简化了线程的处理。

创建一个线程类，有两种形式，一种是定义类，让它实现 Runnable 接口，形式如下：

```kotlin
class MyThread: Runnable{
    override fun run(){
        println("定义实现 Runnable 的类")
    }
}
```

然后，利用这个线程体类对象创建线程对象并启动线程。代码的表现形式如下：

```kotlin
Thread(MyThread()).start()
```

另一种自定义线程的定义方式是，定义一个 Thread 类的子类，然后创建这个线程类的对象，再调用 start() 方法启动线程，形式如下：

```kotlin
Thread {
    override fun run() {
        println("使用对象表达式创建")
    }
}.start()
```

Kotlin 对线程对象的创建和启动进行简化，可以采用下列方式创建和启动线程：

```kotlin
thread {
    println("running from thread(): ${Thread.currentThread()}")
}
```

每个启动的 Android 移动应用都有一个单独的进程。这个进程中有多个线程。这些线程中仅有一个 UI 主线程。Android 应用程序运行时创建的 UI 主线程主要负责控制 UI 界面的显示、更新和控件交互。Android 程序创建之初，一个进程是单线程状态，所有的任务都在主线程中运行，这会导致 UI 主线程的负担过重。一个应用可通过定义多个线程来完成不同的任务,分担主线程的责任,避免主线程阻塞。当主线程阻塞超过规定的时间,会出现 ANR(Application Not Responding,应用程序无响应)问题,导致程序中断运行。

例 5-1 显示计时的应用实例 1,代码如下:

```
//模块 Ch05_01 主活动定义 MainActivity.kt
class MainActivity: ComponentActivity() {
    override fun onCreate(savedInstanceState: Bundle?) {
        super.onCreate(savedInstanceState)
        enableEdgeToEdge()
        setContent {
            Chapter05Theme {var running by remember{mutableStateOf(false)}
                                                            //设置控制线程运行的状态值
                var timer by remember{mutableStateOf(0)}    //计时器状态值
                Column(modifier=Modifier.fillMaxSize(),
                    horizontalAlignment=Alignment.CenterHorizontally,
                    verticalArrangement=Arrangement.Center) {
                    if (running)                            //自定义线程运行显示计时
                        Text("${timer}秒",fontSize=32.sp)
                    else                                    //否则显示"计时器"文本
                        Text("计时器",fontSize=32.sp)
                    Row(modifier=Modifier.padding(20.dp)){
                        Button(onClick={                    //点击"计时开始"按钮
                            running=true
                            thread{                         //创建并启动线程
                                while(running){
                                    Thread.sleep(1000)      //线程休眠 1 秒
                                    timer++                 //计时器加 1
                                }
                            }
                        }){Text("计时开始",fontSize=28.sp}
                        Button(onClick={                    //点击"停止计时"按钮
                            timer=0
                            running=false
                        }){Text("停止计时", fontSize=28.sp) }
                    }
                }
            }
        }
    }
}
```

运行效果如图 5-1 所示。

在 MainActivity 中定义并启动了一个自定义线程。这个线程的主要任务是每秒更新一次文本显示的时间。再通过布尔值 running 控制线程运行。单击"开始计时"按钮,running 为 true,观察日志可以发现,时间是动态显示的。单击"停止计时"按钮,设置 running 为 false,使得线程停止。观察日志可以发现,计时已停止。

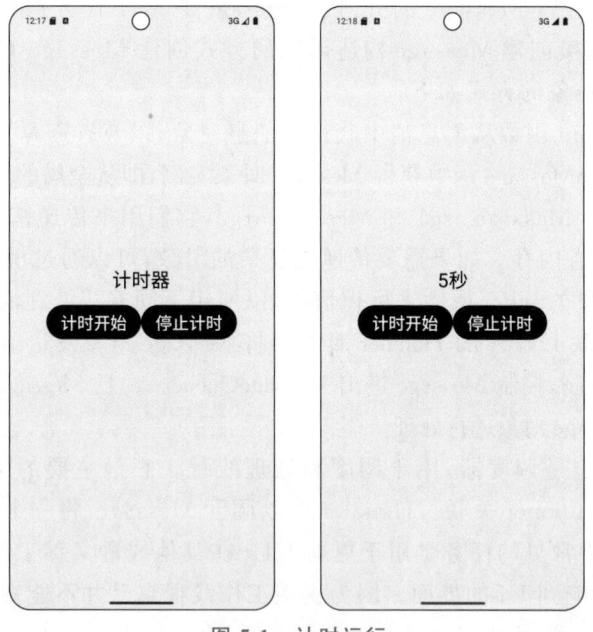

图 5-1　计时运行

5.2　Handler 机制

用于 Message(消息)处理的 Handler 机制是一种异步处理方式。通过 Handler 机制可以实现不同线程之间的通信。Handler 机制相关的类有 Looper、MessageQueue、Message 和 Handler。这些类各司其职,彼此之间的关系如图 5-2 所示。

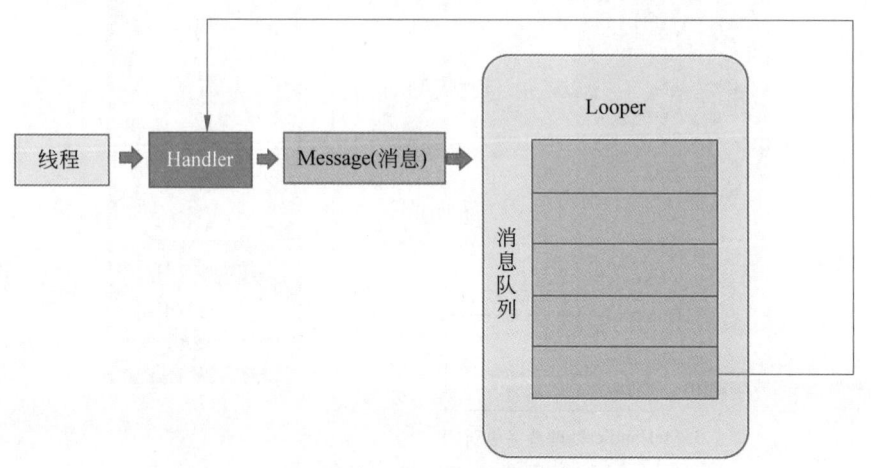

图 5-2　Handler 消息处理机制

其中,Looper(循环体)在线程内部定义,每个线程只能定义一个 Looper。Looper 内部封装了 MessageQueue(消息队列)。Looper 对消息队列进行循环管理。通常所说的 Looper 线程就是循环工作的线程。一个线程不断循环,若有新任务,则执行;执行完毕后,继续等待下一个任务,以这种方式执行任务的线程就是 Looper 线程。

Message(消息)又称任务。有时将 Runnable 线程体对象封装成 Message 对象来处理其中的任务。

Handler 对象发送、接收并进行 Message 的处理。Message 封装了任务携带的信息和处理该任务的 Handler 对象。可以通过直接创建 Message 构造函数的方式创建 Message 对象，但是这种方法并不常用。最常见获取 Message 对象的方式如下。

（1）通过 Message.obtain 函数从消息池中获得空消息对象，以节省资源。

（2）通过 Handler.obtainMessage 函数获得 Message 对象实例，即从全局的消息池中获得消息对象。

Message 对象具有属性 Message.arg1 和 Message.arg2，它们用来传递基本、简单的数据信息，例如数值等。这比用 Bundle 更省内存。如果需要传递更复杂的对象，可以通过消息 Message 对象的 obj 属性来实现。用 Message 对象的 what 属性来标识信息，以便用不同方式处理消息对象。

当产生一个消息时，关联 Looper 的 Handler 对象会将 Message 对象发送给 MessageQueue。Looper 对 MessageQueue 进行管理和控制，控制 Message 进出 MessageQueue。一旦 Message 从 MessageQueue 出列，可以使用 Handler 对这个 Message 对象进行处理。

Handler 既是处理器，也是调度器，用于调度和处理线程。它最主要的工作是发送和接收并处理 Message。Handler 往往与 Looper 关联。Handler 可以调度 Message，将一个任务切换到某个指定的线程中去执行。在 Handler 中常见的任务是用于更新 UI。在其他线程又称工作线程（Work Thread），修改了数据，而这些数据会影响到 UI 的界面。因为这些工作线程自己并不能变更 UI，所以常见的处理方式是将数据封装成 Message，并通过 Handler 对象发送到 MessageQueue 中。最后在 UI 主线程中接收这些数据，对 UI 界面进行变更，如图 5-3 所示。

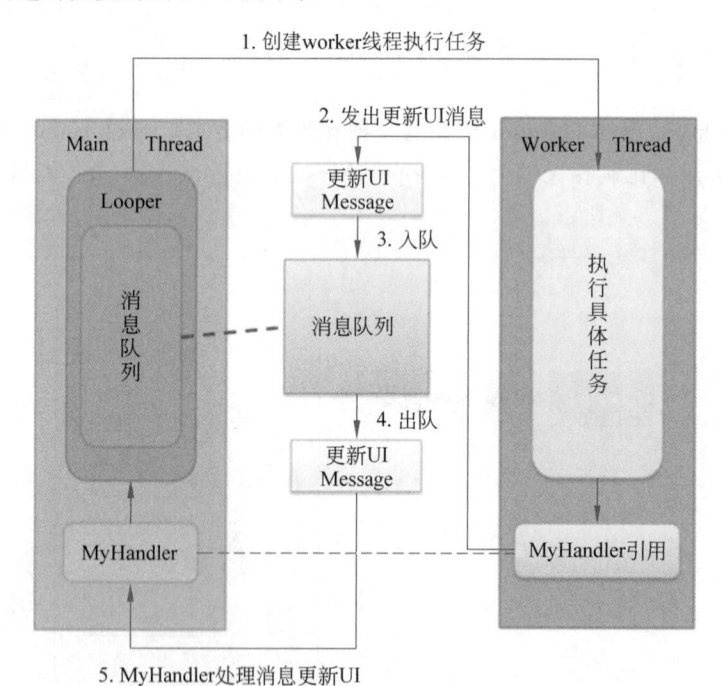

图 5-3　Handler 机制

例 5-2　显示计时的应用实例 2，代码如下：

```
//模块 Ch05_02 主活动的 MainActivity.kt
class MainActivity: ComponentActivity() {
    override fun onCreate(savedInstanceState: Bundle?) {
        super.onCreate(savedInstanceState)
```

```
            enableEdgeToEdge()
            setContent {
                var running by remember{ mutableStateOf(true) }
                var timerTxt by remember{mutableStateOf("")}
                var timer=0                                          //计时器
                val handler=object: Handler(Looper.getMainLooper()) { //定义 Handler 对象
                    override fun handleMessage(msg: Message) {
                        super.handleMessage(msg)
                        if (msg.what==0x123){
                            timerTxt="${msg.arg1}秒"                  //接收数据,并修改文本状态
                        }
                    }
                }
                Chapter05Theme {
                    Column(modifier=Modifier.fillMaxSize(),
                        horizontalAlignment=Alignment.CenterHorizontally,
                        verticalArrangement=Arrangement.Center) {
                        Text("$timerTxt",fontSize=32.sp)
                        Row(modifier=Modifier.padding(20.dp)){
                            Button(onClick={                          //点击按钮创建并启动线程
                                thread{
                                    while(running){
                                        Thread.sleep(1000)
                                        val message=Message.obtain()  //创建消息
                                        message.what=0x123
                                        message.arg1=++timer
                                        handler.handleMessage(message)//发送消息
                                    }
                                }
                            }){
                                Text("计时开始",fontSize=28.sp)
                            }
                            Button(onClick={
                                timer=0
                                running=false
                            }){
                                Text("停止计时", fontSize=28.sp)
                            }
                        }
                    }
                }
            }
        }
    }
}
```

运行结果如图 5-4 所示。

在上述的 MainActivity 中并没有创建 Looper 对象,而是通过 Looper.getMainLooper 函数调用 UI 主线程内置的 Looper 对象。在这个计数器的活动中,可直接利用 UI 主线程的 Looper 控制 MessageQueue。上述代码中,object 表示生成一个匿名的对象。Handler 对象在工作线程(自定义的线程)中,发送消息。在主线程中,Handler 对象对出列的 Message 的 what 来源标识进行判断,确定其是

图 5-4　计时器运行结果

否是从指定工作线程发出的,最后对这个消息进行处理。在本例中,修改了 timerTxt 字符串状态值,从而实现了动态计时功能。

5.3　协　　程

协程是 Kotlin 的特性。协程是一种并发设计模式,在 Android 平台上使用协程可以简化异步执行的代码,有助于管理长时间运行的任务,避免长时间执行的任务阻塞主线程。协程存在于线程中,由程序主动控制切换,开销小,执行效率高。Android JetPack 库包含支持协程的扩展。有些 Android JetPack 库还提供了协程的作用域,用于并发处理。另外,协程可以确保安全地在主线程中调用 suspend 函数。

要在项目中使用 Kotlin 协程并不是 Kotlin 的标准库,因此使用协程需要在项目的 libs.versions.toml 文件配置如下:

```
[versions]
coreKtx="1.10.1"
[libraries]
androidx-core-ktx={ group="androidx.core", name="core-ktx", version.ref="coreKtx" }
```

协程在常规函数的基础上增加 suspend 和 resume 操作。suspend 称为挂起或暂停,用于暂停执行当前的协程,并保存局部变量;resume 用于让已经暂停的协程从暂停处恢复并继续执行。其他函数需要启动协程就必须调用其他挂起函数。协程需要通过协程构造器(如 launch)来启动。

使用 Global.launch 方法是创建协程最简单的方式。Global.launch 方法创建一个顶层全局范围的协程。这种协程在应用程序结束时也会结束。在下列代码中,协程执行的任务非常简单,就是输出 1~500 的整数的字符串,因为执行速度非常快,因此在协程后面增加一个阻塞当前线程 10s 的处理,使得在程序执行的时间中可以打印输出字符串,代码如下:

```kotlin
//模块 Ch05_04 TestCoroutine01.kt
suspend fun test(){                              //定义挂起函数
    for(i in 1..500) {
        delay(1L)                                //延迟协程 1ms
        print("$i ")
    }
}
fun main(){
    GlobalScope.launch {                         //在后台中加载一个协程
        test()
    }
    println("还在处理中……")
    Thread.sleep(10000L)                         //阻塞主线程 10s,保持 JVM 激活状态
    println("处理完毕……")
}
```

运行结果如图 5-5 所示。

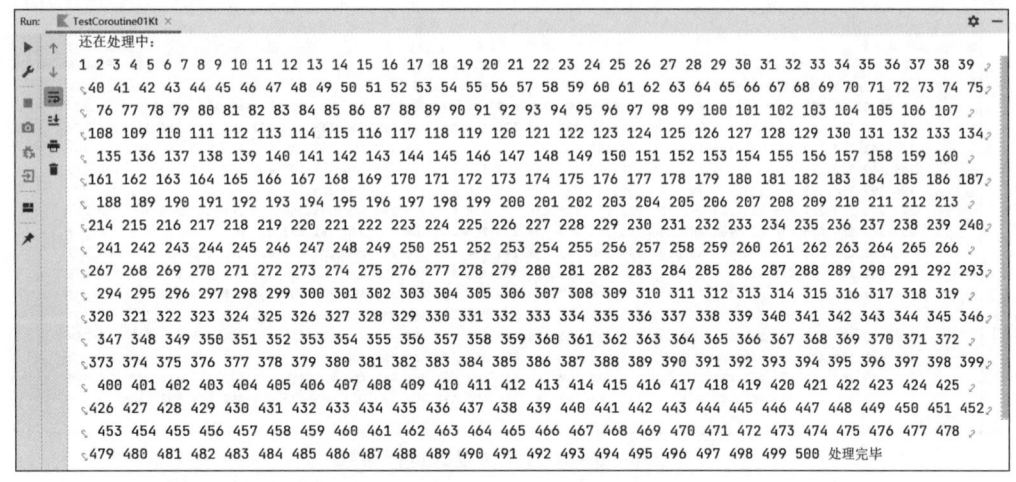

图 5-5 运行结果

在 Android 平台上,要使用协程完成结构化并发处理必须具有 3 个功能。

(1) 取消任务。当某项协程任务不需要的时候可以取消它。

(2) 追踪任务。追踪正在执行的任务。

(3) 发出错误的信号。当协程执行失败,可以发出错误信号表明有错误发生。

因此,为了实现这 3 个功能,在 Kotlin 中使用协程必须指定 CoroutineScope(协程范围)。CoroutineScope 并不运行协程,但是它可以对协程进行追踪。Kotlin 必须在协程范围中启动协程。启动协程有两种方式:一种是用 launch 构建器启动协程,另一种是用 async 构建器启动协程。这两种方式虽然可以启动协程,但是执行情况不同。用 launch 方式启动新协程时,结果不返回调用方,只是执行一个过程;用 async 启动协程,允许使用 await 挂起函数返回结果,代码如下:

```kotlin
//模块 Ch05_04 TestCoroutine02.kt
suspend fun test2(){                             //定义挂起函数
    for(i in 1..500) {
        delay(1L)
```

```
        print("$i ")
    }
}
fun main(){
    println("还在处理中……")
    runBlocking{                                    //在后台中加载一个协程
        val result=async (Dispatchers.IO){
            test2()
            "成功"
        }
        println(result.await())
    }
    Thread.sleep(10000L)                            //阻塞主线程10s,保持JVM激活状态
    println("处理完毕……")
}
```

运行结果如图 5-6 所示。

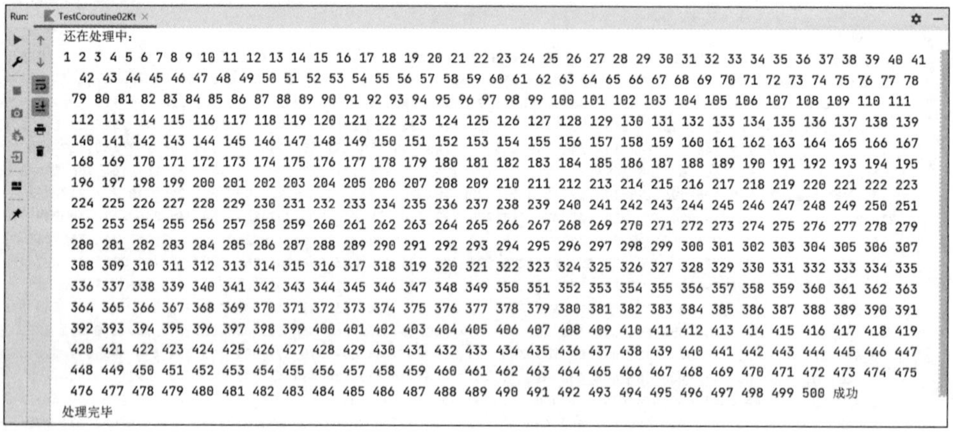

图 5-6　async 启动协程的运行结果

　　TestCoroutine01.kt 和 TestCoroutine02.kt 采用了两种不同的方式启动协程。TestCoroutine01.kt 中,因为执行了 Thread.sleep(10000L) 让主线程休眠 10s,如果在 10s 内不能完成任务,就会被强制中断。因此,要让协程中的所有代码完整执行,在测试场景下可以采用 runBlocking 函数。在 TestCoroutine02.kt 中,采用 runBlocking 函数会创建一个协程的作用域,在这个协程的作用域中所有代码和子协程没有完全执行完之前会阻塞当前线程,直至协程的任务完成为止。另外,TestCoroutine02.kt 中采用 async 启动协程,带有一个返回值。

　　在实际上,GlobalScope.launch 加载顶层协程和 runBlocking 阻塞线程运行协程的两种方式在 Android 环境下并不常用,往往采用如下形式:

```
val job=Job()
val scope=CoroutineScope(job)
scope.launch{
    ...
}
job.cancel()
```

首先，创建 Job 对象表示工作任务，并将 Job 对象作为实参传递给 CoroutineScope 函数（不是 CoroutineScope 类），创建一个协程范围 CoroutineScope 对象 scope。在这个范围启动协程，所有协程都关联在 job 对象中。因此，如果需要取消协程，只需要调用 job.cancel 函数就可以将 job 作用域下的所有协程取消，代码如下：

```kotlin
//模块 Ch05_04 TestCoroutine03.kt
suspend fun test3(){
    for (i in 1..500) {
        delay(1L)
        print("$i ")
    }
}
fun main(){
    println("还在处理中……")
    val job=Job()
    val scope=CoroutineScope(job)
    scope.launch {
        test3()
    }
    Thread.sleep(10000L)             //阻塞主线程 10s,保持 JVM 激活状态
    println("处理完毕……")
}
```

运行结果如图 5-7 所示。

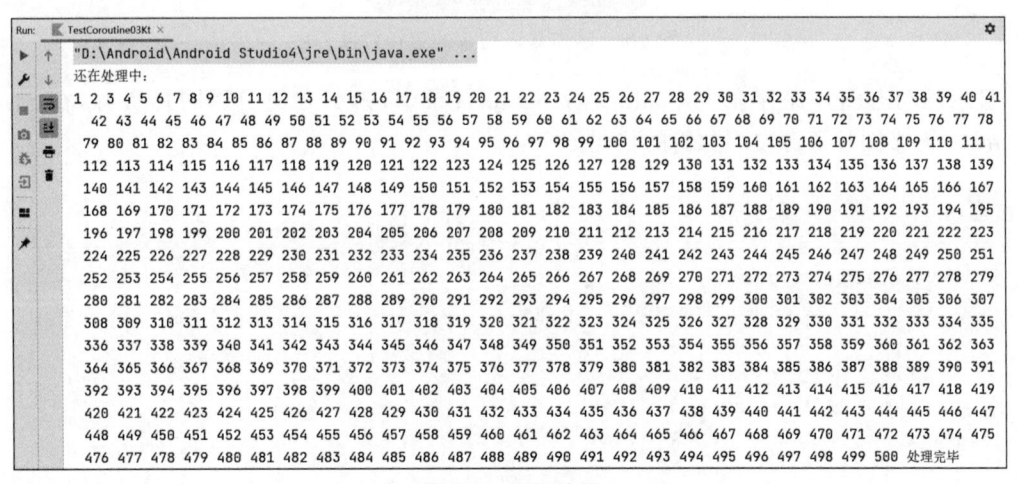

图 5-7　运行结果

上述 3 个例子中都在主线程中使用了协程。但是在实际情况下，可以根据任务的不同指定不同的线程进行调用。为此，Kotlin 提供了 3 个调度器用于指定运行协程：Dispatchers.Main、Dispatchers.IO 和 Dispatchers.Default。

（1）Dispatchers.Main：表示在 Android 主线程中运行协程。可以调用 suspend 函数等用来处理用户界面交互和一些轻量级的任务。

（2）Dispatchers.IO：非主线程，用于磁盘和网络数据的读写的优化。主要用于数据库、文件和网络处理。

（3）Dispatchers.Default：非主线程，对 CPU 的密集型任务进行优化，例如数据排序、处理差异判

断等。

要指定线程运行协程,调用 withContext 函数来创建一个运行的块。例如,withContext(Dispatchers.IO)表示创建一个在 IO 线程中运行的代码块,均由 IO 调度器来执行。

例 5-3 使用协程动态显示图片,运行结果如图 5-8 所示。

图 5-8 动态显示图片的运行效果

(1) 显示的动态图片保存在 res/values/mipmap 目录中,为了方便引用图片资源,在 res/values/strings.xml 中定义图片资源整型数组的代码如下:

```xml
//模块 Ch05_05 的 strings.xml
<resources>
    <string name="app_name">Ch05_04</string>
    <integer-array name="images">
        <item>@mipmap/scene1</item>
        <item>@mipmap/scene2</item>
        <item>@mipmap/scene3</item>
        <item>@mipmap/scene4</item>
        <item>@mipmap/scene5</item>
    </integer-array>
</resources>
```

(2) 定义 MainActivity,代码如下:

```kotlin
//模块 Ch05_05 的 MainActivity.kt
class MainActivity: ComponentActivity() {
    override fun onCreate(savedInstanceState: Bundle?) {
        super.onCreate(savedInstanceState)
        val images=getImages()
        enableEdgeToEdge()
```

```
            setContent {
                Chapter05Theme {
                    ImageScreen(images)
                }
            }
        }
        private fun getImages():List<Int>{              //从资源 stsrings.xml 中获取图片资源,初始化图片
            val imageType:TypedArray=resources.obtainTypedArray(R.array.images)
                                                        //获取类型数组
            val images=mutableListOf<Int>()             //放置图片资源的可变列表
            for (i in 0 until imageType.length()){      //遍历列表
                images.add(imageType.getResourceId(i,R.mipmap.scene1))
                                                        //根据索引获取图片,否则取默认图片 R.mipmap.scene1
            }
            return images
        }
    }
```

（3）动态显示图片的界面,代码如下：

```
//模块 Ch05_05 动态显示图片的 ImageScreen
@Composable
fun ImageScreen(images:List<Int>){                      //定义显示动态图片的界面
    var imageIndex by remember{mutableStateOf(0) }      //控制访问图片的索引号
    var currentImage by remember{mutableStateOf(R.mipmap.scene1) }  //当前访问的图片
    val job=Job()                                       //创建任务
    LaunchedEffect(key1=currentImage) {
        CoroutineScope(job).launch {                    //在协程范围加载
            withContext(Dispatchers.Main) {             //在主线程上下文
                delay(1000)                             //延迟 1s
                imageIndex=(imageIndex1)%images.size    //修改索引值
                if (imageIndex!=0)
                    currentImage=images[imageIndex]     //修改当前图片
            }
        }
    }
    Column(modifier=Modifier.fillMaxSize().background(Color.Black),
        horizontalAlignment=Alignment.CenterHorizontally,
        verticalArrangement=Arrangement.Center) {
        Image(painterResource(id=currentImage), contentDescription="images")
                                                        //绘制图片
        Row{                                            //生成 5 个圆点,显示已经显示图片的进度
            LazyRow{
                items(images){
                    RadioButton(selected=it==currentImage,   //选中
                        onClick={},
                        enabled=false,                       //不可用状态
                        colors=RadioButtonDefaults.colors(   //配置颜色
                            disabledSelectedColor=Color.Green,
                            disabledUnselectedColor=Color.White)
                    )
```

```
                }
            }
        }
    }
```

上述代码调用 CoroutineScope 函数来创建一个协程范围不是一种常见的处理方式。基于 Compose 组件创建的可组合项是通过调用 rememberCoroutineScope 函数获取协程范围。该函数的作用是获取组合感知作用域，以在可组合项外启动协程。因此，上述代码中关于协程处理的部分可以修改成如下形式：

```
val scope=rememberCoroutineScope()                               //创建协程范围
LaunchedEffect(key1=currentImage) {
    scope.launch {                                               //在协程范围加载
        delay(1000)                                              //延迟 1s
        imageIndex=(imageIndex +1)%images.size                   //修改索引值
        if (imageIndex!=0)
            currentImage=images[imageIndex]                      //修改当前图片
    }
}
```

在上述代码中还出现 LanchedEffect 函数。这与 JetPack Compose 中附带效应的内容息息相关，将在 5.4 节详细介绍。

5.4 Compose 的附带效应

Compose 组件定义的可组合函数搭建移动应用界面。理想状态下，移动应用界面的重构和更新是由可组合函数的内部状态的变换造成的。在有些情况下，例如在可组合函数会执行一些耗时的操作时，长时间的执行任务导致卡顿，这就需要在可组合函数中运行一些在可组合函数外部的一些应用状态。而发生在可组合函数作用域外应用状态的变化就是附带效应。有些附带效应是必须，这些操作应从能感知可组合项生命周期的受控环境中调用。但是，可组合函数要控制这些作用域之外的效应并不容易，会造成过度使用。因此，Compose 库中提供了 Effect API 对这些效应进行控制和处理。

5.4.1 附带效应概述

在具体了解 Effect API 之前，通过一个计时器的应用来对附带效应进行初步的理解，代码如下：

```
//模块 Ch05_05 的 MainScreen 的定义
val timer:MutableIntState=mutableIntStateOf(0)                   //全局变量,在 MainScreen 作用域外
@Preview
@Composable
fun MainScreen(){                                                //函数作用域内
    var runningState=remember{mutableStateOf(false)}             //定义内部状态控制计时
    val scope=rememberCoroutineScope()                           //获取协程范围
    Box(modifier=Modifier.fillMaxSize(), contentAlignment=Alignment.Center){
        Column(modifier=Modifier.fillMaxWidth(),
```

```
                horizontalAlignment=Alignment.CenterHorizontally,
                verticalArrangement=Arrangement.Center){
            Text(text="${timer.value}秒",fontSize=32.sp)      //引用外部状态timer的值
            Row(modifier=Modifier.fillMaxWidth(),
                horizontalArrangement=Arrangement.Center){
                Button(onClick={runningState.value=true        //运行状态设置为true
                    scope.launch {                             //启动协程
                        while(runningState.value){
                            delay(1000)                        //暂停1s
                            timer.value+=1                     //修改外部状态值
                        }
                    }
                }){Text("开始计时")}
                Button(onClick={runningState.value=false       //运行状态设置为false
                }){Text("停止计时")}
            }
        }
    }
}
```

运行结果如图 5-9 所示。

图 5-9 计时器的运行结果

在上述代码中,有两种情况需要注意。

(1) 上述代码非常粗暴地在 MainScreen 可组合函数外面定义了一个状态 timer。当 timer 这个外部值变更时,MainScreen 可组合函数构建的 UI 界面会被刷新。MainScreen 函数调用外部状态 timer 的情况就是附带效应。timer 状态在 MainScreen 可组合函数外部,它的变化情况是不可预知的。

(2) rememberCoroutineScope 函数是获取感知应用领域 scope 对象,在可组合项外启动协程。协程可以处理可组合项外的应用状态,如 delay 函数的调用。这也是附带效应的一种形式。另外,scope.launch 已经很好地封装了协程调度过程。在上述代码定义 Button 的 onClick 点击动作处理中,在 scope.launch 的内部调用了 delay 挂起函数,实现不阻塞线程的情况下将协程延迟 1s 的作用。

5.4.2　LaunchedEffect 和 rememberUpdatedState

LaunchedEffect 函数是一个可组合函数，LaunchedEffect 可以在可组合项的作用域内运行挂起函数时，会启动内部的代码块到协程上下文 CoroutineContext。当函数的关键字的值发生变化时，会重构 LaunchedEffect。这时，LaunchedEffect 原来启动的协程会被取消然后又重新启动。当 LaunchedEffect 退出组合项时，协程会被取消。它的函数说明如下：

```
@Composable
@NonRestartableComposable
@OptIn(InternalComposeApi::class)
fun LaunchedEffect(key1: Any?,[key2:Any?,Key3:Any] block: suspend CoroutineScope.()->
    Unit) {
    val applyContext=currentComposer.applyCoroutineContext
    remember(key1) { LaunchedEffectImpl(applyContext, block) }
}
```

LaunchedEffect 函数包含了两类参数。

（1）key1：表示关键字可以是任何类型，如果是可变的状态值，可以根据可变状态值的变化，取消原有协程并启动新的协程。当 key1 为 Unit 或 true 时，LaunchedEffect 函数将与当前重组函数保持一致的生命周期。key2 或 key3 表示可选项，也表示关键字。它们的可变状态的变化，也会取消原有协程并启动新的协程。与 key1 的作用是一致的。

（2）block：表示要调用的挂起函数。需要在协程范围中运行。

例 5-4　用 LaunchedEffect 函数实现计时器，代码如下：

```
//模块 Ch05_06 的 MainScreen
@Preview
@Composable
fun MainScreen(){                                        //函数作用域内
    val timer=remember{mutableIntStateOf(0) }            //计时器状态
    val runningState=remember{mutableStateOf(false) }    //运行状态
    LaunchedEffect(key1=timer.value,key2=runningState.value) {   //处理附带效应
        delay(1000)                                      //暂停 1s
        if (runningState.value)                          //如果运行状态为 true
            timer.value+=1                               //修改计时器的值
    }
    Box(modifier=Modifier.fillMaxSize(),
        contentAlignment=Alignment.Center){
        Column(modifier=Modifier.fillMaxWidth(),
            horizontalAlignment=Alignment.CenterHorizontally,
            verticalArrangement=Arrangement.Center){
            Text(text="${timer.value}秒",fontSize=32.sp)  //引用外部状态 timer 的值
            Row(modifier=Modifier.fillMaxWidth(),
                horizontalArrangement=Arrangement.Center){
                Button(onClick={
                    runningState.value=true              //修改运行状态值 true
                }){Text("开始计时")
```

```
            Button(onClick={
                runningState.value=false                    //修改运行状态值false
            }){Text("停止计时")}
        }
      }
    }
}
```

上述的 LaunchedEffect 函数用于监测 timer 和 runningState 的值是否发生变化，在单击"开始计时"按钮时，runningState 的状态值发生变换，LaunchedEffect 启动新的协程，在 LaunchedEffect 内部暂停 1s 执行了 timer.value++，使得 timer 状态值发生变化。LaunchedEffect 监测到 timer.value 的值发生变化，又会将原来的协程取消，启动新的协程。依次往复，实现了计时的功能。当用户单击"停止计时"时，最后一次启动新协程，但是 timer 状态值没有发生变化，使得后续的 LaunchedEffect 不会启动新协程。整个运行效果就实现了计时停止的功能。这样协程不断切换的处理方式并不可取，可以将上述代码修改为如下形式：

```
//模块 Ch05_06 的 MainScreen
@Preview
@Composable
fun MainScreen(){                                          //函数作用域内
    val timer=remember{mutableIntStateOf(0)}               //计时器状态
    val runningState=remember{mutableStateOf(false)}       //运行状态
    LaunchedEffect(key1=runningState.value) {              //处理附带效应
        while(runningState.value) {                        //如果运行状态为true
            delay(1000)                                    //暂停1s
            timer.value+=1                                 //修改计时器的值
        }
    }
    Box(modifier=Modifier.fillMaxSize(),
        contentAlignment=Alignment.Center){
        Column(modifier=Modifier.fillMaxWidth(),
            horizontalAlignment=Alignment.CenterHorizontally,
            verticalArrangement=Arrangement.Center){
            Text(text="${timer.value}秒",fontSize =32.sp)  //引用外部状态timer的值
            Row(modifier=Modifier.fillMaxWidth(),
                horizontalArrangement=Arrangement.Center){
                Button(onClick={
                    runningState.value=true                //修改运行状态值true
                }){Text("开始计时")}
                Button(onClick={
                    runningState.value=false               //修改运行状态值false
                }){Text("停止计时")}
            }
        }
    }
}
```

通过这种方式，只需要监测 runningState 的状态值，结合循环控制，就实现了开始计时和停止计时功能的控制。LaunchedEffect 函数会根据关键字的值的变化，重启协程。但是，在某些情况下，并不希

望 LaunchedEffect 重启,而是需要 LaunchedEffect 函数中变更的状态的值。可以考虑使用 rememberUpdatedState 函数。

rememberUpdateState 函数是个可组合函数,用于创建可捕获和更新该值的引用。rememberUpdateState 函数的定义如下:

```
@Composable
fun <T> rememberUpdatedState(newValue: T): State<T> = remember {
    mutableStateOf(newValue)
}.apply { value=newValue }
```

修改上述计时器的代码,代码如下:

```
//模块 Ch05_06 的 MainScreen
@Preview
@Composable
fun MainScreen(){                                               //函数作用域内
    var runningState by remember{ mutableStateOf(true) }        //初始设置状态值为 true
    val timer=remember{ mutableIntStateOf(0) }                  //计时器
    val timerState=rememberUpdatedState(newValue=timer)         //记住 timer 状态
    LaunchedEffect(Unit) {
        while(runningState) {
            delay(1000)
            timerState.value.value++                            //修改记住状态值的值
        }
    }
    Box(modifier=Modifier.fillMaxSize(),contentAlignment=Alignment.Center){
        Column(modifier=Modifier.fillMaxWidth(),
            horizontalAlignment=Alignment.CenterHorizontally,
            verticalArrangement=Arrangement.Center){
            Text(text="${timer.value}秒",fontSize=32.sp)        //引用外部状态 timer 的值
            Row(modifier=Modifier.fillMaxWidth(),
                horizontalArrangement=Arrangement.Center){
                Button(onClick={
                    runningState=true
                }){Text("开始计时")}
                Button(onClick={
                    runningState=false
                }){Text("停止计时")}
            }
        }
    }
}
```

在上述代码中,LaunchedEffect 中的关键值的参数为 Unit,表示参数为空。这使得 MainScreen 可组合函数被调用或重组时,才会加载 LaunchedEffect 函数。LaunchedEffect 并不会根据关键字的变化重新加载协程代码。由于 rememberUpdateState 函数的调用,可以通过 timerState 记住 timer 的状态,使得通过 timerState 来获取并变更 timer 的状态值。因为 rememberUpdateState 具有这样的特点,使得上述代码在启动运行后不断修改计时器的值,直至单击"停止计时"按钮使得 runningState 的值为 false,导致 LaunchedEffect 内部的循环终止,才会让计时停止。

5.4.3 DisposableEffect 和 Lifecycle

对于需要在键发生变化或可组合项退出组合后进行清理的附带效应,可以通过 DisposableEffect 来实现。DisposableEffect 可组合函数定义如下：

```
@Composable
@NonRestartableComposable
fun DisposableEffect(
    key1: Any? [,key2: Any?,key3: Any?,…]
    effect: DisposableEffectScope.()->DisposableEffectResult ) {
    remember(key1, key2, key3) { DisposableEffectImpl(effect) }
}
```

DisposableEffect 可组合函数包含两类参数。

(1) key1：表示关键字可以是任何类型,如果是可变的状态值时,可以根据可变状态值的变化,清除可组合项的附带效应。key2 或 key3 表示可选项,也是表示关键字。它们的可变状态的变化,也会清除可组合项的附带效应。与 key1 的作用是一致的。

(2) effect：表示要清除的效应或要调用重新生成效应的操作。当关键字的值发生变化或者 DisposableEffect 离开可组合项,则必须执行撤销或清除操作。

修改 5.4.2 节的计时器代码如下：

```
@Preview
@Composable
fun MainScreen(){                                       //函数作用域内
    var runningState by remember{ mutableStateOf(true) }    //运行状态控制
    val timer=remember{ mutableIntStateOf(0) }              //计时器
    val timerState=rememberUpdatedState(newValue=timer)     //记住计时器状态
    LaunchedEffect(Unit) {                                  //加载效应
        while(runningState) {
            delay(1000)
            timerState.value.value++                        //修改记住的计时器状态值
        }
    }
    DisposableEffect(key1=timer.value){                     //清理效应
        onDispose {                                         //清除操作
            timer.value=0
            runningState=false
        }
    }
    Box(modifier=Modifier.fillMaxSize(),
        contentAlignment=Alignment.Center){
        Column(modifier=Modifier.fillMaxWidth(),
            horizontalAlignment=Alignment.CenterHorizontally,
            verticalArrangement=Arrangement.Center){
            Text(text="${timer.value}秒",fontSize=32.sp)    //引用外部状态 timer 的值
            Row(modifier=Modifier.fillMaxWidth(),
                horizontalArrangement=Arrangement.Center){
```

```
            Button(onClick={
                runningState=true
            }){Text("开始计时")}
            Button(onClick={
                runningState=false
            }){Text("停止计时")}
        }
    }
}
```

运行上述代码,发现运行结果没有任何变化,无论是否点击"开始计时"按钮,计时的状态值一致为 0。这是因为,运行 MainScreen 界面,LaunchedEffect(Unit)在 MainScreen 开始渲染的时候会开始执行一次,修改 timerState 的保存状态的值,即时 timer 的值。在后续的 DisposableEffect 函数调用中,发现了关键字 key1 对应 timer.value 的值发生了变化,立即执行 onDispose 清理效应的操作,让 timer.value 赋值为 0 以及 runningState.value 赋值为 false,这才使得运行效果没有发生任何变化。显然这样运用 DisposableEffect 清理效应是无意义的。

DisposableEffect 往往和 Lifecycle(生命周期)进行关联,当键参数为 LifecycleOwner(生命周期拥有者)对象时,可以利用 DisposableEffect 对 LifecycleOwner 的变化执行撤销或清理的工作。在没有明确说明的情况下,Lifecycle Owner 就是 Activity 或 Fragment 本身。在继续了解清理效应之前,先来了解一下生命周期的相关概念。

如图 5-10 所示,Activity 既是 UI Controller(用户界面控制器)又可作为 LifecycleOwner(生命周期拥有者)拥有 Lifecycle 对象,可对 Lifecycle 的不同状态进行处理。但是这又会带来一些问题:UI Controller 承担了 Lifecycle 变化处理任务,增加了它们的负担。生命周期状态的变化过程产生的对象数据往往在不同的类中存在,导致 Controller 的代码复杂,不得不额外管理和处理 Lifecycle 状态变化产生的对象数据。

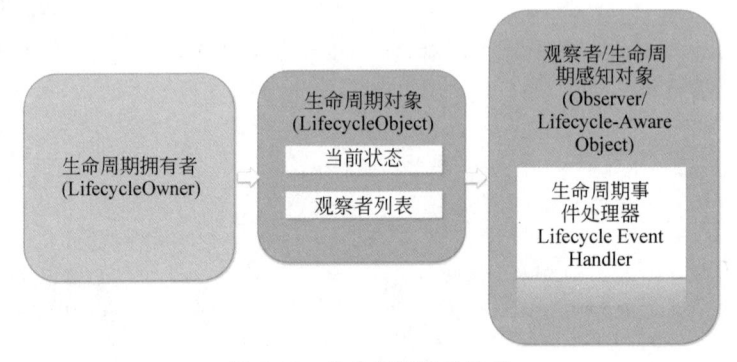

图 5-10 生命周期组件构成

LifecycleOwner 是拥有 Lifecycle 的组件,一般用于实现 LifecycleOwner 接口的对象。Activity 是生命周期拥有者,因为它们本身具有了 Lifecycle,可以对 Lifecycle 进行管理。

如图 5-11 所示,LifecycleOwner 拥有 Lifecycle 对象。一旦 LifecycleOwner 的状态发生变化,它的 Lifecycle 的状态也会发生相应的变化,具体如下:

```
Lifecycle.State.INITIALIZED: 初始化状态
Lifecycle.State.CREATED: 创建状态
Lifecycle.State.STARTED: 启动运行状态
Lifecycle.State.RESUMED: 恢复运行状态
Lifecycle.State.DESTROYED: 销毁状态
```

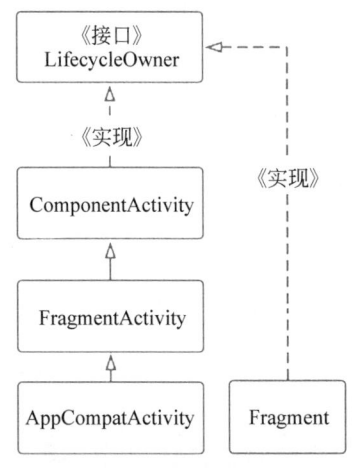

图 5-11 生命周期拥有者

通过 Lifecycle 对象，LifecycleOwner 不但保存了当前状态，而且可以管理 LifecycleObserver 列表或 Lifecycle 感知对象列表。一旦 LifecycleOwner 的 Lifecycle 对象的状态发生了变化，就会让 LifecycleOwner 通知 LifecycleObserver 列表或 Lifecycle 感知对象列表的所有成员，当前状态已经发生变换，可以随之也发生变化。代码如下：

```
public abstract class Lifecycle {
    @MainThread
    public abstract void addObserver(@NonNull LifecycleObserver observer);    //增加观察者
    @MainThread
    public abstract void removeObserver(@NonNull LifecycleObserver observer); //删除观察者
    @MainThread
    @NonNull
    public abstract State getCurrentState();                                  //获得当前状态
}
```

在 Lifecycle 事件切换的过程中，Lifecycle 对象的状态会发生变化，如图 5-12 所示。

图 5-12 LifecycleEvent 中状态的变化

Lifecycle 对象会触发事件给加入 LifecycleObserver 列表中的任何一个 LifecycleObserver 对象。LifecycleObserver 又称 Lifecycle Aware Component(生命周期感知组件)，用于检测和响应移动应用中其他对象的 Lifecycle 状态变化。任何一个类实现了接口 LifecycleObserver，都可以视之为感知 Lifecycle 组件。在 LifecycleObserver 中需要实现如下事件。

(1) Lifecycle.Event.ON_CREATE：创建生命周期。

(2) Lifecycle.Event.ON_START：开始生命周期。

(3) Lifecycle.Event.ON_RESUME：恢复生命周期。

(4) Lifecycle.Event.ON_PAUSE：暂停生命周期。

(5) Lifecycle.Event.ON_STOP：停止生命周期。

(6) Lifecycle.Event.ON_DESTROY：销毁。

(7) Lifecycle.Event.ON_ANY：可以表示上述的任意一个状态，由当时的运行情况决定。

例 5-5 DisposableEffect 函数调用的应用实例。在这个例子中将处理不同 Activity 的切换，从不同的 UI 界面进行跳转，从一个界面跳转到另一个界面前，需要对前一个界面的附带效应进行清理工作。

(1) 定义 MainActivity，代码如下：

```kotlin
class MainActivity: ComponentActivity() {
    override fun onCreate(savedInstanceState: Bundle?) {
        super.onCreate(savedInstanceState)
        enableEdgeToEdge()
        setContent {
            Chapter05Theme {
                HomeScreen(onStart=::onStartCall, onStop=::onStopCall)    //调用 HomeScreen
            }
        }
    }
    private fun onStartCall(timer: MutableState<Int>){                    //启动 start，timer 是状态值
        Log.d("TAG","startCall")                                          //记录日志
        thread{                                                           //创建并启动工作线程
            while(timer.value<100){                                       //判断状态值是否小于 100
                timer.value++                                             //修改状态值
                Thread.sleep(1000)                                        //当前线程暂停 1s
            }
        }
    }
    private fun onStopCall(timer: MutableState<Int>){                     //start 停止
        Log.d("TAG","stopCall")                                           //记录日志
    }
}
```

(2) MainActivity 中的包含的 HomeScreen 界面定义如下：

```kotlin
/**定义 HomeScreen 界面
 * @param lifecycleOwner LifecycleOwner：生命周期拥有者
 * @param onStart Function1<MutableState<Int>, Unit>：函数对象启动动作
 * @param onStop Function1<MutableState<Int>, Unit>：函数对象停止动作 */
@Composable
fun HomeScreen(lifecycleOwner: LifecycleOwner=LocalLifecycleOwner.current,
        onStart:(MutableState<Int>)->Unit,onStop:(MutableState<Int>)->Unit){
    val startAction by rememberUpdatedState(newValue=onStart)             //开始动作记录 onStart 状态
    val stopAction by rememberUpdatedState(newValue=onStop)               //停止动作记录 onStop 状态
    val context=LocalContext.current                                      //当前上下文
    var timer=remember{mutableIntStateOf(0)}                              //计时状态
```

```
DisposableEffect(key1=lifecycleOwner){            //键参数为生命周期拥有者 lifecycleOwner
    val observer=LifecycleEventObserver{_,event->  //创建生命周期观察者
                                                   //观察生周事件的类型
        if (event==Lifecycle.Event.ON_RESUME){     //恢复事件
            startAction(timer)                     //调用 startAction 动作处理
        }else if(event==Lifecycle.Event.ON_STOP){  //停止事件
            stopAction(timer)                      //调用停止动作处理
        }
    }
    /**生命周期拥有者 lifecycleOwner 的生命周期 lifecycle 加入新的观察者 observer,观察者
       可以观察生命周期的变化*/
    lifecycleOwner.lifecycle.addObserver(observer) //生命周期加入生命周期观察者
    onDispose {                                    //当离开可组合项时执行清理处理
        Log.d("TAG","Cleaning is End!")
        timer.value=0                              //计时状态值恢复为 0
        lifecycleOwner.lifecycle.removeObserver(observer)  //移除生命周期观察者
    }
}
Box(contentAlignment=Alignment.Center,
    modifier=Modifier.fillMaxSize()){
    Column{
        Text(text="${timer.value}秒",fontSize=30.sp)
        Button(onClick={                           //跳转到其他活动,即修改了生命周期拥有者对象
            val intent=Intent(context, OtherActivity::class.java)
            context.startActivity(intent)
        }){
            Text("跳转到其他活动")
        }
    }
}
```

(3)定义 OtherActivity,代码如下:

```
class OtherActivity: ComponentActivity() {
    override fun onCreate(savedInstanceState: Bundle?) {
        super.onCreate(savedInstanceState)
        enableEdgeToEdge()
        setContent {
            Chapter05Theme {
                Box(contentAlignment=Alignment.Center,
                    modifier=Modifier.fillMaxSize()){
                    Text("OtherActivity",fontSize=30.sp)
                }
            }
        }
    }
}
```

DisposableEffect 函数中的关键字参数是 LifecycleOwner,表示当前的生命周期拥有者,即为 Activity。观察 Activity 是否发生了变化,通过创建 LifecycleEventObserver 对象获得一个 LifecycleObserver,并将这个

LifecycleObserver 对象加入 LifecycleOwner 中。通过 LifecycleObserver 观察生命周期事件的状态发生什么变换，根据不同的状态做出不同的处理。当生命周期事件的类型是 Lifecycle.Event.ON_RESUME，则启动 startAction 动作；当生命周期事件的类型是 Lifecycle.Event.ON_STOP，则调用 stopAction 动作。

运行结果如图 5-13 所示。

(a) MainActivity　　　(b) OtherActivity　　　(c) 日志

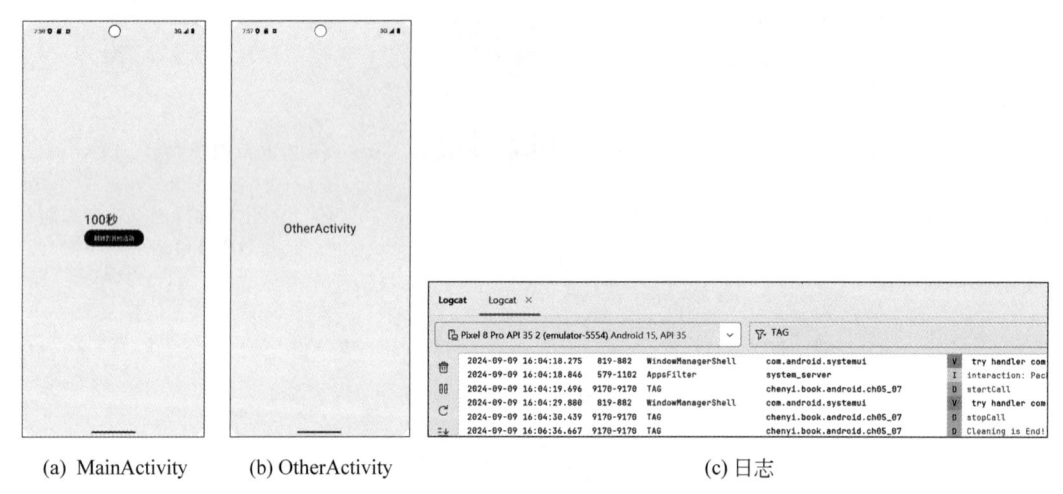

图 5-13　DisposableEffect 运行效果

运行这个模块，MainActivity 处于前台时，会动态显示计时，当单击按钮跳转到 OtherActivity 时，会显示图 5-13(b)的显示效果，日志会显示 stopCall。当彻底从应用退出时，会在日志显示"Cleaning is End!"，如图 5-13(c)所示。这种运行效果与期盼的运行效果有一定差距。这是因为在跳转到 OtherActivity 时，LifecycleOwner 虽然已经切换到 OtherActivity，但是界面 HomeSc,reen 已经在后台，并没有执行 onDispose 的代码块。只有在彻底退出应用，彻底离开可组合函数 HomeScreen 时，可组合函数 HomeScreen 中的 DisposableEffect 中监测到代码生命周期拥有者的对象发生了变化，因此执行 onDispose 代码块。

5.4.4　SideEffect

可组合函数的界面状态值发生变化一般会重组并重新渲染 UI 界面，但是可组合函数进行重组时并不是每次都能成功，这是因为在重组的过程中，一些状态数据又发生了变化，从而导致上次重组没有完全完成。这就使得一些与界面重组无关的数据和代码也会被多次调用，这种情况显然是没有必要的。在这样的前提下，可以使用 SideEffffect。

SideEffect 表示"副作用"，是将 Compose 的状态发布为非 Compose 代码。如需与非 Compose 管理的对象共享 Compose 状态，可使用 SideEffect 中的可组合项，因为只有每次成功重组时才会调用该可组合项。SideEffect 函数的定义内容如下：

```
@Composable
@NonRestartableComposable
@ExplicitGroupsComposable
@OptIn(InternalComposeApi::class)
fun SideEffect(effect: ()->Unit) {
    currentComposer.recordSideEffect(effect)
}
```

运行下列调用 SideEffect 函数的界面,代码如下：

```kotlin
@Preview
@Composable
fun HomeScreen(){
    var timer by remember{ mutableIntStateOf(0) }           //计时状态
    var running by remember{mutableStateOf(false)}          //运行状态
    Log.d("TAG","Log before SideEffect")
    SideEffect{
        Log.d("TAG","run SideEffect Function:running:${running}")
        if (running) {
            timer++                                         //变更计时状态的值
            Thread.sleep(1000)                              //当前线程暂停 1 秒
        }
    }
    Log.d("TAG","Log after SideEffect")
    Box(contentAlignment=Alignment.Center,modifier=Modifier.fillMaxSize()){
        Column{
            Text(text="${timer}秒",fontSize=30.sp)
            Row(horizontalArrangement=Arrangement.Center){
                Button(onClick={
                    running=true
                }){Text("开始计时")}
                Button(onClick={
                    running=false
                    timer=0
                }){Text("停止计时")}
            }
        }
    }
}
```

运行效果如图 5-14 所示。

(a) 运行界面　　　　　　　　　　　　(b) 记录日志

图 5-14　运行效果

运行过程中并没有发生动态更新计时时间的显示。这是因为，running 的值为 false，在条件判断时，使得 timer 状态值无法发生变化，状态值没有变化，使得界面只会在调用时刷新界面。即使点击"开始计时"按钮，修改 running 的值为 true，因为影响界面重组的状态值 timer 并没有变化，因此界面没有发生重组。观察日志，SideEffect 函数内部的 Lambda 代码段中的日志是最后调用的。这是因为可组合函数 SideScreen 重组成功后，才会调用 SideEffect 函数。

例 5-6　SideEffect 的应用实例。每 5 秒记录一个日志，并在 LogCat 中显示日志信息。代码如下：

```kotlin
/**模块 Ch05_08 的实体类 LogInfo 日志信息实体类
 * @property id String: 日志编号
 * @property content String: 日志的内容
 * @property recordedDate LocalDate: 记录日志的日期*/
data class LogInfo(var id:String="0",
    var content:String="empty",
    var recordedDate:LocalDateTime=LocalDateTime.now()){
    override fun toString():String {
        val formater=DateTimeFormatter.ofPattern("yyyy年MM月dd日 HH:mm:ss")
                                                                        //日期格式
        return "编号: $id,内容: $content,记录时间: ${recordedDate.format(formater)}"
                                                                        //字符串表示
    }
}
//模块 Ch05_08 定义的组合函数 rememberLogInfo
@Composable
fun rememberLogInfo(id:String,content:String,recordedTime: LocalDateTime):LogInfo{
    val log=remember{LogInfo()}                              //定义可组合的状态
    SideEffect{                                              //将状态值转换成非可组合项
        log.id=id                                            //修改日志编号数字
        log.content=content
        log.recordedDate=recordedTime
    }
    return log
}
//模块 Ch05_08 定义 SideScreen 界面
@Preview
@Composable
fun SideScreen(){
    var log=rememberLogInfo("0","副效应",LocalDateTime.now())   //非状态的对象
    var running by remember{mutableStateOf(false)}              //运行状态
    LaunchedEffect(running){
        while(running){
            delay(5000)                                         //暂停 5s
            log.id="${log.id.toInt()+1}"
            log.recordedDate=LocalDateTime.now()
            Log.d("TAG","日志: $log 信息")
        }
    }
    Column(modifier=Modifier.fillMaxSize(),
```

```
            verticalArrangement=Arrangement.Center,
            horizontalAlignment=Alignment.CenterHorizontally){
            Text(text="副效应应用示例",fontSize=30.sp)
            Row(horizontalArrangement=Arrangement.Center){
                Button(onClick={
                    running=true
                }){Text("记录日志")}
                Button(onClick={
                    running=false
                }){Text("停止记录")}
            }
        }
    }
```

运行结果如图 5-15 所示。

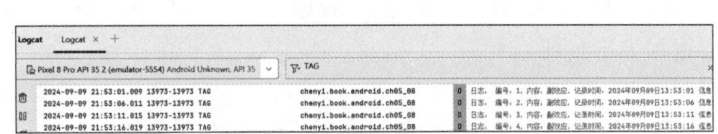

图 5-15 SideEffect 实例的运行效果

从运行结果发现,运行界面没有明显变化。因为在定义的可组合函数中 rememberLogInfo() 函数中将状态值从可组合状态转换成非 Compose 的状态,不会对界面的重组渲染起到作用。但是只要点击"记录日志"按钮,从 LogCat 中观察的结果可以发现每 5 秒就会将修改的日志对象重新显示。SideEffect 副效应往往定义一些分析统计日志等非重组界面的状态进行处理,不会对界面的渲染起到作用。SideEffect 这样的处理方式,会专注于处理与 UI 无关的状态,有效地改善性能,并让代码更易组织和管理。

5.4.5 produceState 和 derivedStateOf

produceState 和 derivedStateOf 都是可组合函数,它们的共同之处就是实现状态的转换。其中,produceState 是将非 Compose 状态转换成 Compose 状态;而 derivedStateOf 是将多个状态转换成其他状态。

1. produceState

produceState 是可组合函数,可将非 Compose 状态转换为 Compose 状态,会在没有定义数据源的

• 193 •

情况下随时间生成可观察的快照状态。produceState 会启动一个协程,在作用域限定范围将生成的值推送到返回的 State 的组合中。produceState 函数定义形式如下:

```
@Composable
fun <T>produceState(initialValue: T,
    [key1:Any?,key2:Any…],
    producer: suspend ProduceStateScope<T>.()->Unit): State<T>{
    val result=remember { mutableStateOf(initialValue) }
    LaunchedEffect(Unit) { ProduceStateScopeImpl(result, coroutineContext).producer() }
    return result
}
```

上述参数说明如下:

initialValue:表示初始值。

key1、key2 等:键值是可选项,表示随着时间的变化,会根据键值生成状态值。

producer:produceState 进入重组界面,加载的 producer 被加载。当重组界面结束后,producer 会被取消。

当 produceState 函数进入重组并渲染界面时会启动 producer 的生成值的操作,创建一个状态值。如果 key1 和 key2 的值发生了变化,正在运行的 producer 会被取消并重新加载创建新的状态源。producer 使用 ProducerStateScope.value 为返回的状态设置新的值。如果 ProducerStateScope.value 混淆了返回状态的新旧值,使得 ProducerStateScope.value 返回的状态的值仍为状态旧值,则不会观察到任何的更改。如果再频繁设置多个新值,那么观察者只会观察到最新的值。

例 5-7　produceState 的应用实例。在这个简单应用实例中实现访问并显示网络中的在线图片,具体如下。

(1) 配置 AndroidManifest.xml 网络权限。为了访问在线图片,需要在 AndroidManifest.xml 配置网络权限,增加如下代码:

```
<uses-permission android:name="android.permission.INTERNET" />
```

(2) 定义图片库。在图片库中加入在线图片资源,代码如下:

```
class ImageRepository constructor(context:Context){
    val context: Context=context
    var id=0
    val imageLst=listOf("https://10.0.2.2:8080/t01df0fad4a860b2327.jpg",
        "https://10.0.2.2:8080/t01166f28ff8d9dc33e.jpg",
        "https://10.0.2.2:8080/t01524438018e990965.jpg",
        "https://10.0.2.2:8080/t011c3679bf35d42d27.jpg",
        "https://10.0.2.2:8080/t0100d9444dcfdf0bc7.jpg")        //定义图片列表
    fun findByUrl(url:String)=imageLst.indexOf(url)             //根据 url 获取列表索引
    suspend fun next():String{                                   //返回下一个 url
        id=(id+1)%imageLst.size
        return imageLst.get(id)
    }
    suspend fun prev():String{                                   //返回上一个 url
```

```
            id=(id-1+imageLst.size)%imageLst.size
            return imageLst.get(id)
    }
    suspend fun loadImageByUrl(url:String): ImageBitmap {          //请求加载在线图片
        val request=ImageRequest.Builder(context).data(url).build() //请求在线图片
        val imageLoader by lazy{ ImageLoader(context) }
        return (imageLoader.execute(request).drawable as BitmapDrawable)
            .bitmap.asImageBitmap().apply{
                delay(1000)                                          //延迟1秒加载图片
            }
    }
}
```

上述代码的图片源于本地服务器localhost(对应于10.0.2.2)。也可以根据需要修改其他图片库的地址。

(3) 定义密封类,表示访问在线资源可能碰到的各种情况,代码如下:

```
sealed class OpResult<T>(){
    object Loading:OpResult<ImageBitmap>()                              //正在加载图片结果
    object Error:OpResult<ImageBitmap>()                                //请求资源错误结果
    data class Success(val image:ImageBitmap):OpResult<ImageBitmap>()   //请求资源成功结果
}
```

(4) 定义可组合函数访问在线图片资源,代码如下:

```
/** 将非Composable状态转换成Composable状态
 * @param imageId Int: 表示在图片仓库中图片列表的序号
 * @param imageRepository ImageRepository 图片仓库
 * @return State<OpResult<ImageBitmap>> */
@Composable
fun loadNetworkImages(imageRepository: ImageRepository):State<OpResult<ImageBitmap>>{
    return produceState<OpResult<ImageBitmap>>( initialValue=OpResult.Loading){
                                                                        //初始值正在加载图片结果
        for (i in 0 until imageRepository.imageLst.size) {
            var image=imageRepository.loadImageByUrl(imageRepository.imageLst[i])
            value=if (image==null) {                                    //图片资源为null
                OpResult.Error                                          //返回请求错误结果
            } else {
                OpResult.Success(image)                                 //返回请求图片成功结果
            }                                                           //并封装图片对象
        }
    }
}
```

因为从网络请求到缓存到本地存在时间差,因此使用produceState处理从图片库请求在线图片资源,且每秒生成一个可观察的快照状态。这个自定义函数会返回一个包含请求资源结果的状态,并根据循环暂停1s,使得图片列表的索引值发生变化,从图库对象imageRepository中要加载的图片。如果图片请求成功,将image图片封装到状态值OpResult.Success(image)中,为后续的界面更新提供保证。只要状态值更新,就会刷新界面。

(5) 定义ImageScreen界面,动态显示在线图片,代码如下:

```
@Composable
fun ImageScreen(imageRepository: ImageRepository) {
```

```
        val result by loadNetworkImages(imageRepository=imageRepository)    //获取状态
        Box(modifier=Modifier.fillMaxSize().background(Color.Black),
            contentAlignment=Alignment.Center) {
            Column(modifier=Modifier.fillMaxWidth().height(300.dp),
                verticalArrangement=Arrangement.Center,
                horizontalAlignment=Alignment.CenterHorizontally) {
                when(result) {
                    is OpResult.Success ->{                                  //请求图片结果成功
                        Image(modifier=Modifier.size(400.dp,300.dp),
                            bitmap=(result as OpResult.Success).image,
                            contentDescription=null)
                    }
                    is OpResult.Error ->{                                    //请求图片结果失败
                        Image(imageVector=Icons.Filled.Warning,contentDescription=null,
                            modifier=Modifier.size(200.dp, 200.dp))
                    }
                    else ->{
                        CircularProgressIndicator()                          //等待加载图片时,显示圆形进度条
                    }
                }
            }
        }
    }
```

（6）在 MainActivity 中加载 ImageScreen 界面，代码如下：

```
class MainActivity: ComponentActivity() {
    override fun onCreate(savedInstanceState: Bundle?) {
        super.onCreate(savedInstanceState)
        enableEdgeToEdge()
        val imageRepository=ImageRepository(this)                //创建图片资源库
        setContent {
            Chapter05Theme {
                ImageScreen(imageRepository)                     //调用 ImageScreen 生成界面
            }
        }
    }
}
```

运行结果如图 5-16 所示。

2. derivedStateOf 函数

derivedStateOf 函数将一个或多个状态对象转换为其他状态,使用此函数可确保仅当计算中使用的状态之一发生变化时才会进行计算。derivedStateOf 函数定义的内容如下：

```
@StateFactoryMarker
fun <T>derivedStateOf( policy: SnapshotMutationPolicy<T>,
    calculation: ()->T,): State<T>=DerivedSnapshotState(calculation, policy)
```

policy 是可选项,表示策略来控制对计算结果的更改何时触发更新。如果调用函数没有设置 policy,会默认为 null。

calculation：函数对象,用来创建一个 State 对象,其 State.value 是计算的结果。

例 5-8 derivedStateOf 函数的应用实例。代码如下：

图 5-16 动态显示在线图片

```
@Preview
@Composable
fun DisplayScreen(){
    val yearState=remember{mutableStateOf(2025)}           //表示年份状态值
    val monthState=remember{mutableStateOf(1)}             //表示月份的状态值
    val calendarState=remember {
        derivedStateOf {
            "${yearState.value}年-${monthState.value}月"    //将两种状态组合生成新的状态值
        }
    }
    Box(contentAlignment=Alignment.Center){
        Column(modifier=Modifier.fillMaxWidth(),
            horizontalAlignment=Alignment.CenterHorizontally){
            Text("${calendarState.value}",fontSize=30.sp)   //显示新状态的值
            Button(onClick={
                monthState.value=(monthState.value)%12      //修改月份状态值
                if (monthState.value==0){
                    yearState.value++                        //修改年份状态值
                }
                monthState.value++
            }){Text("修改月份",fontSize=24.sp) }
        }
    }
}
```

运行结果如图 5-17 所示。点击"修改月份"按钮，使得显示的年月信息发生变化。很明显，derivedStateOf 函数将 yearState 和 monthState 重组生成了一个新的状态值。每次点击按钮，会让 yearState 和 monthState 发生变化，通过 derivedStateOf 将二者组合，重生成一个新状态并保存在内存

中。每产生一个变更的 calculateState 状态，就会使界面发生重组更新。

图 5-17　derivedStateOf 函数实例的运行结果

5.5　RxJava 库实现异步操作

ReactiveX（Reactive eXtension）结合观察者模式、迭代器模式和函数式编程的编程接口，RxJava 是 ReactiveX 的 Java VM 实现。RxJava 库通过可观察的序列来组成异步和基于事件的程序。RxJava 处理异步数据会更加方便。

5.5.1　Observer 模式

Observer（观察者）模式是理解 RxJava 的第一步。现实生活中存在很多多个对象依赖一个对象的行为，例如汽车"红灯停、绿灯行"、每日的报纸发放给订阅者。Observer 模式是为解决对象间存在一对多关系的处理方式。使得一个对象被修改时，则会自动通知依赖它的对象。如果对 Observer 模式有一定了解，可以直接跳过。

如图 5-18 所示，Observer 模式中包含 Subject、ConcreteSubject、Observer 和 ConcreteObserver 4 种角色。

（1）Subject。该角色定义了抽象主题的接口，定义了对 Observer 的管理方法：addObserver()添加 Observer、removeObserver()删除 Observer 和 notifyObservers()通知 Observer 的方法。抽象主题可以包含多个 Observer 对象。

（2）ConcreteSubject。该角色定义了具体的主题，记录主题的相关状态，当主题发生变化时，会通知所有的 Observer 状态发生变化。

（3）Observer。该角色是抽象 Observer 接口，得到主题的通知后会更新自己。

（4）ConcreteObserver。该角色是抽象 Observer 接口的具体实现。当主题状态发生变化时，对应的 Observer 也会随之改变，以与主题的状态保持一致。使用 Observer 模式的代码如下：

```
            Subject                          Observer
   +addObserver(Observer)                   +update()
   +removeObserver(Observer)
   +notifyObservers()


        ConcreteSubject                 ConcreteObserver
   +subjectState                       +observerState
   +addObserver(Observer)              +update()
   +removeObserver(Observer)
   +notifyObservers()
```

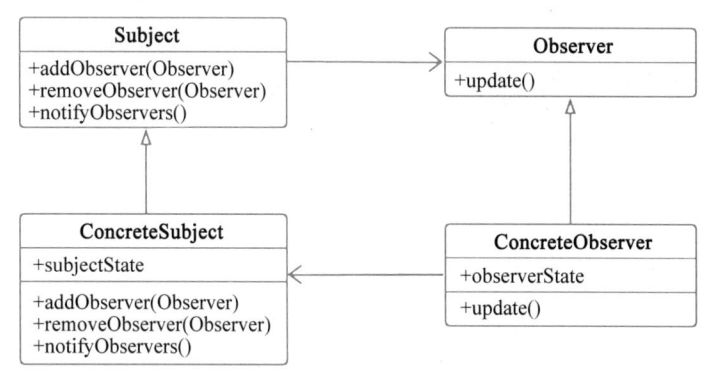

图 5-18 Observer 模式示意图

```kotlin
//模块 Ch05_10 定义抽象主题接口 Subject.kt
interface Subject {
    fun addObserver(observer: Observer)
    fun removeObserver(observer: Observer)
    fun notifyObservers(state: String)
}
```

定义 Observer 接口,代码如下:

```kotlin
//模块 Ch05_10 定义 Observer 接口 Observer.kt
interface Observer {
    fun update(state: String)
}
```

定义 ConcreteSubject 类,代码如下:

```kotlin
//模块 Ch05_10 定义具体主题 ConcreteSubject.kt
class ConcreteSubject:Subject{
    val observers=mutableListOf<Observer>()
    lateinit var state:String
    override fun addObserver(observer: Observer) {
        println("观察者模式-添加观察者${observer.toString()}")
        observers.add(observer)
    }
    override fun removeObserver(observer: Observer) {
        println("观察者模式-移除观察者${observer.toString()}")
        observers.remove(observer)
    }
    override fun notifyObservers(state:String) {
        println("观察者模式-通知所有的观察者")
        observers.forEach {
            print("观察者模式-主题通知${it.toString()}")
            //观察者变更状态
            it.update(state)
        }
    }
    fun changeState(state:String){
```

```
        //修改状态
        this.state=state
        //通知所有的观察者状态变化
        notifyObservers(state)
    }
}
```

定义 ConcreteObserver 类,代码如下:

```
//模块 Ch05_10 定义 ConcreteObserver.kt
data class ConcreteObserver(valname:String):Observer {
    override fun update(state: String) {
        println("观察者模式-观察者${name}接收到主题的${state}")
    }
}
//模块 Ch05_10 测试文件 Test.kt
fun main(){
    val subject=ConcreteSubject()
    for (i in 1..3){
        val observer=ConcreteObserver("观察者${i}")
        subject.addObserver(observer)
    }
    subject.changeState("报纸订阅")
    subject.changeState("电子公告订阅")
}
```

运行结果如图 5-19 所示。

图 5-19　Observer 模式实例的运行结果

在上述的 Test.kt 测试代码中,定义了 3 个 Observer。当主题调用 changeState 函数状态发生变化时,通知所有的 Observer 发生相应的变化,使得运行结果表现了这种变化。两次主题状态变更时,Observer 的两次的输出结果不一致。实现了多个对象对一个对象依赖的关系。

RxJava 基于 Observer 模式,但是与 Observer 模式的不同在于 RxJava 是基于异步的数据流实现的,而不是像 Observer 模式那样根据某些触发器对对象进行更改。发生更改时,将通知每个依赖于主题的 Observer。在 5.5.2 节将对 RxJava 做进一步介绍。

5.5.2　RxJava 的相关概念

要使用 RxJava 库,需要在项目的 libs.versions.toml 文件中增加如下配置:

```
[versions]
rxjava="3.1.9"
rxandroid="3.0.2"
[libraries]
    rxjava3={group="io.reactivex.rxjava3",name="rxjava",version.ref="rxjava"}
    rxandroid={group="io.reactivex.rxjava3",name="rxandroid",version.ref="rxandroid"}
```

在模块的 build.gradle.kts 构建文件中增加如下依赖：

```
dependencies {
    …
    implementation(libs.rxjava3)         //增加 RxJava 库的依赖
    implementation(libs.rxandroid)       //增加在 Android 系统中对 RxJava 库的支持
}
```

首先，了解一下 RxJava 的几个基本概念：Observable（可观察）、Operator（操作符）、Observer（观察者）和 SubScriber（订阅者），如图 5-20 所示。

(1) Observable（可观察）。即被观察者，可以表示任何对象，可以从数据源中获得数据或者状态值。Observable 发送的是数据流。只要有 Observer 接收，Observable 就会发送数据流。Observable 可以有多个 SubScriber。它可以发出零个或多个数据，允许发送错误消息，可以在发出一组数据的同时控制其速度，可以发送有限或无限的数据。Observable 有两种执行类型："非阻塞-异步执行"和阻塞式。"非阻塞-异步执行"的最大特点是，在整个事件流中可以取消订阅，这种方式更为灵活，较容易被接受。"阻塞式"所有的 Observer 的 onNext 调用都是同步的，而且在事件流的过程中不允许取消订阅。

(2) Operator（操作符）。Operator 承担了对 Observable 发出的事件进行修改和变换。实际上，每个 Operator 都是一个方法。作为输入参数，Observable 发送的数据都会在 Operator 的方法中应用，并以 Observable 的形式将结果返回。由于返回的是另外一个 Observable，所以这个 Observable 可以继续向后发送或结束。

图 5-20　RxJava 中 Observable、Operator 和 Observer 之间的关系

(3) Observer（观察者）。Observer 用于订阅 Observable 的序列数据，并对 Observable 的每一项做出反应。Observer 负责处理事件，是事件的消费者。每当关联的 Observable 发送数据时，会通知 Observer，然后 Observer 一个接一个地处理数据。当关联的 Observable 发送数据项时，会激活每个 SubScriber 的 onNext 函数，执行某些特定的任务；当 Observable 完成一些事件处理时，会调用 Observer 的 onComplete 函数；当 Observable 发送的数据有错误时，会调用 Observer 组件的 onError 函数。

使用 RxJava 库时，Operator 可以有若干个，形式如下：

```
dataSource
```

```
.operator1()
.operator2()
.operator3()
```

这些操作符之间构成了上下流的关系。

RxJava 中还有一些基本类必须了解,具体的介绍如表 5-1 所示。

表 5-1　RxJava 中进行可运算处理时常用的基本类

类	说　　明
io.reactivex.rxjava3.core.Flowable	0..N 流,支持响应式流和背压。 按照 onSubscribe onNext(onError 或 onComplete)属性执行,其中 onNext 可以执行多次,onError 和 onComplete 是互斥的
io.reactivex.rxjava3.core.Observable	0..N 流,不支持背压。 按照 onSubscribe onNext(onError 或 onComplete)顺序执行,onNext 可以执行多次,onError 与 onComplete 是互斥的
io.reactivex.rxjava3.core.Single	只有一项或一个错误的单个流。 按照 onSubscribe(onSuccess 或 onError)顺序执行,其中,onSuccess 和 onError 是互斥的
io.reactivex.rxjava3.core.Completable	没有项目但有实现或错误的信号的流。 按照 onSubscribe(onError 或 onComplete)的顺序执行
io.reactivex.rxjava3.core.Maybe	没有项目、只有一项或一个错误的流按照顺序执行 onSubscribe(onSuccess 或 onError 或 onComplete)

Observable 与 Observer 往往在不同的线程处理数据,它们之间是异步的。具体执行线程的切换可以通过线程的 Schedulers(调度器)实现。在表 5-2 中定义了常见的线程调度。

表 5-2　线程调度

线　程　调　度	说　　明
Schedulers.io()	适合输入输出操作和阻塞操作
Schedulers.single()	适合单一线程操作
Schedulers.computation()	适合运行在密集计算的操作
Schedulers.trampoline()	适合按照顺序执行的操作
Schedulers.newThread()	适合为每个任务创建新线程
AndroidSchedulers.mainThread()	Android 的主线程

不同线程中,处理问题所用的时间会随着问题的复杂度不同有所变化,这将导致二者处理数据的速度不同步。如果 Observable 发送数据的速度远远快于 Observer 的数据处理速度,会将数据放入缓存中或者直接放弃。上述两种方法都有不妥之处,因此需要制定背压(Back Pressure)策略来解决二者在异步时 Observable 发送数据和 Observer 处理数据速度不一致问题。通常所说的背压是在异步环境下,控制流速的一种策略。不同于 Observer 模式,采用 Observable 主动通知 Observer 数据发生了变化。在背压策略中,Observer 采用响应式拉取,会根据自身执行的实际情况按需抓取数据,而不是被动接收,间接地让 Observable 发送数据的速度有所减慢。最终达到控制上游 Observable 事件发送速度的目的,实现背压的策略。在 io.reactivex.rxjava3.core.BackpressureStrategy 枚举类中可以指定如表 5-3

所示的背压策略。

表 5-3 背压策略

背压策略	说　　明
MISSING	表示通过 create 方法创建的 Flowable 没有指定背压策略,不会对通过 OnNext 发送的数据做缓存或丢弃处理,下游必须处理操作符
ERROR	发生背压时,会发送 MissingBackpressureException 信号,以免下游不能继续处理
BUFFER	发生背压时,会缓存数据,直至下游完成处理数据
DROP	发生背压时,如果下游不能继续处理数据,将最近发送的值丢弃
LATEST	发生背压时,会缓存最新的数据

例 5-9　RxJava 库的应用实例。动态显示 10 个数据,代码如下:

```
@Preview
@Composable
fun MainScreen(){
    var contentTxt by remember{ mutableStateOf("") }        //显示文本状态
    Column(modifier=Modifier.fillMaxSize(),
        horizontalAlignment=Alignment.CenterHorizontally,
        verticalArrangement=Arrangement.Center){
        Text("$contentTxt",fontSize=36.sp)
        Button(onClick={
            Flowable.create<String>({                        //Flowable 被观察者
                emitter->for (i in 1..10) {
                    emitter.onNext("发送数据=>${i}")          //发送数据
                    Thread.sleep(1000)
                }
                emitter.onComplete()                         //完成发送
            }, BackpressureStrategy.DROP)                    //设置背压策略
                .subscribeOn(Schedulers.io())                //指定被观察者线程处理 I/O 操作
                .observeOn(AndroidSchedulers.mainThread())   //指定观察者的线程为主线程
                .subscribe({
                    it: String->contentTxt="接收并显示: ${it}" //用接受的数据赋值给状态值
                }) { e: Throwable?->
                    e?.printStackTrace()
                }
        }){
            Text("开始",fontSize=24.sp)
        }
    }
}
```

运行结果如图 5-21 所示。

通过测试上述的代码可以发现,点击"开始"按钮后,从 1～10 构成的字符串数据每秒依次发送一次,通过设定背压策略为 DROP,将处理背压的方式设置为丢弃最近发送的数据。subscribeOn 函数指定了 Flowable 的线程,使用 Schedulers.io 线程的 Scheduler 进行线程切换。observeOn 函数指定了 Observer 的线程为 Android 主线程,Android 主线程接收数据。通过这样的异步处理,可使得生成数据

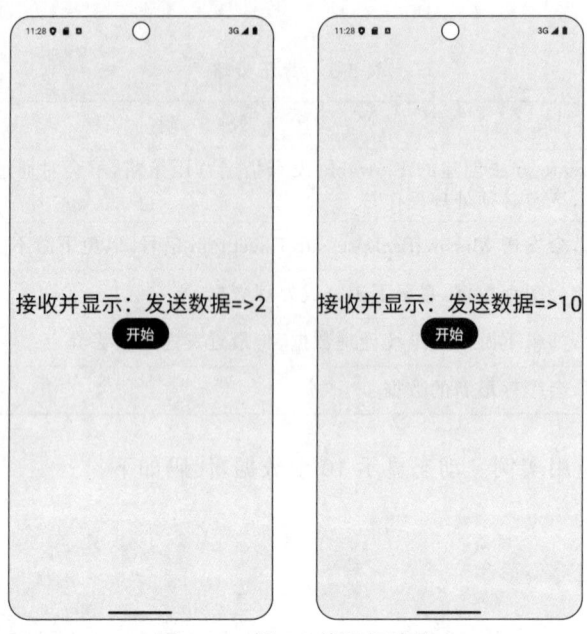

图 5-21 例 5-9 的运行结果

在其他线程实现,而显示数据则在主线程实现。

5.6 歌词同步播放

在 QQ 音乐、网易音乐、酷狗音乐等移动应用播放歌曲时,提供了歌词同步播放的功能。在本节中,将结合 Android 的并发处理,实现该功能。

采用歌词文件为 LRC 格式。LRC 歌词文件是基于文本结合 tag 标记的一种文件格式。LRC 文件包含两种格式标签:定义歌曲的基本信息的标识标签和定义歌词的内容歌词标签。

标识标签:其格式如下:

[标识名:值]

主要包含以下预定义的标签:[ar:歌手名]、[ti:歌曲名]、[al:专辑名]、[by:编辑者(指 lrc 歌词的制作人)]、[offset:时间补偿值]等。

歌词标签的格式主要有 3 种形式。

格式 1(标准格式):

[分:秒.毫秒]歌词

格式 2:

[分:秒]歌词

格式 3:

[分:秒:毫秒]歌词

这 3 种格式的细微差别在于毫秒的处理。格式 1 是用"."将毫秒分隔；格式 2 没有毫秒的定义；格式 3 采用":"来分隔毫秒。实现歌词与音乐同步播放，需要解决以下问题。

（1）歌词的解析。将标识标签和歌词标签分别进行处理，获得有用的信息。这些有用的信息包括歌手名、歌曲名、专辑名和歌词的编辑者等，特别是将歌词按照时间提取成一行行的文本字符串。

（2）将歌词按照时间的定义依次显示在移动终端界面。并伴随时间的流逝，将不同行的歌词从下向上动态显示。为了达到这个目的，本例对指定歌词进行解析。移动模块创建 assets 目录，将歌词和歌曲 mp3 文件放到 assets 目录中。

1. 定义歌词实体类

定义歌词实体类的代码如下：

```kotlin
//模块 LRCPlayer 实体类 LRC.kt
/**定义歌词实体类
 * @param title String 歌曲名
 * @param artist String 歌手
 * @param album String 专辑
 * @param editor String 歌词的编辑者
 * @param offset String 偏移量
 * @param lrcContent MutableMap<Long, String>时间与歌词的映射
 */
data class LRC (var title: String, var artist: String,
    var album: String, var editor: String,
    var offset: String, val lrcContent: MutableMap<Long,String>)
```

2. 歌词解析

进行歌词解析的代码如下：

```kotlin
//模块 LRCPlayer 定义类型转换 TypeConvetor 实用类 TypeConvetor.kt
class TypeConvetor {
    companion object{
        fun strToLong(timeStr: String): Long{           //将时间字符串转换成长整数
            val s=timeStr.split(Regex("[.|:]"))
            val minute=Integer.valueOf(s[0])
            val second=Integer.valueOf(s[1])
            var millSecond=0
            if (s.size>=3)
                millSecond=Integer.valueOf(s[2])
            return minute * 60 * 1000L+ second * 1000L+millSecond * 10L
        }
    }
}
```

TypeConvetor 实用类的主要作用就是将时间字符串，形如"00:55.81"，计算成长整数毫秒值，代码如下：

```kotlin
//模块 LRCPlayer 定义歌词解析 LRCParser 实用类 LRCParser.kt
object LRCParser {                                      //LRC 歌词解析器
    fun parse(lrcName: String): LRC {                   //解析歌词文件
        val assetManager=LRCApp.context.assets          //从 assets 目录读取歌词，
                                                        //歌词文件名从 lrcName 获得
```

```
            val lrcIO=assetManager.open(lrcName)
            val reader=BufferedReader(InputStreamReader(lrcIO,"UTF-8"))
                                                             //读取文件,创建缓存字符流
            val lrc=LRC("","","","",mutableMapOf<Long,String>())
            reader.lines().forEach{line: String->
                parseLine(line,lrc)                          //对每一行的内容进行处理
            }
            return lrc
        }
        fun parseLine(line: String,lrc: LRC){
            when {
                line.startsWith("[ti: ")->lrc.title=line.substring(4,line.length-1)
                                                             //歌曲名
                line.startsWith("[ar: ")->lrc.artist=line.substring(4,line.length-1)
                                                             //歌手
                line.startsWith("[al: ")->lrc.album=line.substring(4,line.length-1)
                                                             //专辑
                line.startsWith("[by: ")->lrc.editor=line.substring(4,line.length-1)
                                                             //歌词编辑者
                else ->{                                     //歌词,定义歌词标签的3种正则表达式
                    val reg="\\[(\\d{1,2}: \\d{1,2}\\.\\d{1,2})\\]|
                        \\[(\\d{1,2}: \\d{1,2})\\]|\\[(\\d{1,2}: \\d{1,2}\\: \\d{1,2})\\]"
                    val pattern=Pattern.compile(reg)
                    val matcher=pattern.matcher(line)
                    while (matcher.find()){                  //匹配正则表达式
                        val groupCount=matcher.groupCount()
                        var curTime: Long =0L
                        val content=pattern.split(line)
                        for (index in 0 until groupCount){
                            val timeStr=matcher.group(index)
                            if (index==0)
                                curTime=TypeConvetor.strToLong(timeStr.substring(1,timeStr
                                    .length-1))
                        }                                    //end for
                        content.forEach{
                            lrc.lrcContent[curTime]=it
                        }
                    }                                        //end while
                }                                            //end else
            }                                                //end when
        }
```

3. 定义显示歌词界面 MainScreen 可组合函数和其他控制函数

MainScreen 可组合函数定义的界面包含播放按钮和停止按钮,以及显示歌词的文本可组合项,代码如下:

```
@Composable
fun MainScreen(){
    val context=LocalContext.current                                    //上下文
    val mediaPlayer=remember{MediaPlayer.create(context,R.raw.yijianmei)} //播放器
```

```kotlin
        val lrcState=remember{mutableStateOf("") }                          //歌词状态
        var playing by remember{mutableStateOf(false) }                     //播放状态
        LaunchedEffect(playing) {                                           //加载效应
            if (playing) {
                withContext(Dispatchers.IO) {                               //IO调度器
                    playMusic("yijianmei.mp3", mediaPlayer)                 //播放音频
                }
                withContext(Dispatchers.Main) {                             //主线程调度器
                    playLRC(lrcState=lrcState)                              //播放歌词
                }
            }
        }
        Box(modifier=Modifier.fillMaxSize(),
            contentAlignment=Alignment.Center){
            Column(modifier=Modifier.fillMaxWidth()) {
                Text (modifier=Modifier.fillMaxWidth().height(400.dp).padding(5.dp),
                    fontSize=28.sp,
                    maxLines=10,
                    textAlign=TextAlign.Start,
                    text=buildAnnotatedString {
                        withStyle(style=ParagraphStyle(lineHeight=36.sp)) {
                            lrcState.value.split("\n").forEach {
                                withStyle(style=SpanStyle(Color.Blue)) {
                                    append("${it}\n")
                                }
                            }
                        }
                    })
                Row(modifier=Modifier.fillMaxWidth().padding(5.dp),
                    horizontalArrangement=Arrangement.Center
                ) {
                    TextButton(colors=ButtonDefaults.buttonColors(containerColor=Color
                        .Blue),onClick={playing=true}) {                    //修改播放状态
                        Text(text="播放", fontSize=24.sp)
                    }
                    TextButton(colors=ButtonDefaults.buttonColors(containerColor=Color
                        .Blue),onClick={ playing=false                      //修改播放状态
                            stopMusic(mediaPlayer)                          //停止播放音频
                    }) {
                        Text("停止", fontSize=24.sp)
                    }
                }
            }
        }
    }
}
```

在 MainScreen 中的点击动作处理需要播放和停止音频以及播放歌词,因此定义如下的函数作为动作的控制器,它们的代码分别如下:

```kotlin
/** 播放歌词 playLRC 函数
 * @param context Context */
fun playLRC(lrcState: MutableState<String>){
    val lrc=LRCParser.parse(LRCApp.context,"yijianmei.lrc")
                                            //解析 assets 目录下的 yijianmei.lrc 歌词获得 LRC 对象 lrc
    val lrcLst=mutableListOf<String>()                              //歌词列表状态
    var range:IntRange
    var currentTimer=0L
    var showLRC:StringBuilder=StringBuilder()
    val lrcContent=lrc.lrcContent                                   //获得时间与歌词的映射
    val lrcTimers=lrcContent.keys
    Flowable.create<String>({
        emitter: FlowableEmitter<String>->lrcTimers.forEach {
            timer:Long->var lrcStr=lrcContent[timer]
            if (lrcStr!=null){
                lrcLst.add(lrcStr)
                range=if(lrcLst.size-10>0) lrcLst.size-10 until lrcLst.size else 0 until
                    lrcLst.size
                showLRC.clear()
                for (i in range){                        //逆序将歌词的最后 10 行加入动态字符串
                    showLRC.append("${lrcLst[i]}\n")
                }
                emitter.onNext(showLRC.toString())
            }
            Thread.sleep(timer-currentTimer)
            currentTimer=timer
        }
        emitter.onNext("")
        emitter.onComplete() },
        BackpressureStrategy.DROP)
        .subscribeOn(Schedulers.io())
        .observeOn(AndroidSchedulers.mainThread())
        .subscribe{it:String->lrcState.value =it   }
}

/** 播放音频 playMusic 函数
 * @param context Context
 * @param musicFileName String
 * @param mediaPlayer MediaPlayer */
fun playMusic(musicFileName:String,mediaPlayer: MediaPlayer){
    val df=LRCApp.context.assets.openFd(musicFileName)
    mediaPlayer.setScreenOnWhilePlaying(true)
    mediaPlayer.start()
}
/** 停止音频 stopMusic 函数
 * @param mediaPlayer MediaPlayer */
fun stopMusic(mediaPlayer: MediaPlayer){
    if (mediaPlayer.isPlaying) {
        mediaPlayer.stop()
        mediaPlayer.release()
    }
}
```

4. MainActivity 调用 MainScreen

代码如下：

```kotlin
//模块 LRCPlayer 定义 MainActivity
class MainActivity: ComponentActivity() {
    override fun onCreate(savedInstanceState: Bundle?) {
        super.onCreate(savedInstanceState)
        enableEdgeToEdge()
        setContent {
            Chapter05Theme {
                MainScreen()
            }
        }
    }
}
```

运行结果如图 5-22 所示。

图 5-22　歌词播放的运行结果

习　题　5

一、选择题

1. 下列创建并启动自定义线程操作正确的是_____。

A.
```
Thread{
    override fun run(){
        print("Hello")
    }
}
```

B.
```
thread{
    override fun run(){
        print("Hello")
    }
}
```

C.
```
thread{
  print("Hello")
}
```

D.
```
Thread{
  print("Hello")
}
```

2. 消息处理机制是一种异步处理方式,通过_____机制可以实现在不同线程之间通信。
    A. Handler　　　　B. AsyncTask　　　C. Looper　　　　D. Coroutine
3. 在消息处理机制中,可以通过调用_____函数从消息池获得消息对象。
    A. Message   B. Message.obtain
    C. Message.obtainMessage()   D. 以上答案均不正确
4. 在消息处理机制中的消息具有属性_____,可用来传递简单的数据信息。
    A. arg1   B. obj
    C. handleMessage   D. dispatchMessage
5. Handler 是消息处理机制的处理器,用于调度和处理线程。通过调用_____函数可以发送消息。
    A. sendMessage   B. handleMessage
    C. handleThread   D. sendThread
6. _____可组合函数用于将 Compose 的状态发布为非 Compose 代码。
    A. SideEffect   B. LaunchedEffect
    C. DisposableEffect   D. 以上答案均不正确
7. Kotlin 协程可以使用_____创建一个顶层全局范围的协程。
    A. Global.launch   B. CoroutineScope.lauch
    C. runBlocking.launch   D. 以上答案均不正确
8. 在测试场景中,常通过_____来创建协程。
    A. runBlocking   B. Global
    C. CoroutineScope   D. 以上答案均不正确
9. 下列调度器可以在 Android 主线程中运行协程的是_____。
    A. Dispatchers.Default
    B. Dispatchers.Main
    C. Dispatchers.IO
    D. 以上答案均不正确

二、填空题
1. 启动协程有_____和_____两种方式。
2. _____是操作系统调度的最小单位。
3. Android 移动应用中单独启动一个_____,它有且仅有一个_____。
4. 调用_____函数并不运行协程,而是追踪协程。

三、上机实践题
1. 分别使用 Kotlin 的协程、Compose 的附带效应和 RxJava 库实现动态显示海报应用,并比较它们的不同。
2. 分别使用消息处理机制、Kotlin 的协程和 Compose 的附带效应实现一个计时器的应用,并比较它们的不同。

3. 结合 5.6 节的歌词同步播放应用实例,为歌词同步播放增加新的功能,除了将歌词按照时间的定义依次显示在移动终端的界面。并伴随时间的流逝,同行歌词从左到右动态显示,将不同行的歌词能从下向上动态显示。

4. 设计和实现一个英文打字移动游戏。要求如下:

(1) 单词从动态从上而下随机位置下降。

(2) 单词还没有触碰到底部,拼写单词成功,加分;如果单词已到达底部还没有拼写成功,则不加分。

(3) 计算积分及拼写正确单词的百分比,并显示和分享统计结果。

第 6 章　Android 的广播机制

广播机制是一种在组件之间实现传播数据的机制。这些组件可以位于不同进程，彼此通过广播机制实现进程间通信。作为 Android 四大组件之一的 BroadcastReceiver（广播接收器）组件是实现广播机制的关键，它可对移动应用的系统群发广播消息（Broadcast Message）做出响应，并将消息发送给其他组件。

6.1　BroadcastReceiver 组件

BroadcastReceiver 组件用于发送广播消息给移动应用或者 Android 系统。这里的广播消息实际是 Event（事件）或者 Intent（意图）。BroadcastReceiver 组件是一种用于对发送出来的广播消息进行过滤、接收和响应的组件。要应用 BroadcastReceiver 组件，需要提前创建和注册。

1. 创建 BroadcastReceiver 组件

要使用 BroadcastReceiver 组件，首先要进行创建，所创建的 BroadcastReceiver 组件必须是 BroadcastReceiver 类的子类，需要对其中的 onReceive 函数进行重新定义，代码如下：

```kotlin
class MyReceiver: BroadcastReceiver() {
    override fun onReceive(context: Context, intent: Intent) {
        …                                                               //处理广播消息
    }
}
```

通过对 onReceive 函数的重新定义，使之能接收指定 Intent 传递的广播消息，并做出相应的处理。

2. 注册 BroadcastReceiver 组件

在移动应用中应用自定义的 BroadcastReceiver 组件，需要对自定义的 BroadactReceiver 组件进行注册。注册 BroadcastReceiver 组件有两种方式：静态注册和动态注册。

静态注册需要在 AndroidManifest.xml 文件中对 BroadcastReceiver 组件进行配置。Android 8.0 及以前的版本需要在 AndroidManifest.xml 系统配置文件中，对 BroadcastReceiver 组件指定关联 BroadcastReceiver 组件的 IntentFilter 对象，实现筛选合适的 Intent，然后将这个 Intent 作为广播消息发送出去，并由 BroadcastReceiver 来处理广播的消息，形式如下：

```xml
<receiver android: name=".MyReceiver"
    android: enabled="true"
    android: exported="true">
    <intent-filter>
        <action android: name="book.android.ch06.Broadcast" />
    </intent-filter>
</receiver>
```

动态注册不是通过配置实现的，无须在应用配置清单文件 AndroidManifest.xml 中配置自定义的 BroadcastReceiver，在代码中直接定义即可。一般情况下，Android 8.0 及以后的版本都需要调用

Activity 的 registerReceiver 函数动态注册 BroadcastReceiver 组件,形式如下:

```
val intentFilter=IntentFilter()
filter.addAction("book.android.ch06.Broadcast")
val receiver=MyReceiver()
registerReceiver(receiver,intentFilter)
```

当已经注册的 BroadcastReceiver 组件不再需要时,可以调用 Activity 的 unregisterReceiver 函数取消注册,形式如下:

```
unregisterReceiver(receiver)
```

例 6-1　检测手机充电状态,代码如下:

```
//模块 Ch06_01 定义 PowerReceiver.kt
class PowerReceiver: BroadcastReceiver() {
    override fun onReceive(context: Context, intent: Intent) {
        when (intent?.action){
            Intent.ACTION_POWER_CONNECTED->showInfo(context,"1: 已经连接充电器")
            Intent.ACTION_POWER_DISCONNECTED->showInfo(context,"2: 充电器已经移除")
        }
    }
    private fun showInfo(context: Context,message: String){             //显示提示信息
        Toast.makeText(context,message,Toast.LENGTH_LONG).show()
    }
}
```

PowerReceiver 扩展于 BroadcastReceiver 类,是一个 BroadcastReceiver 组件。在 onReceive 函数中处理接收的启动 Intent 的两种动作:Intent.ACTION_POWER_CONNTECTED(连接充电器动作)和 Intent.ACTION_POWER_DISCONNECTED(移除充电器的动作)。对这两种动作的处理方式是通过显示信息提示来实现,代码如下:

```
//模块 Ch06_01 定义主活动 MainActivity.kt
class MainActivity: ComponentActivity() {
    private lateinit var receiver:PowerReceiver
    override fun onCreate(savedInstanceState: Bundle?) {                //创建活动
        super.onCreate(savedInstanceState)
        handleReceiver()                                                //注册 PowerReceiver
        enableEdgeToEdge()
        setContent {
            Chapter06Theme {
                Scaffold(modifier=Modifier.fillMaxSize()) {
                    innerPadding ->Box(modifier=Modifier.padding(innerPadding).
                        fillMaxSize(),
                        contentAlignment=Alignment.Center) {
                        Text("监测充电器的状态",fontSize=32.sp)
```

```kotlin
                }
            }
        }
    }
    private fun handleReceiver(){                                          //动态注册
        receiver=PowerReceiver()
        val intentFilter=IntentFilter()
        intentFilter.addAction("android.intent.action.ACTION_POWER_CONNECTED")
                                                                           //增加连接充电器的作用
        intentFilter.addAction("android.intent.action.ACTION_POWER_DISCONNECTED")
                                                                           //增加取消充电器的动作
        registerReceiver(receiver,intentFilter RECEIVER_EXPORTED)          //动态注册广播接收器
    }
    override fun onDestroy() {                                             //销毁活动
        super.onDestroy()
        unregisterReceiver(receiver)                                       //取消注册
    }
}
```

运行结果如图 6-1 所示。

(a) 插入充电器　　　　　　(b) 去掉充电器

图 6-1　插入和去掉充电器的运行结果

在上例中，MainActivity 中定义的 IntentFilter 对象可以监听两种系统的 Intent 动作。

（1）android.intent.action.ACTION_POWER_CONNECTED：连接充电器。

（2）android.intent.action.ACTION_POWER_DISCONNECTED：移除充电器。

当这两个动作发生的时候，一旦监听到这个动作，就会将 Intent 广播发送给 receiver 对象。receiver 对象会调用 onReceive 函数，对 Intent 的动作进行判断，并根据不同的判断结果分别显示提示信息。

6.2 发送广播

发送广播由 3 种方式：标准广播、有序广播和黏性广播。因为黏性广播存在安全性问题，从 Android API 21 开始就不赞成使用。本书对黏性广播不进行介绍。

6.2.1 标准广播

标准广播是一种完全异步执行的广播，在广播发出后所有的广播接收器会在同一时间接收到这条广播，之间没有先后顺序，效率比较高且无法被截断。标准广播需要调用 Activity 的 sendBroadcast 函数来发送 Intent 广播。在例 6-2 中展示了标准广播的处理过程。

例 6-2 标准广播的应用示例。

自定义 MyReceiver01 组件，代码如下：

```kotlin
//模块 Ch06_02 的 MyReceiver01.kt
class MyReceiver01: BroadcastReceiver() {
    override fun onReceive(context: Context, intent: Intent) {
        Log.d("MyReceiver","记录 MyReceiver01 处理意图消息")
    }
}
```

自定义 MyReceiver02 组件，代码如下：

```kotlin
//模块 Ch06_02 的 MyReceiver02.kt
class MyReceiver02: BroadcastReceiver() {
    override fun onReceive(context: Context, intent: Intent) {
        Log.d("MyReceiver","记录 MyReceiver02 处理意图消息")
    }
}
```

在上面分别定义了两个 BroadcastReceiver 类：MyReceiver01 和 MyReceiver02，它们对接收的消息所做的处理是在日志中显示对应的字符串，代码如下：

```kotlin
//模块 Ch06_02 的 MainActivity.kt
class MainActivity: ComponentActivity() {
    private lateinit var myReceiver01: MyReceiver01
    private lateinit var myReceiver02: MyReceiver02
    override fun onCreate(savedInstanceState: Bundle?) {
        super.onCreate(savedInstanceState)
        registerReceivers()                              //注册两个 BroadcastReceiver 组件
        enableEdgeToEdge()
        setContent {
            Chapter06Theme {
                Scaffold(modifier=Modifier.fillMaxSize()) {
                    innerPadding ->Box(modifier=Modifier.padding(innerPadding)
                        .fillMaxSize(),contentAlignment=Alignment.Center){
                        Button(onClick={
```

```
                    sendReceiver()                              //发送广播
            }){Text("发送")}
          }
        }
      }
    }
  }
  private fun registerReceivers(){
      val intentFilter=IntentFilter()
      intentFilter.addAction("chenyi.book.android.ch06_02.MyReceiver")
      myReceiver01=MyReceiver01()
      myReceiver02=MyReceiver02()
      registerMyReceiver(myReceiver01,intentFilter)        //动态注册 myReceiver01
      registerMyReceiver(myReceiver02,intentFilter)        //动态注册 myReceiver02
  }
  private fun registerMyReceiver(myReceiver: BroadcastReceiver, intentFilter:
      IntentFilter){
      if (Build.VERSION.SDK_INT>=Build.VERSION_CODES.Q){
          application.registerReceiver(myReceiver,intentFilter, RECEIVER_EXPORTED)
      } else
          application.registerReceiver(myReceiver,intentFilter)
  }
  private fun sendReceiver(){                              //发送广播的处理
      val intent=Intent()
      intent.action="chenyi.book.android.ch06_02.MyReceiver"
      intent.setPackage("chenyi.book.android.ch06_02")
      sendBroadcast(intent)                                //发送广播
  }
  override fun onDestroy() {
      super.onDestroy()
      unregisterReceiver(myReceiver01)                     //取消注册 myReceiver01
      unregisterReceiver(myReceiver02)                     //取消注册 myReceiver02
  }
}
```

运行结果如图 6-2 所示。

在上述例中,当单击"发送"按钮时,观察日志,可以发现两个 BroadcastReceiver 都处理了接收的广播意图,并分别记录到日志中。

注意：Android 14 调用 registerReceiverI 函数注册 BroadcastReceiver 时,强制要求指定导出标志为 RECEIVER_EXPORTED(导出广播) 或者 RECEIVER_NOT_EXPORTED(不导出广播)。因此,需要对当前的 SDK 版本做一个判断,根据版本采用不同方式进行处理,处理方式如下：

```
if (Build.VERSION.SDK_INT>=Build.VERSION_CODES.Q)
    application.registerReceiver(myReceiver,intentFilter, RECEIVER_EXPORTED)
else
    application.registerReceiver(myReceiver,intentFilter)
```

6.2.2 有序广播

有序广播是一种异步执行的广播。它的传播方式是在广播发出后,同一时刻只有一个 BroadcastReceiver 组件能够接收到,优先级高的 BroadcastReceiver 组件会优先接收,当优先级高的 BroadcastReceiver 组件的

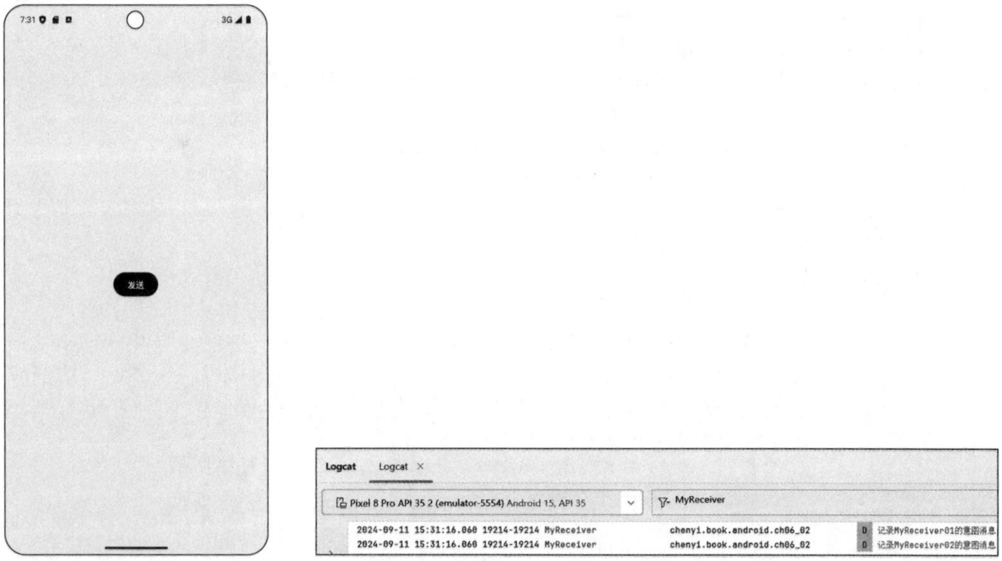

(a) 运行界面　　　　　　　　　　　　(b) 运行日志结果

图 6-2　标准广播

onReceiver 函数运行结束后,广播才会继续传递。前面的 BroadcastReceiver 组件可以选择截断广播,这样后面的 BroadcastReceiver 组件就接收不到这条消息了。调用 Activity 中的 sendOrderedBroadcast 函数可以实现有序广播,使得 BroadcastReceiver 组件依次接收 Intent 广播。

例 6-3　有序广播的示例,代码如下:

```kotlin
//模块 Ch06_03 的 MyReceiver01.kt
class MyReceiver01: BroadcastReceiver() {
    override fun onReceive(context: Context, intent: Intent) {
        val data=intent.getStringExtra("frmMainActivity")
        Log.d("MyReceiver","记录 MyReceiver01 处理意图消息${data}")
        abortBroadcast()                                                    //截断传播
    }
}
```

在 MyReceiver01 这个 BroadcastReceiver 组件对象的 onReceive 函数接收从上下文传来的字符串数据,并记录到日志中。随后调用 abortBroadcast 函数截断 Intent 广播,代码如下:

```kotlin
//模块 Ch06_03 的 MyReceiver02.kt
class MyReceiver02: BroadcastReceiver() {
    override fun onReceive(context: Context, intent: Intent) {
        val data=intent.getStringExtra("frmMainActivity")
        Log.d("MyReceiver","记录 MyReceiver02 处理 Intent 消息${data}")
    }
}
```

调用 MyReceiver02 对象的 onReceive 函数接收从上下文传来的字符串数据,并记录到日志中,代码如下:

```kotlin
//模块Ch06_03的主活动MainActivity.kt
class MainActivity: ComponentActivity() {
    private lateinit var myReceiver01:MyReceiver01
    private lateinit var myReceiver02: MyReceiver02
    override fun onCreate(savedInstanceState: Bundle?) {
        super.onCreate(savedInstanceState)
        registerReceivers()                              //注册BroadcastReceiver组件
        enableEdgeToEdge()
        setContent {
            Chapter06Theme {
                Scaffold(modifier=Modifier.fillMaxSize()) { innerPadding ->
                    Box(modifier=Modifier.padding(innerPadding).fillMaxSize(),
                        contentAlignment=Alignment.Center) {
                        Button(onClick={
                            sendReceiver()               //执行有序广播
                        }){Text("发送")}
                    }
                }
            }
        }
    }
    /**注册两个BroadcastReceiver组件*/
    private fun registerReceivers(){
        myReceiver01=MyReceiver01()                      //创建广播接收器myReceiver01
        myReceiver02=MyReceiver02()                      //创建广播接收器myReceiver02
        val intentFilter1=IntentFilter()                 //定义意图过滤器对象intentFilter1
        intentFilter1.addAction("book.android.ch06_03.Broadcast")   //添加动作
        intentFilter1.priority=200                       //设置优先级为200
        registerMyReceiver(myReceiver01,intentFilter1)   //动态注册myReceiver01

        val intentFilter2=IntentFilter()                 //定义意图过滤器对象intentFilter2
        intentFilter2.addAction("book.android.ch06_03.Broadcast")   //添加动作
        intentFilter2.priority=100                       //设置优先级为100
        registerMyReceiver(myReceiver02,intentFilter2)   //动态注册myReceiver02
    }
    private fun registerMyReceiver(myReceiver: BroadcastReceiver,
        intentFilter: IntentFilter){
        if (Build.VERSION.SDK_INT>=Build.VERSION_CODES.Q){
            application.registerReceiver(myReceiver,intentFilter, RECEIVER_EXPORTED)
        } else {
            application.registerReceiver(myReceiver,intentFilter)
        }
    }
    private fun sendReceiver(){                          //发送广播的处理
        val intent=Intent().apply{
            action="book.android.ch06_03.Broadcast"
            setPackage("chenyi.book.android.ch06_03")
            putExtra("frmMainActivity","来自MainActivity的问候!")
        }
        sendOrderedBroadcast(intent,null)                //发送有序广播
    }
```

```kotlin
    override fun onDestroy() {
        super.onDestroy()
        unregisterReceiver(myReceiver01)                //取消注册 myReceiver01
        unregisterReceiver(myReceiver02)                //取消注册 myReceiver02
    }
}
```

运行结果如图 6-3 所示。

图 6-3　有序广播的运行结果（1）

MainActivity 定义了两个 IntentFilter：intentFilter1 和 intentFilter2，它们的动作相同，均为 book.android.ch06_03.Broadcast，但是它们的优先级不同，intentFilter1 的优先级为 200 高于 intentFilter2 的优先级 100。当 Activity 注册 receiver01 和 receiver02，并分别匹配 intentFilter1 和 intentFilter2 定义 Intent 动作，使得 receiver01 的执行优先级高于 receiver02。在上个程序中，点击"发送"按钮，MyReceiver01 的 receiver01 对象中执行了从 MainActivity 接收消息，并显示日志。由于执行 abortBroadcast()函数，截断了广播的继续传播，使得 MyReceiver02 的 receiver02 没有接收到广播。

如果对上述的 MainActivity.kt 进行修改并设置优先级，代码如下：

```kotlin
//定义 IntentFilter 对象
val intentFilter1=IntentFilter()
intentFilter1.addAction("book.android.ch06_03.Broadcast")
intentFilter1.priority=100
val intentFilter2=IntentFilter()
intentFilter2.addAction("book.android.ch06_03.Broadcast")
intentFilter2.priority=200
```

运行结果如图 6-4 所示。

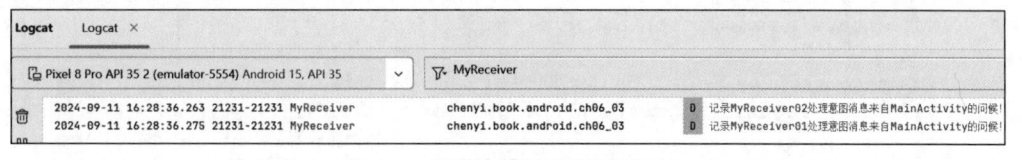

图 6-4　有序广播的运行结果（2）

运行修改后的代码，由于这时 intentFilter2 的优先级为 200，高于 intentFilter1 的优先级 100。使得 MyReceiver02 的 receiver02 对象先执行 onReceive 函数。因为该方法并没有截断，使得 Intent 广播继续向后传播，使得 MyReceiver01 的 receiver01 对象也可以接收到 Intent 广播并做出相应的处理，记录在日志中。

在广播接收器之间可以进行通信，实现数据的传递。BroadcastReceiver 对象中 getResultExtras (true)可以检索由上一个 BroadcastReceiver 组件发送的结果数据包，并发送到下一个接收器中，代码如下：

```kotlin
//检索上一个 BroadcastReceiver 发送的 ResultExtras 对象
```

```kotlin
        val bundle1=getResultExtras(true)
        //获得从上一个 BroadcastReceiver 广播的信息
        val data=bundle1.getString("key")
        //定义数据包
        val bundle2=Bundle()
        bundle2.putString("key","from MyReceiver02")
        //设置向下广播的数据包
        setResultExtras(bundle2)
```

例 6-4 有序广播并传播数据的示例。

定义 MyReceiver01,代码如下:

```kotlin
//模块 Ch06_04 的广播接收器定义 MyReceiver01.kt
class MyReceiver01: BroadcastReceiver() {
    override fun onReceive(context: Context, intent: Intent) {
        val data1=intent.getStringExtra("frmMainActivity")
                                            //接收来自 Activity 的数据
        val receivedBundle=getResultExtras(true)
                                            //处理来自其他 BroadcastReceiver 的数据
        val data2=receivedBundle.getString("frmReceiver")
                                            //处理来自其他 BroadcastReceiver 的数据
        Log.d("MyReceiver","MyReceiver01 来自 MainActivity 的信息: ${data1},
        来自其他 BroadcastReceiver 的信息: ${data2}")
                                            //记录到日志中
        val sendBundle=Bundle()             //发送数据
        sendBundle.putString("frmReceiver","来自 MyReceiver01 的问候!")
                                            //设置键值对
        setResultExtras(sendBundle)         //设置向下传播的数据
    }
}
```

定义 MyReceiver02,代码如下:

```kotlin
//模块 Ch06_04 的定义 MyReceiver02.kt
class MyReceiver02: BroadcastReceiver() {
    override fun onReceive(context: Context, intent: Intent) {
        val data1=intent.getStringExtra("frmMainActivity")
                                            //接收来自 Activity 的数据
        val receivedBundle=getResultExtras(true)  //处理来自其他 BroadcastReceiver 的数据
        val data2=receivedBundle.getString("frmReceiver")
                                            //处理来自其他 BroadcastReceiver 的数据
        Log.d("MyReceiver","MyReceiver02 来自 MainActivity 的信息: ${data1},"
            +"来自其他 BroadcastReceiver 的信息: ${data2}")   //记录到日志中
        val sendBundle=Bundle()             //发送数据
        sendBundle.putString("frmReceiver","来自 MyReceiver02 的问候!")
                                            //设置键值对
        setResultExtras(sendBundle)         //设置向下传播的数据
    }
}
```

MainActivity 执行有序广播,代码如下:

```kotlin
//模块 Ch06_04 的主活动定义 MainActivity.kt
class MainActivity: ComponentActivity() {
```

```kotlin
private lateinit var myReceiver01:MyReceiver01
private lateinit var myReceiver02: MyReceiver02
override fun onCreate(savedInstanceState: Bundle?) {
    super.onCreate(savedInstanceState)
    registerReceivers()                                  //注册 BroadcastReceiver
    enableEdgeToEdge()
    setContent {
        Chapter06Theme {
            Scaffold(modifier=Modifier.fillMaxSize()) { innerPadding ->
                Box(modifier=Modifier.padding(innerPadding).fillMaxSize(),
                    contentAlignment=Alignment.Center) {
                    Button(onClick={
                        sendReceiver()                   //执行有序广播
                    }) {
                        Text("发送")
                    }
                }
            }
        }
    }
}
private fun registerReceivers(){
    myReceiver01=MyReceiver01()                          //创建广播接收器 myReceiver01
    myReceiver02=MyReceiver02()                          //创建广播接收器 myReceiver02
    val intentFilter1=IntentFilter()                     //定义意图过滤器对象 intentFilter1
    intentFilter1.addAction("book.android.ch06_4.Broadcast")   //添加动作
    intentFilter1.priority=200                           //设置优先级为 200
    registerMyReceiver(myReceiver01,intentFilter1)       //动态注册 myReceiver01

    val intentFilter2=IntentFilter()                     //定义意图过滤器对象 intentFilter2
    intentFilter2.addAction("book.android.ch06_4.Broadcast")   //添加动作
    intentFilter2.priority=100                           //设置优先级为 100
    registerMyReceiver(myReceiver02,intentFilter2)       //动态注册 myReceiver02
}
private fun registerMyReceiver(myReceiver: BroadcastReceiver,
    intentFilter: IntentFilter){
        if (Build.VERSION.SDK_INT>=Build.VERSION_CODES.Q){
            application.registerReceiver(myReceiver,intentFilter, RECEIVER_EXPORTED)
        } else {
            application.registerReceiver(myReceiver,intentFilter)
        }
}
private fun sendReceiver(){                              //发送广播的处理
    val intent=Intent().apply{
        action="book.android.ch06_4.Broadcast"
        setPackage("chenyi.book.android.ch06_04")
        putExtra("frmMainActivity","来自 MainActivity 的问候!")
    }
    sendOrderedBroadcast(intent,null)                    //发送有序广播
}
override fun onDestroy() {
```

```
        super.onDestroy()
        unregisterReceiver(myReceiver01)        //取消注册 myReceiver01
        unregisterReceiver(myReceiver02)        //取消注册 myReceiver02
    }
}
```

运行结果如图 6-5 所示。

图 6-5 数据传递的运行结果

在 MyReceiver01 和 MyReceiver02 中定义了类似的功能，要求处理从 MainActivity 接收的数据和从其他 BroadcastReceiver 中获得的数据，并将这些数据记录在日志中。同时，将字符串信息向下传播。

因为 MainActivity 设置 MyReceiver01 的 receiver01 对象的执行优先级高于 MyReceiver02 的 receiver02 对象的执行优先级，MainActivity 发送 Intent 广播时采用有序广播的方式。因此，receiver01 对象可以接收 MainActivity 发送的 Intent 传播，并没有接收其他 BroadcastReceiver 传来的信息，receiver01 将 Intent 广播继续向后传递，MyReceiver02 的 receiver02 对象接收传来的 Intent 广播，以及上一个 MyReceiver01 发送的数据。

习 题 6

一、选择题

1. BroadcastReceiver 是 Android 的四大组件之一，静态注册该类型的组件需要在项目的_____文件中进行配置，才可以使用。

  A. build.gradle.kts   B. setting.gradle.kts

  C. AndroidManifest.xml   D. gradle.properties

2. 在定义 BroadcastReceiver 组件中，需要重写_____函数，它可以接收指定 Intent 传递的广播消息。

  A. onReceive   B. sendBroadcast

  C. onCreate   D. onStart

3. 要使用 BroadcastReceiver 组件，也可以通过动态注册来实现。如果在一个 Activity 中动态注册 BroadcastReceiver 组件，则必须调用_____函数来实现。

  A. sendBroadcast   B. registerReceiver

  C. onReceive   D. sendStickyBroadcast

4. 假设有一个 BroadcastReceiver 组件，要求在一个 Activity 中采用标准广播发送，这时需要调用_____函数。

  A. sendBroadcast   B. sendStickyBroadcast

  C. sendOrderedBroadcast   D. 以上答案均不正确

5. 假设 Activity 中定义两个 BroadcastReceiver 组件对象分别是 myReceiver01 和 myReceiver02，注册它们的动作同为 test.ACTION，myReceiver01 优先级别设置为 100，而 myReceiver02 的优先级别

设置为 200。如果在 Activity 中采用有序广播方式,则_____组件先接收到消息。

    A. myReceiver01　　　　　　　　B. myReceiver02

    C. 同时接收到消息　　　　　　　　D. 按照先注册的组件先接收消息

二、思考题

1. BroadcastReceiver 组件的静态注册如何实现?
2. BroadcastReceiver 组件的动态注册如何实现?
3. 发送广播有标准广播方式、有序广播方式和黏性广播方式 3 种方式,比较它们的异同。

三、上机实践题

1. 使用 BroadcastReceiver 组件实现对移动终端充电状态的检测,并显示当前的状态。
2. 使用 BroadcastReceiver 组件实现对移动终端联网状态的检测,并显示连接网络的状态。
3. 使用 BroadcastReceiver 组件实现对移动终端飞行模式的检测,并显示是否处于飞行模式状态。

第 7 章　执行后台任务

移动应用中有大量的后台任务需要处理。本章主要介绍如何使用 Service 组件和 WorkManager 组件处理后台任务。

Service(服务)组件是 Android 的四大组件之一，是在后台运行的组件。Service 组件与 Activity 组件最大的不同是 Service 组件没有界面，只在后台为移动应用提供支持。Service 组件不受 Activity 组件生命周期的限制。一般情况下，服务应用于运行时间长、需要重复执行的任务。Service 组件并不独立，需要依赖创建它的移动应用的进程。一旦移动应用的进程终止，则 Service 组件也会销毁。

WorkManager 组件是 Android JetPack 库中的重要组成，用于执行后台的异步任务。因为与 Service 组件有类似的功能，因此也在此介绍。

7.1　Service 组件

Service 组件在后台执行移动应用指定的操作。应用 Service 组件时，首先要定义 Service 子类，然后才能在 AndroidManifest.xml 文件中进行配置，最后才可以在需要的上下文中启动。

1. Service 的定义

Service 组件必须是 android.app.Service 的子类。Service 类中定义了一些回调函数，必须根据要求对这些回调函数进行重写，赋予自定义的 Service 组件的功能。这些回调函数如表 7-1 所示。

表 7-1　Service 组件的常见函数

函　　数	说　　明
onCreate	Service 组件第一次创建后将立即回调该函数
onBind(Intent)	返回的 IBinder 应用程序与 Service 组件通信
onUnbind(Intent)	当 Service 组件上绑定的所有客户端都断开连接时将会回调该函数
onStartCommand(Intent,int,int)	调用 startService(Intent) 方法启动 Service 组件时回调该函数
onDestroy	在该 Service 被关闭之前会回调该函数

Service 的定义可以有两种方式。

方式 1：

```kotlin
class MyService: Service() {
    override fun onCreate() {
        super.onCreate()
    }
    override fun onStartCommand(intent: Intent?, flags: Int, startId: Int): Int {
        return super.onStartCommand(intent, flags, startId)
```

```kotlin
    }
    override fun onDestroy() {
        super.onDestroy()
    }
    override fun onBind(intent: Intent): IBinder {
        TODO("Return the communication channel to the service.")
    }
}
```

在这种方式定义的 Service 组件中必须重写 onStartCommand( )函数。这是与通过调用 startService( )函数启动这类服务组件的调用方式相关联的。

方式 2：

```kotlin
class MyService: Service() {
    override fun onCreate() {
        super.onCreate()
    }
    override fun onDestroy() {
        super.onDestroy()
    }
    override fun onBind(intent: Intent): IBinder {
        TODO("Return the communication channel to the service.")
    }
    override fun onUnbind(intent: Intent?): Boolean {
        return super.onUnbind(intent)
    }
}
```

用这种方式定义的 Service 组件中没有 onStartCommand 函数，但是必须对 onBind 函数进行重写。这种方式创建的 Service 组件可以进行通信，但是必须在 Activity 中通过 bindService 函数启动。

2. 配置 Service

要让 Service 组件可以使用，就必须将其注册到 AndroidManifest.xml 文件，形式如下：

```xml
<service
    android: name=".MyService"
    android: enabled="true"
    android: exported="true" />
```

在上述配置中，android：enabled 表示允许本移动应用中使用 Service 组件，android：exported 则表示其他的移动应用也允许使用定义的 Service 组件。由于移动应用的资源相对有限，Android 系统进行内存管理配置时，对于一些优先级别较低的 Service，在资源不足的情况下会杀死。如果希望降低 Service 被销毁的可能性，可以在配置 Service 时，通过 android：priority 来提高 Service 的优先级。优先级的取值范围为−1000～1000。数字越高，优先级别越高，数字越低，优先级别越低，代码如下：

```xml
<service
    android: name=".MyService"
    android: enabled="true"
    android: priority="500"
    android: exported="true" />
```

3. 启动 Service

Service 由移动应用中的 Activity 组件或 BroadcastReceiver 组件等内部调用 startService 函数和 bindService 函数进行启动。这两种启动方式分别对应了 Service 的两种类型。

（1）当通过 startService 函数启动服务时，会调用 Service 的 onStartCommand 函数。使用这种启动方式时，即使启动 Service 的 Activity 组件已经销毁，Service 组件仍然可以在后台独立运行。这种启动 Service 的方式并不鼓励，这是因为，即使移动应用的其他组件销毁了，但是后台仍占据存储空间并消耗资源。通过 startService 函数启动的 Service 可以通过 stopService 函数来关闭，也可调用 Service 的 stopSelf 函数关闭本身。

例 7-1　使用 startService 方式启动播放音乐的服务，代码如下：

```kotlin
//模块 Ch07_01 自定义服务 MusicService.kt
class MusicService: Service() {
    lateinit var musicPlayer: MediaPlayer
    override fun onCreate() {
        super.onCreate()
        musicPlayer=MediaPlayer.create(this,R.raw.music)    //创建音乐播放器
    }
    override fun onStartCommand(intent: Intent?, flags: Int, startId: Int): Int {
        musicPlayer.start()
        musicPlayer.setOnCompletionListener {
            musicPlayer.release()                            //释放与 musicPlayer 关联的资源
            stopSelf()                                       //关闭当前 Service
        }
        return super.onStartCommand(intent, flags, startId)
    }
    override fun onDestroy() {
        super.onDestroy()
        if (musicPlayer.isPlaying) {
            musicPlayer.stop()                               //停止播放
            musicPlayer.release()                            //释放资源
        }
    }
    override fun onBind(intent: Intent): IBinder {
        TODO("Return the communication channel to the service.")
    }
}
```

上述的 MusicService 的 onCreate 函数通过 raw 目录的 music.mp3 音频文件创建了一个 MediaPlayer 对象，然后在 onStartCommand 函数中启动媒体播放器播放音频文件。监视 MediaPlayer 对象是否播放完毕；如果播放完毕，则调用 stopSelf 函数来结束当前的 Service；如果 Service 结束，则调用 onDestroy 函数，检测 MediaPlayer 对象是否播放完毕；如果没有播放完毕，则调用 MediaPlayer 对象的 stop 函数停止播放，以及 release 函数释放空间，代码如下：

```kotlin
//模块 Ch07_01 启动服务的 MainActivity.kt
class MainActivity : ComponentActivity() {
    override fun onCreate(savedInstanceState: Bundle?) {
        super.onCreate(savedInstanceState)
```

```
enableEdgeToEdge()
val musicIntent=Intent(this,MusicService::class.java)              //创建意图
setContent {
    Chapter07Theme {
        Scaffold(modifier=Modifier.fillMaxSize()) {
            innerPadding ->Column(modifier=Modifier.padding(innerPadding)
            .fillMaxSize(),
            verticalArrangement=Arrangement.Center,
            horizontalAlignment=Alignment.CenterHorizontally){
            Text("音乐播放",fontSize=36.sp)
            Row{
                Button(onClick={ startService(musicIntent) }){ Text("播放") }
                                                                  //启动服务
                Button(onClick={ stopService(musicIntent) }){ Text("停止") }
                                                                  //停止服务
            }
          }
        }
      }
    }
}
```

运行结果如图 7-1 所示。

图 7-1　运行界面

运行上述的模块 ch07_01。点击"开始"按钮,播放音乐。这是因为在点击按钮时,通过 startService

启动 Service，即调用了 Service 组件的 onStartCommand()函数，则开始播放音乐。点击"停止"按钮，停止播放音乐。在"停止"按钮的事件处理中，执行了关闭 Service 自身的操作。但是当音乐正在播放时，按用户手机终端的返回键，即使 MainActivity 已经退出，也可以听到播放的音乐。这说明 MusicService 仍在后台默默地运行。这种类型的 Service 与 Activity 并没有太强的关联。

（2）当通过 bindService 启动 Service 时，启动 Service 的其他组件和 Service 组件就构成了一种"服务器-客户端"关联。绑定的 Service 提供"客户端-服务器"接口，使得其他组件与 Service 组件进行交互。一旦启动 Service 的其他组件销毁，Service 组件也会随之销毁，释放资源。这种启动方式可以让 Service 组件与其他组件进行交互通信，进行数据的传递。通过 bindService()函数启动的 Service，可以通过 unBind()函数进行关闭。

例 7-2 使用 bindService 启动播放音乐的 Service，代码如下：

```kotlin
//模块 Ch07_02 自定义服务 MusicService.kt
class MusicService: Service() {
    lateinit var mediaPlayer: MediaPlayer
    override fun onCreate() {
        super.onCreate()
        mediaPlayer=MediaPlayer.create(this,R.raw.music)
    }
    override fun onBind(intent: Intent): IBinder? {
        mediaPlayer.start()
        mediaPlayer.setOnCompletionListener {
            mediaPlayer.release()
        }
        return null
    }
    override fun onUnbind(intent: Intent?): Boolean {
        if (mediaPlayer!=null&&mediaPlayer.isPlaying) {
            mediaPlayer.stop()
            mediaPlayer.release()
        }
        return super.onUnbind(intent)
    }
}
```

同样，上述 MusicService 在 onCreate 函数中利用 values/raw 目录的 music.mp3 文件创建了一个 MediaPlayer 对象。在 onBind 函数中启动 MediaPlayer 实现音乐的播放。在 onUnbind 函数中进行解除绑定的处理。执行对 MediaPlayer 对象状态的检测，如果仍处于播放音乐状态，则先停止播放，然后释放资源，代码如下：

```kotlin
//模块 Ch07_02 启动服务的 MainActivity.kt
class MainActivity: ComponentActivity() {
    override fun onCreate(savedInstanceState: Bundle?) {
        super.onCreate(savedInstanceState)
        val musicIntent=Intent(this,MusicService::class.java)
        val conn=object:ServiceConnection{                    //构建与 Service 连接的对象
            override fun onServiceConnected(name: ComponentName?, service: IBinder?) {
```

```kotlin
            }
            override fun onServiceDisconnected(name: ComponentName?) {
            }
        }
        enableEdgeToEdge()
        setContent {
            Chapter07Theme {
                Scaffold(modifier=Modifier.fillMaxSize()) {
                    innerPadding ->Column(modifier=Modifier.padding(innerPadding)
                        .fillMaxSize(),
                        verticalArrangement=Arrangement.Center,
                        horizontalAlignment=Alignment.CenterHorizontally){
                        Text("音乐播放",fontSize=36.sp)
                        Row{
                            Button(onClick={ bindService(musicIntent, conn,
                                Context.BIND_AUTO_CREATE) }){Text("播放") }
                                                                        //绑定服务
                            Button(onClick={unbindService(conn) }){Text("停止") }
                                                                        //停止服务
                        }
                    }
                }
            }
        }
    }
}
```

运行上述的模块 Ch07_02。当点击"开始"按钮时,会播放音乐。这是因为点击按钮会调用 bindService 创建并启动服务。但是当用户手机终端的"返回"键时,MainActivity 已经退出,音乐也会停止播放,说明 MusicService 也已经退出。在 MainActivity.kt 中定义的 conn 对象是 ServiceConnection 的对象实例,用于构建 MainActivity 与 MusicService 之间通信的中介。具体详细内容将在 7.3 节定义。

7.2 Service 的生命周期

每个 Service 具有生命周期。根据 Service 启动方式的不同,生命周期可分为两种类型。在图 7-2 中展示了 Service 组件的不同生命周期的过程。

当在上下文中通过 startService 启动服务时,Service 会调用 onCreate 函数进行初始化操作,然后在 onStartCommand 函数中定义 Service 核心处理。当前 Service 处理完毕后,会调用 onDestroy 函数关闭 Service。

如果在上下文(可以是其他组件,也可以是使用 Service 的客户端)中通过 bindService 启动服务时, Service 会调用 onCreate 函数进行初始化操作。然后通过调用 onBind 函数使得客户端与 Service 绑定。 当客户端请求 unBindService 时,会调用 Service 的 onUnbind 函数,然后再调用 onDestroy 函数关闭 Service。

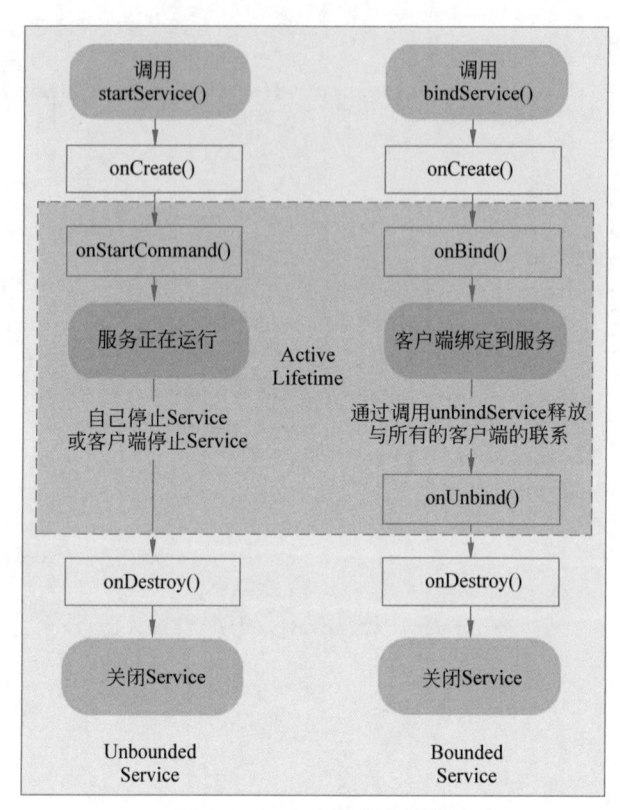

图 7-2 Service 的生命周期

7.3 Activity 和 Service 的通信

例 7-2 中使用 bindService 让 Activity 组件与 Service 组件构建了联系,但是它们并没有进行数据的传递。实际上,Activity 组件和 Service 组件是可以进行通信和数据传递的。之所以能实现通信,是因为 ServiceConnection 对象发挥了作用。ServiceConnection 对类有两个函数:onServiceConnected 和 onServiceDisConnected。onServiceConnected 函数用于获得 IBinder 类型的对象,通过 IBinder 对象从 Service 中获取数据。

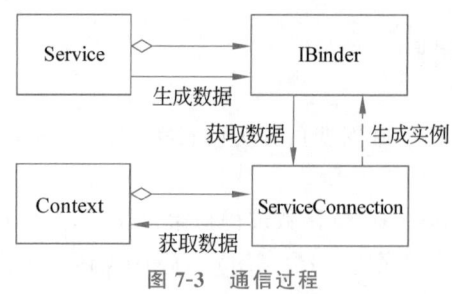

如图 7-3 所示,在 Service 中会定义一个 IBinder 对象,通过 IBinder 对象设置在 Service 中不断变更的数据。Android 的 Binder 类实现了 IBinder 接口,所以可以创建 Binder 子类对象实际上就是一个 IBinder 对象。ServiceConnection 是 Service 和上下文(例如 Activity 组件)之间的中介。Service 组件中包含了 ServiceConnection 的对象。ServiceConnection 对象可以获得 Service 中 IBinder 对象的数据,从而获取了 Service 对象的数据。

图 7-3 通信过程

例 7-3 音乐播放并显示音乐进度,代码如下:

```
//模块 Ch07_03音乐播放控制的服务 MusicService.kt
class MusicService: Service() {
```

· 230 ·

```kotlin
        lateinit var mediaPlayer:MediaPlayer
        var running=true
        var time=0
        var musicProgress=0
        inner class ProgressBinder(): Binder() {        //定义内部类记录音乐播放进度
            fun getMusicProgress()=musicProgress        //获得音乐当前播放的进度,取值范围为0~100
            fun getTime()=time                          //播放的时间
        }
        override fun onBind(intent: Intent): IBinder {
            mediaPlayer=MediaPlayer.create(this,R.raw.music)
            mediaPlayer.start()
            mediaPlayer.setOnCompletionListener {
                mediaPlayer.release()
                time=0
            }
            thread{
                while(running){
                    Thread.sleep(1000)                  //休眠1秒
                    time++                              //增加时间
                    musicProgress=(100 * mediaPlayer.currentPosition)/mediaPlayer.duration
                                                        //计算音乐播放进度
                    if (musicProgress>=100)             //判断音乐是否已经播放完毕
                        running=false                   //中断线程
                }
            }
            return ProgressBinder()
        }
        override fun onUnbind(intent: Intent?): Boolean {
            if (mediaPlayer.isPlaying){
                mediaPlayer.stop()
            }
            return super.onUnbind(intent)
        }
    }
```

在上述代码的 MusicServiceService 组件中,onBind 函数中创建并启动了一个线程。这个线程通过(100 * mediaPlayer.currentPosition)/mediaPlayer.duration 计算音乐播放的进度。并通过 ProgressBinder 获取音乐播放进度。通过这种方式,为 Activity 与 Service 之间的数据通信提供了支持。

下面定义 MainScreen 的可组合函数,定义音乐播放和停止的按钮以及音乐播放进度的 Slider 组件。通过传递不同的状态值以及用户交互控制按钮,调用不同的动作,达到更新界面的目的。代码如下:

```kotlin
//模块 Ch07_03 MainScreen
/* * 播放音乐界面
 * @param modifier Modifier: 修饰符
 * @param timerState MutableState<String>: 播放时间状态
 * @param progressState MutableState<Float>: 进度状态
 * @param playAction Function0<Unit>: 播放音频动作
 * @param stopAction Function0<Unit>: 停止播放音频动作 */
@Composable
```

```kotlin
fun MainScreen(modifier:Modifier,timerState:MutableState<String>,
    progressState: MutableState<Float>,playAction:()->Unit, stopAction:()->Unit){
    Column(modifier=modifier.fillMaxSize()
        .background(colorResource(id=android.R.color.holo_orange_light)),
        verticalArrangement=Arrangement.Center,
        horizontalAlignment=Alignment.CenterHorizontally) {
        Text("音乐播放器",fontSize=36.sp)
        Slider(value=progressState.value,
            onValueChange={it:Float->progressState.value=it },
            colors=SliderDefaults.colors(activeTrackColor=Color.Green))
        Text(text="${timerState.value}",fontSize=24.sp)
        Row(modifier=Modifier.fillMaxWidth().wrapContentHeight(),
            horizontalArrangement=Arrangement.Center) {
            TextButton(onClick={playAction.invoke() }) {            //调用播放动作
                Row(verticalAlignment=Alignment.CenterVertically) {
                    Icon(painter=painterResource(id =R.mipmap.play),
                        contentDescription=null, tint=Color.Unspecified)
                    Text("播放",fontSize=24.sp)
                }
            }
            TextButton(onClick={ stopAction.invoke() }) {           //调用停止播放
                Row(verticalAlignment=Alignment.CenterVertically) {
                    Icon(painter=painterResource(id=R.mipmap.pause),
                        contentDescription=null, tint=Color.Unspecified)
                    Text("停止",fontSize=24.sp)
                }
            }
        }
    }
}
```

在 MainActivity 中调用 MainScreen，代码如下：

```kotlin
//模块 Ch07_03 主活动 MainActivity.kt
class MainActivity: ComponentActivity() {
    lateinit var musicIntent: Intent
    lateinit var conn:ServiceConnection
    var running=false
    override fun onCreate(savedInstanceState: Bundle?) {
        super.onCreate(savedInstanceState)
        musicIntent=Intent(this,MusicService::class.java)
        enableEdgeToEdge()
        setContent {
            val timerState:MutableState<String>=remember{mutableStateOf("") }
            val progressState:MutableState<Float>=remember{ mutableFloatStateOf(0.0F) }
            conn=object: ServiceConnection {                        //创建 ServiceConnection 对象
                override fun onServiceConnected(name: ComponentName?,
                    service: IBinder?) {val binder=service as MusicService.ProgressBinder
                    thread{                                          //定义工作线程
```

```kotlin
                    while(running){
                        progressState.value=binder.getMusicProgress().toFloat()/100
                                                                //修改播放进度状态
                        timerState.value=convertStr(binder.getTime())
                                                                //修改进度显示文本状态
                    }
                }
            }
            override fun onServiceDisconnected(name: ComponentName?) {
                TODO("Not yet implemented")
            }
        }
        Chapter07Theme {
            Scaffold{
                innerPadding->MainScreen(modifier=Modifier.padding(innerPadding),
                    timerState=timerState,
                    progressState=progressState,
                    playAction=::startMusic,stopAction=::stopMusic)
            }
        }
    }
}
private fun startMusic(){
    running=true                                        //设置线程体循环运行 true
    bindService(musicIntent,conn, Context.BIND_AUTO_CREATE)  //绑定服务
}
private fun stopMusic(){
    running=false                                       //设置线程体循环运行 false
    unbindService(conn)                                 //停止绑定
}
private fun convertStr(seconds:Int):String{             //将时间秒转换成字符串
    var m=seconds/60
    var s=seconds%60
    var ms=if(m<10)"0$m" else "$m"
    var ss=if(s<10) "0$s" else "$s"
    return "${ms}:${ss}"
}
}
```

运行结果如图 7-4 所示。

在 MainActivity 中定义了一个 ServiceConnection 对象。在这个 ServiceConnection 对象中的 onServiceConnected 函数中通过 service as MusicService.ProgressBinder 获得 MusicService.ProgressBinder 对象。通过这个对象的 getMusicProgress 函数可以获取到 MusicService 播放音乐的进度。得到进度数据后，修改 progressState 的状态值，再将 progressState 传递给 MainScreen 界面的 Slider 可组合项，使得播放进度发生相应的变化，从而达到 Activity 与 Service 的通信，使得界面的 SeekBar 伴随音乐的播放进行变更。

图 7-4 音乐播放的运行界面

7.4　Notification 通知和前台服务

7.4.1　Notification

Notification(通知)是在移动应用的通知栏中提供给用户的消息提示。当移动应用不在运行时或者在后台状态时,可通过发布 Notification 的方式,提醒用户进行某些操作。从 Android 13 开始,必须在项目模块的 AndroidManifest.xml 中配置发送许可,代码如下:

```
< uses-permission android:name="android.permission.POST_NOTIFICATIONS" />
```

在图 7-5 中展示了 Notification 在顶部状态栏的情况。可以通过下拉操作显示具体内容,如图 7-6 所示。

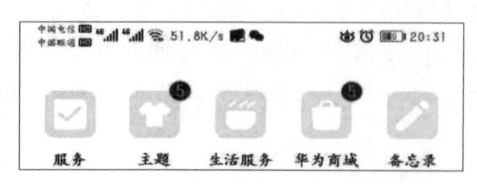

图 7-5　显示在通知栏中的通知

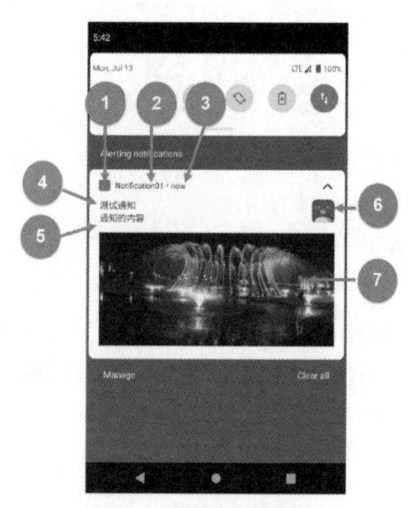

图 7-6　通知的下拉显示

· 234 ·

由图7-6可知,Notification一般由如下元素构成。

(1) 小图标:必要图标,通过setSmallIcon()设置。
(2) 应用名称:由系统提供。
(3) 时间戳:由系统提供,可以通过setWhen()进行替换或使用setShowWhen(false)将其隐藏。
(4) 标题:可选内容,通过setContentTitle()设置。
(5) 文本:可选内容,通过setContentText()设置。
(6) 大图标:可选图标(通常仅用于联系人照片),通过setLargeIcon设置。
(7) 样式:通过样式设置大图片。

Android 8.0提出了NotificationChannel(通知渠道)概念。NotificationChannel可以理解Notification的类别,所有的Notification必须分配相应的Channel。移动应用中的每个NotificationChannel都有独一无二的标识。还需要定义NotificationChannel的名称和描述。将通知加入NotificationChannel并进行分类管理。NotificationChannel也设定了优先级别,根据优先级别的高低,对同一个NotificationChannel的Notification进行不同的处理,如表7-2所示。

表7-2 通知渠道的优先级

通知渠道的优先级	说　　明
NotificationManager.IMPORTANCE_HIGH	紧急:发出声音并显示为提示通知
NotificationManager.IMPORTANCE_DEFAULT	高:发出声音
NotificationManager.IMPORTANCE_LOW	中等:没有声音
NotificationManager.IMPORTANCE_MIN	低:无声音并且不会出现在状态栏中

要创建NotificationChannel,需要通过NotificationManager(通知管理器)实现,代码如下:

```
//定义NotificationManager
val notificationManager=getSystemService(Context.NOTIFICATION_SERVICE)
    as NotificationManager
//定义NotificationChannel的标识
val channelId="chenyi.book.android.ch07"
//定义NotificationChannel的名称
val name="Android移动应用开发"
//定义NotificationChannel
val channel=NotificationChannel(channelId,name,NotificationManager
    .IMPORTANCE_DEFAULT)
//定义NotificationChannel的描述
channel.description="Android移动应用开发的NotificationManager"
//创建并配置NotificationChannel
notificationManager.createNotificationChannel(channel)
```

在NotificationChannel的前提下,创建一个Notification,需要设置标题、内容、时间、显示的图标,可以修改Notification显示的样式,也可以增加大图的显示或者文本的显示,代码如下:

```
//创建Notification对象
val notification=Notification.Builder(this,channelId).apply {
    setContentTitle("通知示例")                              //设置标题
    setContentText("欢迎使用通知!")                          //设置内容
```

```
            setWhen(System.currentTimeMillis())                    //设置时间
            setShowWhen(true)                                      //允许显示时间
            setSmallIcon(android.R.drawable.ic_Dialog_Info)        //设置小图标
            setStyle(Notification.BigPictureStyle()
                .bigPicture(BitmapFactory.decodeResource(resources,R.mipmap.scene)))
                                                                   //设置样式为大图样式
        }.build()                                                  //创建 Notification 对象
        val notificationID=1                                       //定义 Notification 标识
        notificationManager.notify(notificationID,notification)
                                                                   //发布 Notification 到通知栏中
```

发布 Notification 到通知栏时,可以发现它的图标为灰色显示。即使设置的小图标为彩色,显示的内容仍是灰色。这是因为从 Android 5.0 开始,对通知栏图标的设计进行了修改。所有应用程序的通知栏图标只支持带 alpha 图层的图片,不支持 RGB 图层,这使得通知栏图标失去了原本的色彩。如果希望保留色彩,必须进行设置,代码如下:

```
        setColorized(true)                                         //可以设置彩色
        setColor(getColor(R.color.teal_200))                       //设置指定的颜色
```

Notification 可以通过特定的操作实现特定的某些功能。这时可以利用 android.app.PendingIntent 实现。

PendingIntent(待定意图)并不是 Intent 的子类,它表示执行 Intent 和目标动作的描述。PendingIntent 本身是系统维护令牌(Token)的引用,该 Token 用于检索原始数据。这表明,即使所属的应用进程被终止,PendingIntent 也可以在为其分配的其他进程中继续使用。如果以后创建的应用程序重新检索同样类别的 PendingIntent(例如,同样的操作、同样的意图动作、数据、类别和组件以及同样的标记),它将接收这个相同 Token 的 PendingIntent,也可以调用 cancel()函数删除它。PendingIntent 支持 3 种待定意图:启动 Activity、启动 Service 和发送广播。

(1) getActivity(context:Context!, requestCode:Int, intent:Intent!, flags:Int):返回 PendingIntent 对象,用于检索将启动新 Activity 的 PendingIntent。

(2) getService(context:Context!, requestCode:Int, intent:Intent, flags:Int):返回 PendingIntent 对象,检索将启动 Service 的 PendingIntent。

(3) getBroadcast(context:Context!, requestCode:Int, intent:Intent!, flags:Int):返回 PendingIntent 对象,检索将执行广播的 PendingIntent。

下面,以调用 PendingIntent.getActivity()函数从活动中获得一个 PendingIntent 的对象为例,了解 PendingIntent 的使用,形式如下:

```
        val intent=Intent(this,SomeActivity::class.java)                      //某个意图对象
        val pendingIntent=PendingIntent.getActivity(activity,requestCode,intent,flags)
```

生成 pendingIntent 对象的相关 4 个参数。

(1) activity:表示上下文。

(2) requestCode:表示请求码,用数字表示。

(3) intent:表示要加载的意图。

(4) flags：表示标记，可以取值为 FLAG_ONE_SHOT（获取的 PendingIntent 只能使用一次）、FLAG_NO_CREATE（获取 PendingIntent，若描述的 Intent 不存在，则返回 NULL 值）、FLAG_CANCEL_CURRENT（使用新的 Intent 取消当前的描述）、FLAGUPDATE_CURRENT（使用新的 Intent，更新当前的描述）。

在 NotificationCompat.Builder 中调用 addAction 函数将配置的 Notification 与某个 pendingIntent 对象描述的任务进行关联，并执行描述中指定的意图，形式如下：

```
val notificationBuilder=NotificationCompat.Builder(this,"chenyi.book.android.ch07_04")
    .apply{
    ...                                                        //其他配置略
    addAction(R.drawable.play,"按钮文本",pendingIntent)
    }
```

7.4.2 前台服务

Service 在后台运行时，若系统资源不足，为了释放空间，就会被销毁，导致需要完成的任务提前结束。为了避免被提前销毁，可以使用以下 4 种方法。

1. 提高 Service 的优先级

在 AndroidManifest.xml 中设置 Service 的优先级（取值范围为 -1000～1000）为较大的数字，例如 android：priority="1000"，数字越大，优先级别越高。

2. 设置 Service 为系统服务

在 AndroidManifest.xml 中设置 Service 组件的 android：persistent 属性为 true，配置 Service 为系统服务，则可使其在内存不足时不被终止，形式如下：

```
<service
    android: name=".MusicService"
    android: enabled="true"
    android: exported="true"
    android: persistence="true"  />
```

这种做法不鼓励，这是因为，若自定义 Service 配置为系统服务太多，将严重影响系统的整体运行。同理，在应用中设置 android：persistent 属性为 true，则该应用为系统应用。应用程序通常不应设置此标志；持久模式仅适用于某些系统应用程序，形式如下：

```
<application
    android: fullBackupContent="true"
    android: icon="@mipmap/ic_launcher"
    android: label="@string/app_name"
    android: persistent="true"
    android: largeHeap="true"
    android: supportsRtl="true"
    android: theme="@style/AppTheme1">
</application>
```

3. 利用 Android 的系统广播

利用 Android 的 Intent.ACTION_TIME_TICK 系统广播检查 Service 的运行状态，如果 Service 被

终止,就重新启动 Service 组件。系统广播可以每分钟发送一次,检查一次 Service 的运行状态,如果已经终止,就重新启动 Service。

以上 3 种方式都存在不足。最佳的处理方式是设置 Service 为前台服务。此时,会在状态栏显示一个 Notification。通知用户界面与 Service 进行关联。设置 Service 为前台服务时,需要在 Service 组件的定义中使用 startForeground(int,Notification)函数。如果不想显示 Notification,只要把参数里的 int 设为 0 即可。

从 Android 9.0 以后版本,Service 在前台运行时,必须在 AndroidManifest.xml 中进行权限声明,形式如下:

```
<use-permission android: name="android.permission.FOREGROUND_SERVICE" />
```

例 7-4 前台服务控制音乐播放的应用实例。

(1) 在 AndroidManifest.xml 配置除了通知的相关权限许可和前台服务的权限许可,因为涉及播放音频还需要设置 android.permission.FOREGROUND_SERVICE_MEDIA_PLAYBACK,代码如下:

```xml
<uses-permission android:name="android.permission.POST_NOTIFICATIONS" />
<uses-permission android:name="android.permission.ACCESS_NOTIFICATION_POLICY" />
<uses-permission android:name="android.permission.FOREGROUND_SERVICE" />
<uses-permission android:name="android.permission.FOREGROUND_SERVICE_MEDIA_PLAYBACK" />
```

(2) 定义控制音乐播放的 MusicService,代码如下:

```kotlin
//模块 Ch07_04 的 MusicService.kt
class MusicService: Service() {
    lateinit var mediaPlayer: MediaPlayer
    lateinit var notificationBuilder:NotificationCompat.Builder
                                                        //定义访问通知对象的辅助类的构建器
    var running=true                                    //音乐进度播放线程的运行控制
    var timer=0                                         //播放时间
    var musicProgress=0                                 //播放进度
    inner class ProgressBinder: Binder(){               //定义内部类 ProgressBinder
        fun getMusicProgresss()=musicProgress           //获取音乐播放进度
        fun getRunning()=running                        //获取线程运行状态
    }
    override fun onCreate() {
        super.onCreate()
        createNotificationBuilder()

    }
    override fun onBind(intent: Intent): IBinder {
        postNotification()
        playMusic()
        return ProgressBinder()
    }
    override fun onUnbind(intent: Intent?): Boolean {
        stopMusic()
```

```kotlin
        return super.onUnbind(intent)
    }
    private fun playMusic(){                                    //播放音乐
        running=true
        musicProgress=0
        mediaPlayer=MediaPlayer.create(this,R.raw.song3)
        mediaPlayer.setOnPreparedListener {
            mediaPlayer.start()
        }
        mediaPlayer.setOnCompletionListener {
            mediaPlayer.release()
            running=false
        }
        changeProgress()
        postNotification()
    }
    private fun stopMusic(){                                    //停止播放音乐
        running=false
        if (mediaPlayer.isPlaying)
            mediaPlayer.stop()
    }
    private fun changeProgress(){                               //改变播放进度
        thread{                                                 //定义工作线程
            while (running){
                Thread.sleep(1000)
                timer ++                                        //计算音乐播放进度
                musicProgress=(100 * mediaPlayer.currentPosition)/mediaPlayer.duration
                if (musicProgress>=100){
                    running=false
                }
            }
        }
    }
    private fun createNotificationBuilder(){                    //创建通知构造器,配置通知
        val playPendingIntent=getPlayPendingIntent()            //定义播放音乐待处理的意图
        val stopPendingIntent=getStopPendingIntent()            //定义停止播放音乐待处理意图
        notificationBuilder=NotificationCompat.Builder(this,
            "chenyi.book.android.ch07_04").apply {              //配置通知
                setOngoing(true)
                setOnlyAlertOnce(true)
                setWhen(System.currentTimeMillis())
                setContentTitle("播放音乐")
                setContentText("一剪梅")
                setSmallIcon(R.mipmap.ic_launcher)
                setColorized(true)
                color=resources.getColor(R.color.purple_200,null)
                setDefaults(Notification.DEFAULT_SOUND or Notification.DEFAULT_LIGHTS
                    or Notification.DEFAULT_VIBRATE)
                setAutoCancel(true)                             //禁止自动停止
                addAction(R.drawable.play,"播放",playPendingIntent)   //增加播放动作
```

```kotlin
            addAction(R.drawable.pause,"停止",stopPendingIntent)
                                                            //增加停止播放动作
        }
    }
    private fun postNotification(){                         //发布通知
        NotificationManagerCompat.from(this).apply{
            notificationBuilder.setProgress(100,0,false)    //设置通知的进度条100
            startForeground(1,notificationBuilder.build())  //启动前台服务

            thread{                                         //定义工作线程修改通知中的进度条
                var currentProgress=0
                while (running){
                    notificationBuilder.setProgress(100,currentProgress,false)
                    Thread.sleep(1000)
                    currentProgress=musicProgress           //修改播放进程
                    startForeground(1,notificationBuilder.build())   //启动前台服务
                }
                notificationBuilder
                    .setContentText("一剪梅已经播放完毕")
                    .setProgress(0,0,false)
                startForeground(1,notificationBuilder.build()) //启动前台服务
            }
        }
    }
    private fun getPlayPendingIntent():PendingIntent{       //获取播放意图
        val intentFilter=IntentFilter()                     //创建意图过滤器
        intentFilter.addAction("PLAY_ACTION")               //增加意图过滤器的动作
        val playActionReceiver=object:BroadcastReceiver(){  //创建播放音乐广播接受器
            override fun onReceive(context: Context?, intent: Intent?) {
                playMusic()                                 //播放音频
            }
        }
        registerMyReceiver(playActionReceiver,intentFilter) //注册播放音乐广播器
        val intent=Intent("PLAY_ACTION")                    //创建播放意图
        return PendingIntent.getBroadcast(this, 0,intent,
            PendingIntent.FLAG_IMMUTABLE or PendingIntent.FLAG_UPDATE_CURRENT)
    }
    private fun getStopPendingIntent():PendingIntent{
        val intentFilter=IntentFilter()                     //创建意图过滤器
        intentFilter.addAction("STOP_ACTION")               //增加意图过滤器的动作
        val stopActionReceiver=object:BroadcastReceiver(){  //创建停止音乐播放接收器
            override fun onReceive(context: Context?, intent: Intent?) {
                stopMusic()                                 //停止播放
            }
        }
        registerMyReceiver(stopActionReceiver,intentFilter) //注册停止播放音乐广播器
        val intent=Intent("STOP_ACTION")                    //创建停止播放音乐的意图
```

```kotlin
        return PendingIntent.getBroadcast(this,0,intent,
            PendingIntent.FLAG_IMMUTABLE or PendingIntent.FLAG_UPDATE_CURRENT)
    }
    private fun registerMyReceiver(receiver:BroadcastReceiver,
        intentFilter:IntentFilter){
                                                    //根据不同版本注册 BroadcastReceiver
        if(Build.VERSION.SDK_INT>=Build.VERSION_CODES.Q){    //判断是否是 SDK 34 及以上版本
            registerReceiver(receiver,intentFilter, RECEIVER_EXPORTED)
        }else {
            registerReceiver(receiver, intentFilter)
        }
    }
}
```

在上述的 MusicService 中,有 3 处需要注意。

- Service 与 Notification 绑定时,除了在 AndroidManifest.xml 中配置相关的权限,还需要硬代码对通知权限许可进检查和授权,如 postNotification 函数前部分代码所示。
- 定义了 getPlayPendingIntent 函数和 getStopPendingIntent 函数,分别获取两个 PendingIntent 待定意图分别控制播放音频和停止播放音频的动作。在 Service 中,通过 PendingIntent 对象启动 BroadcastReceiver 组件实现不同操作的处理。
- 在 MusicService 中启动 Notification 时,也可以通过调用 notificationBuilder 的 notify(notifyId, notification)函数来实现,作用是一致的。

(3) 定义控制音乐播放主界面 MainScreen,代码如下:

```kotlin
/**模块 Ch07_04 的主界面 MainScreen
 * @param progressState:MutableState<Float>:音乐播放进度状态
 * @param playAction Function0<Unit>播放音频
 * @param stopAction Function0<Unit>停止播放 */
@Composable
fun MainScreen(modifier:Modifier,progressState: MutableState<Float>,
    playAction:()->Unit, stopAction:()->Unit){
    Column(modifier=modifier.fillMaxSize()
        .background(colorResource(id=android.R.color.holo_orange_light)),
        horizontalAlignment=Alignment.CenterHorizontally,
    verticalArrangement=Arrangement.Center) {
        Text("音乐播放器",fontSize=36.sp)
        Slider(value=progressState.value,onValueChange={        //音乐播放的进度条
            it:Float->progressState.value=it
        })
        Row{
            TextButton(onClick={playAction.invoke() }) {        //调用 playAction 函数对象
                Row(verticalAlignment=Alignment.CenterVertically){
                    Icon(painter=painterResource(id=R.drawable.play),
                        contentDescription="play",tint=Color.Unspecified)
                    Text("播放音乐",fontSize=24.sp)
```

```
            }
        }
        TextButton(onClick={ stopAction.invoke() }) {
                                                        //调用 stopAction 函数对象
            Row(verticalAlignment=Alignment.CenterVertically){
                Icon(painter=painterResource(id=R.drawable.pause),
                    contentDescription="pause",tint=Color.Unspecified)
                Text("暂停播放",fontSize=24.sp)
            }
        }
    }
  }
}
```

(4) 定义 MainActivity,代码如下：

```
class MainActivity: ComponentActivity() {
    lateinit var musicIntent:Intent
    lateinit var conn: ServiceConnection
    override fun onCreate(savedInstanceState: Bundle?) {
        super.onCreate(savedInstanceState)
        enableEdgeToEdge()
        requestNotificationPermission()
        musicIntent=Intent(this,MusicService::class.java)        //创建启动服务的意图
        setContent {
            val progressState=remember {mutableFloatStateOf(0.0F)}//显示播放进度的状态
            val runningState=remember {mutableStateOf(false)}    //音乐播放状态

            LaunchedEffect(Unit) {                                //加载效应
                conn=object:ServiceConnection{                    //确保只有一个 conn
                    override fun onServiceConnected(name: ComponentName?, service: IBinder?) {
                        val binder=service as MusicService.ProgressBinder
                        thread{                                    //定义工作线程追踪音乐播放的进度
                            do{
                                runningState.value=binder.getRunning()
                                                                    //从服务获取运行状态
                                progressState.value=binder.getMusicProgresss().toFloat()/100
                                                                    //从服务获取进度状态
                            }while(progressState.value<1&&runningState.value)
                        }
                    }
                    override fun onServiceDisconnected(name: ComponentName?) {
                        progressState.value=0F
                        runningState.value=false
                    }
                }
            }
            Chapter07Theme {
```

```
                    Scaffold(modifier=Modifier.fillMaxSize()) { innerPadding ->
                        MainScreen(modifier=Modifier.padding(innerPadding),
                            progressState,
                            playAction=::playMusic,
                            stopAction=::stopMusic)
                    }
                }
            }
    }
    private fun playMusic(){
        bindService(musicIntent,conn, Context.BIND_AUTO_CREATE)     //绑定服务
    }
    private fun stopMusic(){
        unbindService(conn)                                         //解绑服务
    }
    private fun requestNotificationPermission(){
        if (ActivityCompat.checkSelfPermission(
            applicationContext,
            Manifest.permission.POST_NOTIFICATIONS
            ) !=PackageManager.PERMISSION_GRANTED
        ) {
            if (Build.VERSION.SDK_INT>=Build.VERSION_CODES.TIRAMISU) {
                ActivityCompat.requestPermissions(
                    this,arrayOf(android.Manifest.permission.POST_NOTIFICATIONS),0
                )
            }
        }
    }
}
```

运行结果如图 7-7 所示。

(a) 播放音频的运行结果　　(b) 播放音频的运行结果

图 7-7　前台服务控制音乐播放的运行结果

7.5 WorkManager 组件

大部分移动应用存在后台执行任务的需求,例如从网络上传下载数据、对应用的用户数据进行备份等。这些需要后台执行的任务往往会消耗有限的设备资源,例如内存和电池的电量。这些后台任务可能会缩短设备的续航时间或者设备的卡顿等问题,进而影响用户体验。

自从 Android M(Android 6.0)版本开始,谷歌公司致力于解决后台运行耗电多的问题,对电池的使用进行了优化。Android 6.0 引入了 Doze 模式和 App Standby 待命模式。通过这两种省电技术,降低了功耗,延长了电池的使用时间。从 Android 8.0 开始,对后台任务的执行限制地越来越严格,例如,对 Service 要求使用与 Activity 关联的 Service 要工作在前台。此外,谷歌公司还提出了很多不同的后台任务调度的解决方案来执行后台异步任务,例如 Firebase JobDispatcher 方式、Job Scheduler 方式、Alarm Manager 与 BroadcastReceiver 相结合的处理方式。但是,从这些任务调度的后台任务方案中选择合适的解决方式往往让程序员无所适从。基于上述原因,谷歌公司在 Android JetPack 中加入了 WorkManager 组件来实现对任务的调用。当需要执行一个后台任务时,将后台任务提交给 WorkManager 组件。由 WorkManager 组件根据需要决定使用哪种具体的任务调度程序实现后台任务的执行,从而达到降低功耗的目的。

WorkManager 组件的本质是一个安排异步任务的程序。虽然这些任务不需要立即完成,允许在一段时间,甚至设备重启后仍可以执行,但是最终必须完成,只是完成的时间限制较为不严格。WorkManager 组件兼容范围广泛,最低兼容到 Android 1.4 版本,这使得程序开发人员可以更加关注任务的定义和执行,而不是关注移动设备的性能。WorkManager 组件可以创建并配置任务,在收到提交的任务后,WorkManager 会选择合适的任务调度程序在合适的时间执行任务。让异步任务的运行更加容易。在图 7-8 中,展示了 WorkManager 组件的工作原理。

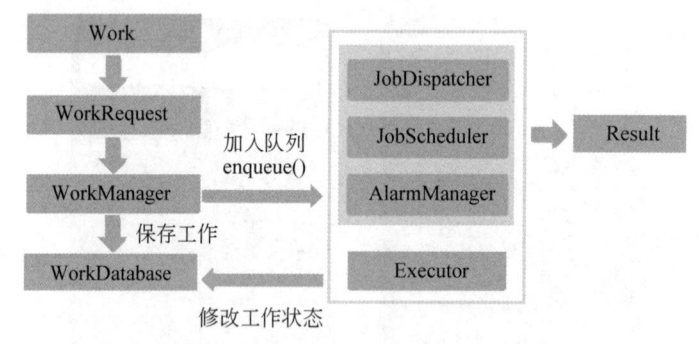

图 7-8 WorkManager 组件工作原理

(1) Work:用于定义后台执行的任务,指定需要执行的具体任务。

(2) WorkRequest:用于任务请求,对 Worker 任务进行配置包装,一个 WorkRequest 配置一个 Worker 类。具体配置包括设置任务触发的约束条件、设置退避策略等。

(3) WorkManager:主要管理任务请求和任务队列,调用 enqueue 函数将包含任务的 WorkRequest 任务请求加入队列。通过 WorkManager 组件来调度任务,以分散系统资源的负载。WorkManager 组件调度任务,无论采用哪种调度方案,最终的任务都会交付 Executor 来执行的。

(4) WorkDatabase:用于持久化存储任务的状态、约束条件等信息。当 Executor 执行任务后,会将修改的 WorkStatus 发送给 WorkDatabase 保存。

（5）WorkPolicy：定义了任务调度的策略。

（6）WorkResult：任务一旦执行，便会返回执行结果。执行的结果分为 3 种状态。

① Result.success()：任务执行成功。

② Result.fail()：任务执行失败。

③ Result.retry()：任务需要重新执行。

7.5.1　WorkManager 的基本使用方法

要使用 WorkManager 组件，需要模块依赖"androidx.work:work-runtime-ktx：$work_verion"。因此，在项目的 libs.versions.toml 文件中配置如下内容：

```
[versions]
work="2.9.1" # 当前$work_version 版本
[libraries]
androidx-work={group="androidx.work",name="work-runtime-ktx",version.ref="work"}
```

在应用模块的构建配置文件 build.gradle.kts 中增加下列依赖：

```
dependencies {
    implementation(libs.androidx.work)
    ...
}
```

在本节中通过实现一个访问本地图片资源，生成带水印的图片副本并将其保存在本地文件中的简单应用，了解 WorkManager 组件的基本应用。

1. 定义 Work 任务

定义实用类 FileUtils 类，用于实现图片文件的读取操作，代码如下：

```
//模块 Ch07_05 文件操作的实用类定义 FileUtils.kt
object FileUtils {
    fun saveFile(context: Context,bitmap: Bitmap, fileName: String){
                                                            //将 bitmap 对象保存到文件中
        try {var outputStream=context.openFileOutput(fileName, Context.MODE_PRIVATE)
            if (bitmap.compress(Bitmap.CompressFormat.PNG,100,outputStream)){
                outputStream.flush()
                outputStream.close()
            }
        }catch(e: IOException){
            e.printStackTrace()
        }
    }
    fun readImageFile(context: Context,fileName: String): Bitmap? {
                                                            //读取文件生成 Bitmap 图片
        var bitmap: Bitmap?=null
        if (isFileExist(context,fileName)) {
            try {val inputStream=context.openFileInput(fileName)
```

```kotlin
                inputStream.use {
                    bitmap=BitmapFactory.decodeStream(inputStream)
                }
            } catch (e: IOException) {
                e.printStackTrace()
            }
        }
        return bitmap
    }
    private fun isFileExist(context: Context,fileName: String): Boolean{      //判断文件是否存在
        val files=context.fileList()                                           //获得指定目录下的文件名
        files.forEach {
            if (it.contains(fileName))
                return true
        }
        return false
    }
}
```

下列 ImageUtils 实用类处理指定的图片 Bitmap 对象,创建添加水印图片副本。水印字符串从外界参数传入,代码如下:

```kotlin
//模块 Ch07_05 图片操作的实用类定义 ImageUtils.kt
object ImageUtils {
    fun addWorkMark(bitmap: Bitmap, text: String): Bitmap {                    //给图片增加水印
        val newBmp=Bitmap.createBitmap(bitmap.width, bitmap.height,
            Bitmap.Config.ARGB_8888)
                                                                               //创建一样大小的图片
        val canvas=Canvas(newBmp)                                              //创建画布
        canvas.drawBitmap(bitmap,0f,0f,null)                                   //绘制图片
        canvas.save()
        canvas.rotate(45f)                                                     //顺时针旋转 45°
        val paint=Paint(Paint.ANTI_ALIAS_FLAG)                                 //定义笔刷
        /*配置笔刷属性*/
        paint.typeface=Typeface.DEFAULT_BOLD                                   //设置字体
        paint.color=Color.WHITE                                                //设置文字颜色
        paint.textSize=bitmap.width.toFloat()/10                               //设置文字的大小
        paint.isDither=true                                                    //设置防抖动
        paint.isFilterBitmap=true                                              //设置抗锯齿
        /*得到 text 占用宽高*/
        val rectText=Rect()
        paint.getTextBounds(text,0,text.length,rectText)
        /*角度值 45°是 1.414*/
        val beginX=(canvas.height /2 -rectText.width()/2) * 1.4
        val beginY=(canvas.width /2 -rectText.width()/2) * 1.4
        canvas.drawText(text, beginX.toFloat(), beginY.toFloat(), paint);
        canvas.restore()
        return newBmp
    }
}
```

假设定义一个实现给图片增加水印的任务。这时需要定义一个扩展于 Worker 类的子类 WaterMarkWorker，在这个类的 doWork() 函数实现为图片添加水印的功能，并将图片保存到当前应用模块中。定义 WaterMarkWorker 类的代码如下：

```kotlin
//模块 Ch07_05 添加水印任务的定义 WaterMarkWorker.kt
class WaterMarkWorker(val context: Context, params: WorkerParameters): Worker
    (context,params) {
    override fun doWork(): Result {
        val bitmap=BitmapFactory.decodeResource(context.resources, R.drawable.sunflower)
                                                            //获得原始图片
        val waterMarkBitmap=ImageUtils.addWorkMark(bitmap,"梵高：向日葵")
                                                            //创建添加水印的新图片
        FileUtils.saveFile(context,waterMarkBitmap,"waterMark.png")
                                                            //将添加水印的图片保存到文件中
        val data=Data.Builder().putString("waterMarkFileName","waterMark.png").build()
                                                            //传递文件名
        return Result.success(data)
    }
}
```

WaterMarkWorker 类的 doWork() 函数用于定义后台任务。在这里定义了创建一个添加水印的图片并保存到当前应用模块中，保存在移动终端的路径为 data/data/chenyi.book.android.ch07_05/files。

Worker 可以通过 Data 来传递数据，在上述代码中，利用

```kotlin
val data=Data.Builder().putString("waterMarkFileName","waterMark.png").build()
```

通过 Data.Builder 函数构建一个 Data 对象，该对象包含一组键值对，即关键字 waterMarkFileName 对应的取值 watermark.png。要将数据 Data 数据传递给 WorkManager，需要通过 Result.success(data) 函数，将 Data 对象（即 data）作为参数进行传递。值得注意的是，Data 对象只能用于传递一些基本数据类型的数据，并且数据大小不能超过 10KB。

如果定义 Worker 类不需要传递参数，则只需要返回的 Result.success 函数即可，即 Result.success 函数中的没有任何参数。

2. 利用 WorkRequest 配置任务

定义好任务后需要使用 WorkRequest 来配置任务 Worker。WorkRequest 类是一个抽象类，它有两种实现方式：OneTimeWorkRequest 和 PeriodicWorkRequest。

OneTimeWorkRequest 定义一次执行的任务，如果不需要配置后台任务，可以直接执行如下代码：

```kotlin
val request=OneTimeWorkRequest.from(MyWork:: class.java)
```

如果需要利用 WorkRequest 来配置后台任务，则需要采用下列形式来创建：

```kotlin
val request=OneTimeWorkRequestBuilder<WaterMarkWorker>().build()
```

一次性任务执行完毕后，就彻底结束了后台任务。

PeriodicWorkRequest 定义了周期性执行的任务，定义的形式如下：

```kotlin
val request=PeriodicWorkRequestBuilder<WaterMarkWorker>(15, TimeUnit.MINUTES).build()
```

PeriodicWorkRequest定义的周期性任务会按照设定的时间周期运行。上述定义周期性任务请求对象表示在15min周期性执行任务,这个周期性任务的时间间隔不能超过15min。

3. 配置后台任务

由于移动应用的需求存在不同,往往触发这些后台任务的执行需要满足一定条件,因此需要创建触发任务的条件约束。这些触发条件可以是联网、是否正在充电或者存储空间大小处在允许的范围等。下面展示约束构造器的setRequires×××函数和setRequired×××函数是所有约束需要调用的对应方法,在实际情况下并不是所有的设置都需要。代码如下:

```
val constraints: Constraints=Constraints.Builder()
    .setRequiresBatteryNotLow(true)                             //电量充分
    .setRequiresStorageNotLow(true)                             //存储空间充分
    .setRequiresCharging(true)                                  //需要充电
    .setRequiredNetworkType(NetworkType.CONNECTED)              //需要联网
    .setRequiresDeviceIdle(true)                                //需要设备闲置
    .build()
```

定义触发约束条件后,可以在任务请求调用setContraints函数配置约束,形式如下:

```
OneTimeWorkRequestBuilder<WaterMarkWorker>().setContraints(contraits).build()
```

有时并不需要任务立即执行,可以通过配置WorkRequest对象的setInitialDelay函数设置后台任务执行的延迟时间,代码如下:

```
OneTimeWorkRequestBuilder<WaterMarkWorker>()
    .setInitialDelay(5,TimeUnit.SECONDS)                        //延迟5s再执行后台任务
    .build()
```

在执行任务时可能会出现异常的情况,但又希望任务能重试。针对这种情况,可以为Worker后台任务的doWork函数返回Result.retry函数。这样一来,系统会提供默认的指数退避策略BackoffPolicy.EXPONENTIAL重试后台任务的执行。当然,也可以通过调用WorkRequest对象的setBackoffCriteria函数配置后台任务的退避策略。代码如下:

```
OneTimeWorkRequestBuilder<WaterMarkWorker>()
    .setBackoffCriteria(BackoffPolicy.LINEAR,                   //设置线性增加重试时间的退避策略
OneTimeWorkRequest.MIN_BACKOFF_MILLIS,TimeUnit.MILLISECONDS)
    .build()
```

此外,通过设置WorkRequest对象的tag标签,可以对多个后台任务进行分组管理及批量化处理。可以调用addTag函数为后台任务增加标签。WorkManager组件可以对具有相同标签的后台任务进行批量化执行相同的操作。代码如下:

```
OneTimeWorkRequestBuilder<WaterMarkWorker>().addTag("ImageHandlerTag")
    .build()                                                    //增加标签
```

4. 执行和取消任务

配置WorkRequest的后台任务后,调用WorkManager.enqueue函数,将WorkRequest对象提交给WorkManager组件,交付系统来执行后台任务。代码如下:

```
WorkManager.getInstance(this).enqueue(request)
```

也可以通过以下方法取消任务,代码如下:

```
WorkManager.getInstance(this).cancelAllWork()                              //取消所有任务
WorkManager.getInstance(this).cancelAllWorkByTag("ImageHandlerTag")
                                                                //取消执行 Tag 标记的任务
```

5. Worker 类与 WorkManager 组件之间数据的传递

在上述定义的 WaterMarkWorker 中定义返回 Result.success(data) 成功返回并附有数据。WorkManager 组件可以通过 LiveData 获得从 Worker 成功返回的 Data 对象,代码如下:

```
WorkManager.getInstance(this).getWorkInfoByIdLiveData(request.id)
    .observe(MainActivity@this) {
    if (it!=null&&it.state==WorkInfo.State.SUCCEEDED) {          //检查 WorkInfo 的状态
        val data=it.outputData                                    //接收任务发送的数据
        ...
    }
}
```

在 MainActivity 中创建并配置 WaterMarkWorker 为后台任务,实现添加水印的图片并显示功能,代码如下:

```
//Ch07_05模块主活动的定义 MainActivity.kt
class MainActivity: ComponentActivity() {
    override fun onCreate(savedInstanceState: Bundle?) {
        super.onCreate(savedInstanceState)
        enableEdgeToEdge()
        setContent {
            val imageState:MutableState<ImageBitmap?>=remember{mutableStateOf(null)}
                                                                       //显示图片状态
            Chapter07Theme {
                Scaffold(modifier=Modifier.fillMaxSize()) { innerPadding ->
                    Column(modifier=Modifier.fillMaxSize().padding(innerPadding),
                        horizontalAlignment=Alignment.CenterHorizontally,
                        verticalArrangement=Arrangement.Center){
                        Text("添加图片水印应用",fontSize=42.sp)
                            imageState.value?.let{                      //添加水印
                                Image(bitmap=imageState.value!!, contentDescription=null)
                            }?:run{                                     //未加水印的图片
                                Image(modifier=Modifier.size(500.dp,610.dp),
                                    painter=painterResource(R.drawable.sunflower),
                                    contentDescription="sunflower")
                            }
                            Button(onClick={
                                handleWork(imageState)
                            }){
```

· 249 ·

```kotlin
                    Text("添加水印",fontSize=24.sp)
                }
            }
        }
    }
}

    private fun handleWork(imageState: MutableState<ImageBitmap?>){
        val constraints: Constraints=Constraints.Builder()
            .setRequiresBatteryNotLow(true)
            .setRequiresStorageNotLow(true)
            .build()                                    //定义触发约束条件
        val request=OneTimeWorkRequestBuilder<WaterMarkWorker>()
            .setConstraints(constraints).build()        //定义 WorkRequest 配置后台任务
        WorkManager.getInstance(this).enqueue(request)  //加入队列
        WorkManager.getInstance(this)
            .getWorkInfoByIdLiveData(request.id)
            .observe(MainActivity@this) {
                if (it!=null&&it.state==WorkInfo.State.SUCCEEDED){
                    val fileName=it.outputData.getString("waterMarkFileName")
                                                        //接收传递的数据
                    imageState.value=FileUtils.readImageFile(MainActivity@this,
                        fileName!!)?.asImageBitmap()
                }
            }
    }
}
```

运行结果如图 7-9 所示。

(a) 开始界面 (b) 显示水印图片界面

图 7-9 运行结果

7.5.2 任务链

移动应用往往需要同时执行多个后台任务。可以使用 WorkManager 组件创建任务链,并将任务链加入队列。可以利用 WorkManager 组件通过任务链指定这些后台任务的运行顺序。

创建任务链的形式如下:

```
WorkManager.beginWith(workRequestA)
    .then(workRequestB)
    .enqueue()
```

或

```
WorkManager.beginWith(listOf(workRequestA,workRequestB))
    .enqueue()
```

例 7-5 将资源的图片灰度化并添加水印在图片后,发布消息到通知栏。将处理的图片显示在移动终端中。

(1) 定义实用类。实用类包括 FileUtils 类和 ImageUtils 类。它们分别处理对文件和图像的相关处理。此处 FileUtils 的定义与 7.5.1 节定义的 FileUtils 类代码相同,在此不再说明。

ImageUtils 类增加的图像的灰度的处理,代码如下:

```kotlin
//模块 Ch07_06 的 ImageUtils.kt
object ImageUtils {
    fun grayImage(bitmap:Bitmap):Bitmap{                       //将图片转换为灰度
        val newBitmap=Bitmap.createBitmap(bitmap.width,bitmap.height,
            Bitmap.Config.RGB_565)                             //创建目标灰度图片
        val canvas=Canvas(newBitmap)                           //创建 newBitmap 的画布
        val painter=Paint()                                    //创建画笔
        val colorMatrix=ColorMatrix()                          //颜色矩阵
        colorMatrix.setSaturation(0F)                          //设置颜色饱和度
        val filter=ColorMatrixColorFilter(colorMatrix)         //创建颜色过滤器
        painter.setColorFilter(filter)                         //设置画笔的颜色过滤器
        canvas.drawBitmap(bitmap,0F,0F,painter)                //在画布绘制图片
        return newBitmap
    }
    fun addWorkMark(bitmap: Bitmap, text:String): Bitmap {     //给图片增加水印
        val newBmp=Bitmap.createBitmap(bitmap.width,bitmap.height,
            Bitmap.Config.ARGB_8888)                           //创建一样大小的图片
        val canvas=Canvas(newBmp)                              //创建画布
        canvas.drawBitmap(bitmap,0f,0f,null)                   //绘制图片
        canvas.save()
        canvas.rotate(45f)                                     //顺时针旋转 45°
        val paint=Paint(Paint.ANTI_ALIAS_FLAG)                 //定义笔刷
        /*配置笔刷属性*/
        paint.typeface=Typeface.DEFAULT_BOLD                   //设置字体
        paint.color=Color.WHITE                                //设置文字颜色
```

```kotlin
        paint.textSize=bitmap.width.toFloat()/10           //设置文字的大小
        paint.isDither=true                                //设置防抖动
        paint.isFilterBitmap=true                          //设置抗锯齿
        /*得到 text 占用宽高*/
        val rectText=Rect()
        paint.getTextBounds(text, 0, text.length, rectText)
        /*45°的角度值是 1.414*/
        val beginX=(canvas.height/2 - rectText.width()/2) * 1.4
        val beginY=(canvas.width/2 - rectText.width()/2) * 1.4
        canvas.drawText(text, beginX.toFloat(), beginY.toFloat(), paint);
        canvas.restore()
        return newBmp
    }
}
```

（2）定义后台任务。GrayImageWorker 类定义将图片的内容灰度化，并将处理的临时文件 tmp.png 保存在本地的 data/data/chenyi.book.android.ch07_06/files 目录中。将新建灰度的图片的文件名生成 Data 对象，伴随 Result 传递给 WorkManager。代码如下：

```kotlin
//模块 Ch07_06 的 GrayImageWorker.kt
class GrayImageWorker(val context: Context, params: WorkerParameters):Worker(context,
    params) {
    override fun doWork(): Result {
        val bitmap=BitmapFactory.decodeResource(context.resources,
            R.drawable.sunflower)                          //获得原始图片
        val grayBitmap=ImageUtils.grayImage(bitmap)
        FileUtils.saveFile(context,grayBitmap,"tmp.png")   //创建并保存已经灰度化的新图片
        val data=Data.Builder().putString("tmpImage","tmp.png").build()
                                                           //记录文件名到数据中
        return Result.success(data)                        //传递文件名
    }
}
```

WaterMarkWorker 类根据临时文件创建一个带有文字时间戳的水印，并添加新图片中，以 PNG 格式保存在本地中。将新建带水印的图片的文件名生成 Data 对象，伴随 Result 传递给 WorkManager。代码如下：

```kotlin
//模块 Ch07_06 的 WaterMarkWorker.kt
class WaterMarkWorker(val context: Context, params: WorkerParameters): Worker(context,
    params) {
    override fun doWork(): Result {
        val bitmapFileName=inputData.getString("tmpImage")   //获得灰度化的图片的名称
        val bitmap=FileUtils.readImageFile(context,bitmapFileName!!)
                                                             //加载已经灰度化的图片
        val waterMarkBitmap=ImageUtils.addWorkMark(bitmap!!,
            "梵高：向日葵${LocalDate.now()}")
        FileUtils.saveFile(context,waterMarkBitmap,"tmp.png")//创建并保存添加水印的新图片
        val data=Data.Builder().putString("tmpImage","tmp.png").build()
```

```
            return Result.success(data)                          //将添加水印的图片文件记录到数据中
                                                                 //传递文件名
    }
}
```

NotificationWorker 类执行发布通知,将相关的执行情况发布到通知栏中,提示用户操作情况。代码如下:

```
//模块 Ch07_06 的 NotificationWorker.kt
class NotificationWorker (val context: Context, params: WorkerParameters): Worker(context,
    params) {
    override fun doWork(): Result {
        val bitmapFileName=inputData.getString("tmpImage")  //获得添加水印的图片的名称
        val data=Data.Builder().putString("tmpImage","tmp.png").build()
                                                            //将添加水印的图片文件记录到数据中
        showNotification(bitmapFileName)
        return Result.success(data)
    }
    private fun showNotification(imageFileName:String?){
        val notificationManager: NotificationManager=context.getSystemService(Context.
            NOTIFICATION_SERVICE) as NotificationManager //定义通知管理器
        val channelId="chenyi.book.chenyi.ch07"            //定义通知渠道的标识
        val channelName="Android 移动应用开发"              //定义通知渠道的名称
        val channel=NotificationChannel(channelId,channelName,
            NotificationManager.IMPORTANCE_DEFAULT)
                          //定义通知渠道:指定通知渠道的标识、名称和通知渠道的重要级别
        notificationManager.createNotificationChannel(channel)
                                                            //创建并配置通知渠道
        val notification=Notification.Builder(context,channelId).apply{   //创建通知
            setContentTitle("Ch07_06 的通知")              //设置通知标题
            setContentText("$imageFileName 图片已经灰度化并加水印处理")  //设置通知内容
            setWhen(System.currentTimeMillis())            //设置通知时间
            setSmallIcon(R.drawable.sunflower)             //设置通知的小图标
        }.build()
        val notificationID=1                               //创建通知标记
        notificationManager.notify(notificationID,notification)   //发布通知到通知栏
        Log.d("TAG","notification published")
    }
}
```

(3) 定义视图模型。定义视图模型 WorkViewModel 类处理核心的后台任务链,并修改图片的状态,达到更新界面的目的。在 WorkViewModel 中定义了私有的 MutableFlowState<ImageBitmap?>对象_imageState,处理核心业务时会修改_imageState 状态值。WorkViewModel 提供了 output 这个 StateFlow<ImageBitmap?>对象,为界面获取_imageState 保存的状态值,从而实现不同的_imageState 的值的变化,更新用户界面。

```kotlin
//模块 Ch07_06 的 WorkViewModel.kt
class WorkViewModel: ViewModel() {
    private var _imageState:MutableStateFlow<ImageBitmap?>=MutableStateFlow(null)
    val output: StateFlow<ImageBitmap?>=_imageState.asStateFlow()
    fun handleWorks(context: Context){                              //处理任务链
        //定义约束条件
        val constraints: Constraints=Constraints.Builder()
            .setRequiresBatteryNotLow(true)
            .setRequiresStorageNotLow(true)
            .build()                                                //定义触发约束条件
        //定义将图片灰度化
        val grayImageRequest=OneTimeWorkRequestBuilder<GrayImageWorker>()
            .setConstraints(constraints)
            .build()                                                //定义添加水印任务请求
        val waterMarkRequest=OneTimeWorkRequestBuilder<WaterMarkWorker>()
            .setConstraints(constraints)
            .setInitialDelay(100, TimeUnit.MICROSECONDS)
            .build()                                                //定义发送通知请求
        val notificationRequest=OneTimeWorkRequestBuilder<NotificationWorker>()
            .setConstraints(constraints)
            .setInitialDelay(100, TimeUnit.MICROSECONDS)
            .build()                                                //创建任务链并按照顺序执行
        WorkManager.getInstance(context)
            .beginWith(grayImageRequest)                            //执行图片灰度化
            .then(waterMarkRequest)                                 //添加水印
            .then(notificationRequest)                              //发送通知
            .enqueue()
        WorkManager.getInstance(context)
            .getWorkInfoByIdLiveData(waterMarkRequest.id)
            .observe(context as MainActivity) {
                if (it!=null&&it.state==WorkInfo.State.SUCCEEDED){
                    val fileName=it.outputData.getString("tmpImage")_imageState
                        .value=FileUtils.readImageFile(context,
                        fileName!!)?.asImageBitmap()
                }
            }
    }
}
```

（4）定义界面。通过可组合函数 MainScreen 定义展示给用户的界面，代码如下：

```kotlin
//模块 Ch07_06 的可组合函数 MainScreen
@Composable
fun MainScreen(modifier:Modifier,viewModel:WorkViewModel){
    val context=LocalContext.current as MainActivity
    val imageState=viewModel.output.collectAsState()
    Column(modifier=modifier.fillMaxSize(),
        horizontalAlignment=Alignment.CenterHorizontally,
        verticalArrangement=Arrangement.Center){
        Text("简单处理图片应用",fontSize=42.sp)
```

```
        if (imageState.value!=null){
            Image(bitmap=imageState.value!!, contentDescription=null)    //添加水印的图片
        } else {
            Image(modifier=Modifier.size(500.dp,610.dp),painter=painterResource(R.drawable
                .sunflower),contentDescription="sunflower")              //未添加水印的图片
        }
        Button(onClick={ viewModel.handleWorks(context) }){
            Text("处理图片",fontSize=24.sp)
        }
    }
}
```

（5）定义 MainActivity。MainActivity 检查是否具有发布通知的权限,如果没有就申请,代码如下:

```
//模块 Ch07_06 的 MainActivity.kt
class MainActivity: ComponentActivity() {
    override fun onCreate(savedInstanceState: Bundle?) {
        super.onCreate(savedInstanceState)
        val viewModel:WorkViewModel=ViewModelProvider(this)
            .get(WorkViewModel::class.java)
        checkNotificationPermission()                                    //请求通知权限
        enableEdgeToEdge()
        setContent {
            Chapter07Theme {
                Scaffold(modifier=Modifier.fillMaxSize()) {
                    innerPadding->MainScreen(modifier=Modifier.padding(innerPadding),
                        viewModel=viewModel )                            //生成界面
                }
            }
        }
    }

    private fun checkNotificationPermission(){                           //检查通知权限
        if (ActivityCompat.checkSelfPermission(this,Manifest.permission
            .POST_NOTIFICATIONS) !=PackageManager.PERMISSION_GRANTED) {
            requestNotificationPermission(this)                          //请求通知许可
        }
    }

    private fun requestNotificationPermission(context: Context){         //请求通知权限
        if (Build.VERSION.SDK_INT>=Build.VERSION_CODES.TIRAMISU) {
            ActivityCompat.requestPermissions(context as Activity,
                arrayOf(android.Manifest.permission.POST_NOTIFICATIONS),0
            )
        }
    }
}
```

上述任务链是先执行图片灰度化,然后再是延迟 0.1s(100ms)后执行添加水印的任务,再延迟 0.1s

（100ms）发布通知到通知栏。确保 3 个任务按照顺序执行。

运行结果如图 7-10 所示。

    (a) 显示原始图片　　　　(b) 显示处理后的图片　　　　(c) 显示通知

图 7-10　任务链执行结果

  上述例子展示了任务链的任务按照顺序依次执行，如果多个任务没有先后关系，可以将它们并行放入 WorkManager 的 beginWith 函数中。假设有如下代码，其中 workRequestA、workRequestB 和 workRequestC 分别配置任务 WorkerA、WorkerB 和 WorkerC 的任务请求，代码如下：

```
val workerRequestA=OneTimeWorkRequestBuilder<WorkerA>().build()
val workerRequestB=OneTimeWorkRequestBuilder<WorkerB>().build()
val workerRequestC=OneTimeWorkRequestBuilder<WorkerC>().build()
WorkManager.getInstance(this).beginWith(workRequestA,
    workRequestB).then(workRequestC).enqueue()
```

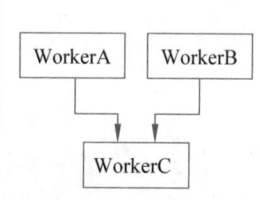

执行任务的顺序如图 7-11 所示。

  注意：如果要执行连接网络的后台任务，那么需要为该任务设置联网的触发约束条件。代码如下：

图 7-11　任务链执行流程

```
val constraints:Constraints=Constraints.Builder().setRequiredNetworkType(NetworkType
    .CONNECTED).build()                                          //定义联网约束
val workRequest=OneTimeWorkRequestBuilder<SomeWorker>().setConstraints(constraints)
    .build()                                                     //为任务请求配置约束
```

  在连网测试这样的任务时，必须使用真机进行测试。这是因为在 Android 模拟器上运行时，ConnectivityManager.activeNetwork 不具有 NetworkCapabilities.NET_CAPBILITY_VALIDATED 功能，模拟器不能满足网络约束，会导致模拟器运行联网后台任务失败。

习 题 7

一、选择题

1. 自定义的 Service 组件必须在 AndroidManifest.xml 中进行配置,其中 service 元素设置_____属性表示是否可以让其他应用调用该组件。
   A. android：enabled   B. android：exported
   C. android：priority   D. android：name

2. 配置 Service 的优先级的取值范围是_____。
   A. －1000～1000   B. 0～1000   C. －100～100   D. 0～100

3. 自定义服务组件是 Service 类的子类,_____函数在 Service 组件第一次创建后会被立即调用。
   A. onCreate   B. onStartCommand
   C. onBind   D. onUnbind

4. 在 Activity 中通过_____函数启动自定义 Service,该 Service 必须定义_____方法。
   A. startService onCreate   B. startService onStartCommand
   C. startService onBind   D. bindService onStartCommand

5. 要实现 Service 组件和 Activity 组件通信,必须在 Activity 中调用_____函数来启动 Service 组件。
   A. startService   B. bindService
   C. sendService   D. 以上答案均不正确

6. Service 组件和 Activity 组件进行通信时,必须在 Activity 组件中创建_____对象。
   A. Ibinder   B. Binder
   C. ServiceConnection   D. 以上答案均不正确

7. _____用于为用户提供消息提示,可以在移动系统的通知栏中显示。
   A. Toast   B. Notification   C. Snackbar   D. AlertDialog 前台

8. Service 必须在 AndroidManifest.xml 配置_____权限。
   A. android.permission.INTERNET
   B. android.permission.WRITE_EXTERNAL_STORAGE
   C. android.permission.READ_EXTERNAL_STORAGE
   D. android.permission.FOREGROUND_SERVICE

9. Android JetPack 中加入了_____组件来实现对任务的调用。
   A. WorkManager   B. Compose
   C. Service   D. Work

10. 调用 WorkMananger 的_____函数,将任务请求 WorkRequest 对象提交给 WorkManager 组件,交付系统来执行后台任务。
    A. setConstraints   B. getInstance
    C. enqueue   D. observe

二、填空题

1. Service 组件有两组生命周期。当在上下文中通过 startService 启动 Service 时,在 Service 第一次创建后,会调用_____函数进行初始化操作,然后执行_____函数进行定义。当前的 Service 处理完毕后会调用_____函数关闭 Service。

2. 如果在上下文（也是其他组件，也可以是使用 Service 的客户端）中通过 bindService 启动 Service 时，在服务第一次创建后，会调用_____函数进行初始化操作，然后执行_____函数使得客户端与 Service 绑定，一旦客户端请求 unBindService，就会调用 Service 的_____函数，然后调用_____函数关闭 Service。

3. Service 在后台运行时，若系统资源不足，就会被销毁，释放空间，导致需要完成的任务提前结束。要避免 Service 被提前销毁，可以有 4 种方式进行解决这个问题：_____、_____、_____和_____。

三、上机实践题

自选一个歌曲专辑，设计一个可以播放该专辑的移动应用，具体功能如下。

（1）只显示某个特定专辑的歌曲列表。

（2）根据专辑的歌曲列表，选择一首歌曲进行播放，播放时显示 Notification 用于控制歌曲的播放。界面自行设计。

第8章 Android 的网络应用

目前,移动互联网已经可以提供高速的数据传输,这使得许多基于移动互联网的移动应用成为人们日常生活的必备,如微信、支付宝、QQ、高德地图等。通过互联网,移动应用可以实现与云端的数据下载与上传,进行特定的数据访问和处理。Android 已内置了网络服务支持,可以方便地实现网络应用。Android 还提供了支持网络访问的 GUI 控件——WebView,以及 HttpURLConnection 用于 HTTP 访问方式支持,特别对云端的 XML、JSON 等多种格式的数据提供了解析和处理。有些第三方 Android 库(如 Retrofit 库)也对 HTTP 访问数据提供了支持。本章对基于 Android 的网络应用进行介绍。

8.1 网络访问相关配置

在开发基于互联网的移动应用时,只有在移动应用中设置相应的属性和取值,才能顺利访问网络。

1. 设置网络的访问权限

网络的访问权限一般在移动应用的系统配置清单文件 AndroidManifest.xml 中进行配置,形式如下:

```xml
<uses-permission android: name="android.permission.INTERNET" />
```

2. 设置明文访问

Android 9.0(Android API 28)以后,即使已经设置了网络访问权限,默认情况下也不能直接通过 URL 链接明文访问网络资源。Android 限制了明文流量的网络请求,对未加密流量不再信任,会被直接放弃。在进行访问时,会提示 net:ERR_CLEARTEXT_NOT_PERMITTED 错误。解决明文禁止访问,常见的方式有两种。

方式1:在移动应用的系统配置文件 AndroidManifest.xml 中配置 application 元素增加 android:usesCleartextTraffic 属性,并设定该属性为 true,表示允许使用明文访问在线资源,形式如下:

```xml
<manifest …>
    <application …
        android: usesCleartextTraffic="true"
        …>
        …
    </application>
</manifest>
```

方式2:在移动应用项目的 res 目录下新建 xml 目录,然后创建一个 xml 文件,例如命名为 network_security_config.xml,指定文件的网络安全配置允许设计明文访问,形式如下:

```xml
<?xml version="1.0" encoding="utf-8"?>
<!--network_security_config.xml-->
<network-security-config>
    <base-config cleartextTrafficPermitted="true" />
</network-security-config>
```

为了让这个网络安全配置文件发挥作用,还需要在移动应用项目的系统配置文件 AndroidManifest.xml 中,将 android：networkSecurityConfig 配置指定为网络安全配置文件,代码如下：

```xml
<manifest …>
    <application …
        android: networkSecurityConfig="@xml/network_security_config"
        …>
        …
    </application>
</manifest>
```

上述两种方式都能实现让 Android 移动应用对网络进行明文访问。

8.2 WebView 组件

WebView 组件是一种基于 Android 的 Webkit 引擎展现 Web 页面的控件。通过 WebView 控件,可以方便地显示和渲染 Web 页面,直接使用 HTML 文件(网络上或本地 assets 中)作为显示的布局。通过 WebView 控件,还可以实现与 JavaScript 交互调用。在表 8-1 中列出了 WebView 控件常见的函数。

表 8-1 WebView 控件常见的函数

函　　数	说　　明
canGoBack()	加载的网页是否可以后退,是则返回 true,否则返回 false
goBack()	后退网页
canGoForward()	加载的网页是否可以前进,是则返回 true,否则返回 false
goForward()	前进网页
goBackOrForward(steps)	以当前的 index 为起始点前进或者后退到历史记录中指定的 steps
clearCache(true)	清除网页访问留下的缓存
clearHistory()	清除当前 WebView 组件访问的历史记录
clearFormData()	清除自动完成填充的表单数据
loadUrl(url)	加载 url 指定的页面
pauseTimers()	停止所有的布局、解析以及 JavaScript 计时器
resumeTimers()	恢复 pauseTimers 时的所有操作

要加载网页页面,并实现 JavaScript 与页面的交互,代码如下：

```
//WebView 必须设置支持 JavaScript
webView.settings.javaScriptEnabled=true
//创建 WebViewClient 实例,处理各种通知和请求事件
webView.webViewClient=WebViewClient()
//加载页面
webView.loadUrl(link)
```

因为 WebView 是一种传统的 Android 控件,并不是 Compose 内置的可组合项。因此,WebView

往往是嵌入传统的布局 xml 文件中才被使用。为了在 Compose 中调用 WebView 控件，可以利用 AndroidView 这个 UI 可组合函数包装 WebView，从而达到在 Compose 的可组合界面中使用 WebView 访问 HTML 页面的功能，形式如下：

```
AndroidView(factory={
    ctx: Context->WebView(ctx).apply{          //初始化 WebView
        ...                                     //略
    }
})
```

例 8-1 WebView 控件的应用示例。通过 WebView 访问 www.baidu.com 网站。

首先配置清单文件 AndroidManifest.xml 设置 Internet 访问权限，然后在 application 元素中用 android:networkSecurityConfig 属性设置指定网络安全配置属性引用 xml 目录的 network_security_config.xml 文件配置，即采用明文方式访问网络资源。具体内容同上。

定义应用界面的可组合函数 MainScreen，代码如下：

```
//模块 Ch08_01 的 MainScreen
@Composable
fun MainScreen(modifier:Modifier){
    var webView: WebView? =null
    val scope=rememberCoroutineScope()
    Box(modifier=modifier.fillMaxSize()){
        AndroidView(factory={
            ctx: Context->WebView(ctx).apply{          //初始化 WebView
                webView=this
                settings.javaScriptEnabled=true        //允许 JavaScript 交互
                webViewClient=WebViewClient()          //创建客户端
                loadUrl("http://www.baidu.com")        //加载百度网站
            }
        })
    }
    BackHandler {                                       //处理后退动作
        scope.launch {
            webView?.goBack()                           //返回上一个网页
        }
    }
}
```

在 MainScreen 可组合函数中，利用 AndroidView 包装了传统 UI 控件 WebView，对 WebView 进行初始化，配置 WebView 为允许 JavaScript 交互，创建 WebView 的 WebViewClient 客户对象，最后成功加载网站 http://www.baidu.com。因为加载网页后希望能实现网页的后退处理，这样的操作不可能同步操作。因此在协程范围中允许执行 webView.goBack 函数。

在 MainActivity 中调用 MainScreen 可组合函数，生成移动界面，代码如下：

```
//模块 Ch08_01 的主活动 MainActivity.kt
class MainActivity: ComponentActivity() {
    override fun onCreate(savedInstanceState: Bundle?) {
```

· 261 ·

```
        super.onCreate(savedInstanceState)
        enableEdgeToEdge()
        setContent {
            Chapter08Theme {
                Scaffold(modifier=Modifier.fillMaxSize()) {
                    innerPadding->MainScreen(modifier=Modifier.padding(innerPadding))
                }
            }
        }
    }
```

如图 8-1 所示,在 WebView 控件中显示了百度的网站,通过百度首页提供的链接可以进入下一个页面。当按模拟器的 Back 键时,会访问百度的前一个页面。

(a) 百度首页　　　　　(b) 单击链接进入下一个页面

图 8-1　WebView 的运行界面

如图 8-2 所示,在 Activity 中使用 WebView 控件时,一旦 Activity 启动调用 onResume 函数,WebView

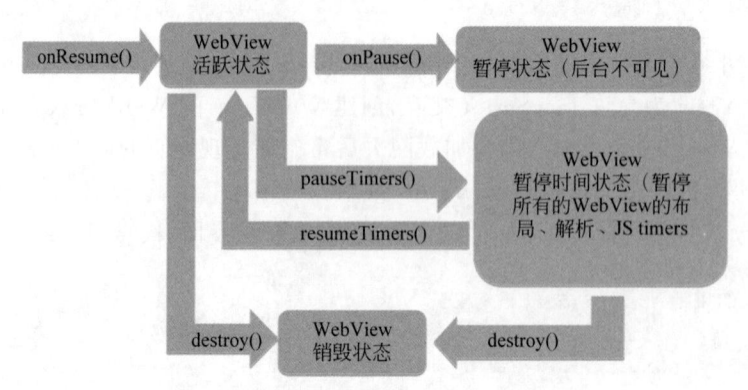

图 8-2　WebView 控件在 Activity 中的运行

就会进入活跃状态。WebView 控件可通过调用 pauseTimers 函数暂停自己的时间状态（包括 WebView 的布局、解析和 JavaScript 的定时器），也可以通过 resumeTimers 函数实现恢复所有 pauseTimers 函数暂停的操作。当 Activity 暂停时，WebView 控件也进入暂停状态。只有提前销毁 WebView 控件，才能销毁 Activity 并释放空间。

8.3 使用 HttpURLConnection 访问网络资源

HttpURLConnection 是 Android 自带的一种使用 HTTP 访问网络资源的轻量级 API。通过它可以用比较简单的方式对网络资源进行访问。一般情况下，HttpURLConnection 基于 HTTP 对网络资源进行访问可分为 5 个步骤。

（1）创建 HttpURLConnection 连接对象。

（2）配置连接的相关参数，例如设置 HTTP 访问所使用方法的 GET（从指定的资源请求数据）或 POST（向指定的资源提交要处理的数据）方式，建立连接需要时间，读取在线资源的限定时间，等等。

（3）根据连接对象的不同，创建当前连接的输入输出流对象。

（4）通过输入输出流实现读取资源。

（5）关闭连接。

代码如下：

```kotlin
var connection: HttpURLConnection?=null
val response=StringBuilder()
val url=URL(urlStr)                                          //创建 URL 链接对象
try {
    //1.创建连接
    connection=url.openConnection() as HttpURLConnection
    //2.配置连接
    connection.requestMethod="GET"                           //设置请求方式为 GET
    connection.connectTimeout=8000                           //设置请求建立连接时间
    connection.readTimeout=8000                              //设置读取数据限定的时间
    //3.利用连接的输入流创建缓冲输入流对象
    val reader=BufferedReader(InputStreamReader(connection.inputStream))
    //4.读取资源操作
    reader.use {
        reader.forEachLine {                                 //从输入流中读取每一行
            response.append(it)                              //加入到动态字符串 response 中
        }
    }
    val result=response.toString()
}catch(e:IOException){
    e.printStackTrace()
}finally{
    //5.关闭连接
    connection?.disconnect()
}
```

例 8-2 利用 HttpURLConnection 访问 QQ 新闻（https://news.qq.com/）。

首先，在 AndroidManifest.xml 文件中配置 android.permission.INTERNET 访问权限；设置

application 的 android：userCleartextTraffic 属性为 true，指定采用明文方式访问在线资源；然后，定义通过基于 HTTP 访问在线资源的通用实用类，代码如下：

```kotlin
//模块 Ch08_02 HttpUtils.kt
class HttpUtils {
    companion object{
        fun getWebContent(urlStr: String): String {
            lateinit var connection: HttpURLConnection
            lateinit var inputStream: BufferedReader
            val url=URL(urlStr)
            val response=StringBuilder()
            try {
                connection=url.openConnection() as HttpURLConnection    //1.创建连接
                connection.requestMethod="GET"                          //2.配置连接
                connection.connectTimeout=8000
                connection.readTimeout=8000
                inputStream=BufferedReader(InputStreamReader(connection.inputStream))
                                                                        //3.建立输入流
                inputStream.use{                                        //4.读取信息
                    it.forEachLine {                                    //对每一行记录进行处理
                        response.append(it)
                        response.append("\n")
                    }
                }
            } catch (e: IOException) { e.printStackTrace() } finally {//5.断开连接
                connection.disconnect()
            }
            return response.toString()
        }
    }
}
```

上述的 HttpUtils 类在伴随对象中定义了 getWebContent 函数，用于采用 GET 方式访问网络资源，使得可以通过类名直接访问伴随类内定义的函数。具体的操作步骤如下。

（1）按照给定的 URL 创建 HttpURLConnection 对象。

（2）配置连接，设置请求资源的方式采用 GET 方式，连接的时间限定为 8s，读取数据的时间限定为 8s。

（3）创建输入流。首先从连接 HttpURLConnection 对象中获得字节流，为了处理方便，最终打包成缓存字符输入流。

（4）对字符输入流的每行字符串进行处理并添加到 StringBuffer 对象中。

（5）当访问结束后，调用连接对象的 disconnect 函数断开连接。

定义了一个通用的 HttpUtils 类，通过它实现对网络资源的获取。这样处理的好处是可以重新反复使用这个类，即使访问的 URL 地址字符串不同，也可以访问网络资源，代码如下：

```kotlin
//模块 Ch08_02 MainActivity.kt
class MainActivity: ComponentActivity() {
    override fun onCreate(savedInstanceState: Bundle?) {
        super.onCreate(savedInstanceState)
```

```
enableEdgeToEdge()
setContent {
    var data by remember{mutableStateOf("")}            //获取网络资源数据
    val scope=rememberCoroutineScope()                  //协程
    Chapter08Theme {
        Scaffold(modifier=Modifier.fillMaxSize()) {
            innerPadding ->Column(modifier=Modifier.padding(innerPadding)
                .fillMaxSize(),
                horizontalAlignment=Alignment.CenterHorizontally,
                verticalArrangement=Arrangement.Top) {
                Text("$data",fontSize=12.sp, maxLines=35,minLines=35)
                                                //设置最大最小行数为 35 行

                Button(onClick={
                    scope.launch {
                        withContext(Dispatchers.IO){    //在 IO 线程中执行协程
                            data=HttpUtils.getWebContent("https://news.qq.com/")
                                                //访问网络资源
                        }
                    }
                }){Text("获取数据")}
            }
        }
    }
}
```

运行结果如图 8-3 所示。

(a) 初始界面　　(b) 单击按钮显示界面

图 8-3　获取内容

MainActivity 中对 HttpUtils 类通过调用 getWebContent(urlStr)函数,获得 urlStr 字符串指定的 url 链接的资源。由于访问在线资源有些耗时,因此采用 Kotlin 协程的方式进行处理。首先通过 Dispatcher.IO 将协程处理访问网络资源发送给后台线程来执行。当访问资源成功后,再让主线程执行修改 UI 界面的内容。最后实现在主线程中在文本可组合项中显示获取资源的内容。

上述的 HttpUtils 类定义了 HTTP 请求的 GET 方式访问网络资源。如果需要通过 HTTP 请求的 POST 方式提交表单信息,可以设置成如下形式:

```
//设置 HTTP 请求方式为 POST
connection.requestMethod="POST"
//根据连接的输出流创建数据输出流对象(也可以是其他输出流)
val writer=DataOutputStream(connection.outputStream)
//通过输出流写入数据
writer.writeBytes("username=guest&&password=123456")
```

例 8-3 应用 HttpURLConnection 实现 HTTP POST 访问"Bing 搜索"。"Bing 搜索"提供了检索关键词功能,检索的连接形式为 https://www.bing.com/search? q=关键词。

(1) 网络配置,代码如下:

```
//模块 Ch08_03 网络安全配置 res/xml/network_security_config.xml
<?xml version="1.0" encoding="utf-8"?>
<network-security-config>
    <base-config cleartextTrafficPermitted="true" />
</network-security-config>
```

为了能用明文访问在线资源,在 res 目录下创建 xml 目录,定义 net_security_config.xml 文件设置明文访问为 true,代码如下:

```
//模块 Ch08_03 应用配置 AndroidManifest.xml
<?xml version="1.0" encoding="utf-8"?>
<manifest xmlns: android="http://schemas.android.com/apk/res/android"
    package="chenyi.book.android.ch08_3">
    <uses-permission android: name="android.permission.INTERNET" />
    <application
        android: allowBackup="true"
        android: icon="@mipmap/ic_launcher"
        android: label="@string/app_name"
        android: roundIcon="@mipmap/ic_launcher_round"
        android: supportsRtl="true"
        android: theme="@style/AppTheme"
        android: networkSecurityConfig="@xml/network_security_config">
        <activity
            android: name=".MainActivity"
            android: exported="true">
            <intent-filter>
                <action android: name="android.intent.action.MAIN" />
                <category android: name="android.intent.category.LAUNCHER" />
            </intent-filter>
        </activity>
    </application>
</manifest>
```

在 AndroidManifest.xml 中定义网络访问权限,并设置 android:networkSecurityConfig 属性为引用 xml 目录的 network_security_config.xml 文件,实现明文访问。

（2）POST 方式访问处理，代码如下：

```kotlin
//模块 Ch08_03 HTTP 访问实用类 HttpUtils.kt
class HttpUtils {
    companion object{
        fun searchByPost(keyword:String):String{
            val url=URL("https://www.bing.com/")
            val response=StringBuilder()
            lateinit var connection:HttpURLConnection
            try {
                //1.创建连接
                connection=url.openConnection() as HttpURLConnection
                //2.配置连接
                connection.requestMethod="POST"
                connection.connectTimeout=8000
                connection.readTimeout=8000
                //3.创建输入输出流
                val output=DataOutputStream(connection.outputStream)
                //4.写入数据
                output.writeBytes("search? q=${keyword}}")
                //5.发送数据给服务器
                val input=BufferedReader(InputStreamReader(connection.inputStream))
                //6.接收从服务器获取的信息
                input.use{
                    input.forEachLine {
                        response.append(it)
                        response.append("\n")
                    }
                }
            }catch (e:IOException){
                e.printStackTrace()
            }finally {
                //7.关闭连接
                connection.disconnect()
            }
            return response.toString()
        }
    }
}
```

上述的 HttpUtils 类是一个实用类，定义了采用 POST 方式访问在线资源。searchByPost 函数定义在伴随对象中定义，这样就可以通过类名来直接访问。在对在线资源进行访问时，这个函数需要以下步骤进行定义。

（1）通过 URL 创建连接。
（2）对创建的连接进行配置，配置请求的方式为 POST，访问网络的时间限制为 8s，读取网络资源的时间限制为 8s。

(3) 利用连接 HttpURLConnection 创建输出流，为了写入数据方便，打包为数据字节输出流。

(4) 将需要提交给服务器的数据写入输出流。

(5) 利用连接 HttpURLConnection 对象获得输入字节流，最后打包成缓冲字符输入流。

(6) 在字符输入流中逐行读取字符串，实现网络资源的获取。

(7) 一旦网络访问完成，调用连接对象的 disconnect() 函数断开连接。

在 MainActivity 的由多个可组合项构建界面中定义了文本输入框用于接收关键字的输入，定义了按钮用于提交网络访问请求，定义了文本用于显示最后的访问资源的结果，代码如下：

```kotlin
//模块 Ch08_03 主活动 MainActivity.kt
class MainActivity: ComponentActivity() {
    override fun onCreate(savedInstanceState: Bundle?) {
        super.onCreate(savedInstanceState)
        enableEdgeToEdge()
        setContent {
            var data by remember{ mutableStateOf("") }            //获取网络资源数据
            var keyword by remember{ mutableStateOf("人工智能") }  //关键字
            val scope=rememberCoroutineScope()                     //协程
            Chapter08Theme {
                Scaffold (modifier=Modifier.fillMaxSize()) {
                    innerPadding ->Column(modifier=Modifier.padding(innerPadding)
                        .fillMaxSize(),horizontalAlignment=Alignment.CenterHorizontally,
                        verticalArrangement=Arrangement.Top) {
                        Row(modifier=Modifier.fillMaxWidth()){
                            TextField(value= keyword, onValueChange = {it:String-> keyword =
                                it})
                            Button(onClick={
                                scope.launch {
                                    withContext(Dispatchers.IO){
                                        Log.d("TAG","${keyword}")//在 IO 线程中执行协程
                                        data=HttpUtils.searchByPost(keyword) //访问网络资源
                                    }
                                }
                            }){
                                Row{
                                    Icon(Icons.Filled.Search,
                                    contentDescription="search")Text("检索")
                                }
                            }
                        }
                        Text("$data",fontSize=12.sp, maxLines=35,minLines=35)
                                                        //设置最大最小行数为 35 行
                    }
                }
            }
        }
    }
}
```

运行结果如图 8-4 所示。

(a) 初始界面　　　　(b) 单击"检索"按钮显示界面

图 8-4　以 HTTP POST 方式提交

在 MainActivity 中调用 HttpUtils 的 searchByPost()函数获得访问在线资源。在 MainActivity 中采用了 Kotlin 的协程实现资源的访问和显示。首先通过 Dispatchers.IO 执行协程中发送的关键字和获取在线资源的信息。然后再通过主线程更新 UI 界面的操作。

8.4　JSON 数据的解析

如果只是显示网页的内容，可以采用 8.2 节介绍的 WebView 组件显示。但是在现实中，往往需要对获取的数据进行解析重组，再以自定义界面设计的方式重新展现，以适应不同终端的设计的要求。因此，需要在网络上传递中间格式的数据。在网络上传递数据的常见格式有 XML 和 JSON 两种。XML 格式的数据表达规范完整。与之相比，JSON 格式的表达形式灵巧简单、易于读写、占用的体积更小、节省带宽。因此，在许多网站中都采用 JSON 格式进行数据的传递。在本章中侧重 JSON 数据的解析处理。

8.4.1　JSON 格式

JSON 数据的最大特点是采用键值对的方式，形式如下：

```
"关键字": "取值"
```

关键字是独一无二的，取值的数据类型可以是字符串、数字、数组、Boolean 布尔类型、对象 Object 和 null。JSON 的键值对可以构建成 JSON 对象和 JSON 数组对象这两种常见的数据形式。在下面的代码中展示一个简单的 JSON 对象，JSON 对象就是"键-值"构成的内容，形式如下：

```
{"name": "张三","gender": "男","age": 20}
```

JSON 数组可以表示一组 JSON 对象,形式如下:

```
[{"name": "张三","gender": "男","age": 20},
 {"name": "李四","gender": "男","age": 19},
 {"name": "王五","gender": "男","age": 20}
]
```

常见的 JSON 格式的数据往往用 JSON 对象和 JSON 数组组合表示,形式如下:

```
{   "student": [
        {"name": "张三","gender": "男","age": 20},
        {"name": "李四","gender": "男","age": 19},
        {"name": "王五","gender": "男","age": 20}
    ],
    "teacher": [
        {"name": "张老师","gender": "男","age": 40,"major": "语文"},
        {"name": "吴老师","gender": "男","age": 31,"major": "数学"}
    ]
}
```

JSON 数据作为一种中间数据,根据需求将 JSON 数据转换成多种不同形式,以满足不同的要求。例如,JSON 数据可以解析处理成 XHTML、HTML5 以及自定义格式的数据,通过不同的界面展示出来。

8.4.2 JSONObject 解析 JSON 数据

JSONObject 是 Android 系统官方解析 JSON 数据的 API,可根据不同类型的 JSON 数据提供不同的解析方式。可以通过 JSON 对象的 JSONObject 的 getXxx(key)函数进行解析。其中,key 表示关键字,getXxx(key)表示获取关键字所对应的值。在表 8-2 中展示了 JSONObject 常见的解析函数。

表 8-2 JSONObject 常见的解析函数

解析函数	说明
get(name: String)	返回 name 映射的值,返回的类型为 Any。如果映射不存在,则抛出异常
getBoolean(name: String)	返回 name 映射的 Boolean 值。如果映射不存在或映射的值不是布尔类型,则提示异常
getDouble(name: String)	返回 name 映射的 Double 值。如果映射不存在或映射的值不是 Double 类型,则提示异常
getInt(name: String)	返回 name 映射的 Int 值。如果映射不存在或映射的值不是整数类型,则提示异常
getLong(name: String)	返回 name 映射的 Long 值。如果映射不存在或映射的值不是长整数类型,则提示异常
getString(name: String)	返回 name 映射的 String 值。如果映射不存在或映射的值不是字符串类型,则提示异常
isNull(name: String)	如果 name 映射不存在或映射的值为 NULL,则返回 true,否则为 false

续表

解 析 函 数	说　明
getJSONArray(name: String)	返回 name 映射的 JSONArray 值。如果映射不存在或映射的值不是 JSONArray 类型，则提示异常
getJSONObject(name: String)	返回 name 映射的 JSONObject 值。如果映射不存在或映射的值不是 JSONObject 类型，则提示异常

在表 8-2 所列的解析函数中，通过 JSONObject 对象的 getJSONArray(key) 函数可以获取到一个 JSONArray 对象，即获得 JSON 数组。JSONArray 对象包括了一组数据，可以再调用 JSONArray 对象的 get(i) 函数获取数组元素，其中 i 为数据的索引值。

例 8-4 要求在日志中记录 2000—2010 年中国肉类消费量（原始数据来源于经济合作与发展组织农业统计数据，并加工成 JSON 数据格式）。

首先将数据预处理成 JSON 数据格式，通过本地 Web 服务器 http://127.0.0.1:5000/json/meat_consumption.json 进行访问，使用浏览器访问的 JSON 数据如图 8-5 所示。也可以自行利用 Python＋Flask（默认为 5000 端口）或 Tomcat（默认为 8080 端口）或 nginx（默认为 80 端口）等来搭建本地 Web 服务器。

图 8-5　使用浏览器访问要处理的 JSON 数据

上述 JSON 数据包括了 JSON 对象和 JSON 的数组。在解析这样的 JSON 数据时，需要按照不同类型进行处理。另外，解析后的数据往往处理成具体的实体类对象，方便后续的操作和处理，代码如下：

```
//模块 Ch08_04 实体类 MeatComputicon.kt
data class MeatConsumption(val location: String,
    val meatConsumptionList: List<MeatConsumptionData>) {
```

```
    data class MeatConsumptionData(val subject: String,val measure: String,
    val time: Int,val value: Double)
}
```

MeatConsumption 是一个数据实体类,定义了 location 属性表示国家和地区、meatConsumptionList 表示 2000—2010 年的肉类消费记录。每项肉类消费记录通过 MeatConsumption 类的内部定义 MeatConsumptionData 类来表示,具体的属性包括类型名称、重量单位、时间和肉类消费量,代码如下:

```
//模块 Ch08_04 访问网络资源 HTTPUtils.kt
class HttpUtils {
    companion object{
        fun getWebContent(urlStr: String): String {           //获取 urlStr 指定网址的内容
            lateinit var connection: HttpURLConnection
            lateinit var inputStream: BufferedReader
            val url=URL(urlStr)
            val response=StringBuilder()
            try { connection=url.openConnection() as HttpURLConnection
                                                              //1.创建连接
                connection.requestMethod="GET"                //2.配置请求方式
                connection.connectTimeout=8000
                connection.readTimeout=8000
            inputStream=BufferedReader(InputStreamReader(connection.
                inputStream))                                 //3.建立输入流
                inputStream.use{                              //4.读取信息
                    it.forEachLine {                          //5.对每一行记录进行处理
                        response.append(it)
                        response.append("\n")
                    }
                }
            } catch (e: IOException) {
                e.printStackTrace()
            } finally {                                       //6.关闭连接
                connection.disconnect()
            }
            return response.toString()
        }
    }
}
```

HttpUtils 处理了采用 GET 方式访问本地服务器的 JSON 数据,将 JSON 数据按行保存到 StringBuilder 对象中,最后返回包括所有 JSON 数据的字符串,代码如下:

```
//模块 Ch08_04 JSON 解析的实用类 JSONUtils.kt
class JSONUtils {
    companion object{
        fun parseJSON(jsonData: String): MeatConsumption{
            val jsonObject=JSONObject(jsonData)      //将 jsonData 字符串生成 JSONObject 对象
            val location=jsonObject.getString("location")
                                                     //获得 location 关键字的值
```

```kotlin
            val meatConsumptionList=mutableListOf<MeatConsumption
                .MeatConsumptionData>()
                                        //创建保存 MeatConsumptionData 的列表
            val jsonArray=jsonObject.getJSONArray("meat_consumption")
                                        //获得 meat_consumption 关键字对应的数组
            for (i in 0 until jsonArray.length()){
                                        //对数组依次进行访问
                var jsonObj=jsonArray.getJSONObject(i)
                                        //获得索引为 i 的 JSONObject
                var subject=jsonObj.getString("subject")
                var measure=jsonObj.getString("measure")
                var time=jsonObj.getInt("time")
                var value=jsonObj.getDouble("value")
                meatConsumptionList.add(MeatConsumption.MeatConsumptionData(
                    subject,measure,time,value))
                                        //添加到肉类消费的列表中
            }
            return MeatConsumption(location,meatConsumptionList)
        }
    }
}
```

JSONUtils 类实现了对 JSON 数据的解析。在 JSONUtils 的伴随对象中定义的 parseJSON 函数包含 JSON 数据的字符串解析成 JSON 对象。具体的做法是，结合 JSONObject 将 JSON 字符串生成 JSON 对象。因为对象 meat_comsumption 属性中包含了 JSON 数组，再通过 getJSONArray 函数把数组数据提取出来，按照数组的索引值，依次读取 JSON 数组的每个数据生成 MeatConsumption.MeatConsumptionData 对象，保存到列表中。通过这种函数，获得一个 MeatConsumption 对象，它包含国家地区名和一组肉类消费的记录，代码如下：

```kotlin
//模块 Ch08_04 使用解析后的数据 MainActivity.kt
class MainActivity: ComponentActivity() {
    override fun onCreate(savedInstanceState: Bundle?) {
        super.onCreate(savedInstanceState)
        enableEdgeToEdge()
        setContent {
            var meatConsumption:MutableState<MeatConsumption?>=remember{
                mutableStateOf(null)
            }
            LaunchedEffect(Unit) {
                withContext(Dispatchers.IO){
                    val jsonData=HttpUtils.getWebContent("http://10.0.2.2:5000/json/
                        meat_consumption.json")              //获得 JSON 数据字符串
                    meatConsumption.value=JSONUtils.parseJSON(jsonData)
                                                             //解析 jsonData
                    Log.d("TAG","${meatConsumption.value?.location}消费情况")
                    meatConsumption.value?.meatConsumptionList?.forEach {
                        Log.d("TAG","${it.subject}在${it.time}消费: ${it.value}${it.
                            measure}")
```

```kotlin
                    }
                }
            }
            Chapter08Theme {
                Scaffold(modifier=Modifier.fillMaxSize()) {
                    innerPadding ->Column(modifier=Modifier.padding(innerPadding)
                        .fillMaxSize(),
                        horizontalAlignment=Alignment.CenterHorizontally,
                        verticalArrangement=Arrangement.Center){
                        Text("2000-2010年肉类消费情况",fontSize =24.sp)
                    }
                }
            }
        }
    }
}
```

用计算机的浏览器访问本地 Web 服务器的资源一般是通过 localhost 或 127.0.0.1 来实现的。Android 模拟器无法通过 localhost 或 127.0.0.1 进行访问,这是因为 Android 模拟器将自己视为服务器,占用了 127.0.0.1 地址。如果 Android 模拟器要访问本地服务器,可通过 10.0.2.2 实现。MainActivity 调用 HttpUtils 的 getWebContent 函数访问 http://10.0.2.2:5000/json/meat_consumption.json 获得 JSON 格式的数据字符串后,可借助 JSONUtils 实用类的 parseJSON 函数获得最后的 MeatConsumption 对象,即指定国家地区的一组肉类消费记录。最后遍历肉类消费记录,并输出到日志中,如图 8-6 所示。

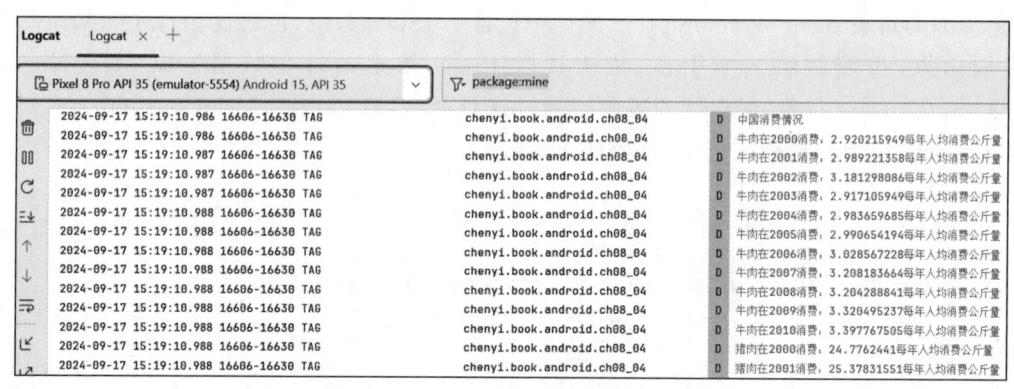

图 8-6　日志显示肉类消费量

8.4.3　GSON 解析 JSON 数据

GSON 库是解析 JSON 数据常用的开源 Java 库。用它可以非常容易地实现对象的序列化和反序列化,即实现两个方向的转换。它可以将 JSON 数据绑定成特定类型的对象,也可以将特定类型的数据转换成 JSON 数据。GSON 库提供了 fromJson 和 toJson 函数用于实现上述功能。这样就可按照参数规格要求进行调用,使操作简单便利。

假设有一个 JSON 数据如下所示:

```
{"id": "60001223","name": "张三","gender": "男","age": 20}
```

将 JSON 数据转换成对象的具体做法如下:

(1) 配置 GSON 库。要多个项目模块使用 GSON 库,需要在 libs.versions.toml 文件中增加如下配置:

```
[versions]
gson="2.10"
[libraries]
gson={module="com.google.code.gson:gson",version.ref="gson"}
```

(2) 在模块的构建文件 build.gradle.kts 中增加依赖,形式如下:

```
dependencies {
    implementation(libs.gson)
    ...                                              //略
}
```

(3) 定义与 JSON 数据对应的实体类 Student,形式如下:

```
data class Student(val id: String,val name: String,val gender: String,val age: Int)
```

注意:使用 GSON 解析成对应的实体对象的过程中,实体类的属性名必须与 JSON 关键字的名字保持一致。如果实体类的属性与 JSON 的关键字不一致,可以通过@SerializedName 标注该属性指定对应的 JSON 关键字。这时,可以写成

```
import com.google.gson.annotations.SerializedName
data class Student(@SerializedName("id") val no: String,val name: String,
    val gender: String,val age: Int)
```

上述代码表示在序列化和反序列化时采用 no 转换 id 对应的 JSON 数据。

(4) 创建并调用 GSON 对象的 fromJson 函数将 JSON 数据转换为生成对象,代码如下:

```
val gson=Gson()
val student=gson.fromJson(jsonData,Student::class.java)
```

jsonData 为一个包含 JSON 数据的字符串,按照 Student 类的方式进行转换,可得到 student 对象。
如果 JSON 数据的形式比较复杂(如 JSON 数组形式),则包含一组数据。例如,JSON 数据如下所示:

```
[{"id": "60001223","name": "张三","gender": "男","age": 20},
 {"id": "60001224","name": "李四","gender": "男","age": 19},
 {"id": "60001113","name": "王五","gender": "男","age": 21}]
```

可以借助 TypeToken 进行数据转换。TypeToken 用于存储通用对象的类型,按照存储的数据类型进行转换,操作的代码可以写成下面的形式:

```
val types=object: TypeToken<List<Student>>(){}.type
val students=gson.fromJson<List<Student>>(jsonData,types)
```

例 8-5 要求结合 GSON 库解析例 8-3 提供的 2000—2010 年中国肉类消费量的 JSON 数据(原始

数据来源于经济合作与发展组织农业统计数据),并对数据格式进行加工处理。

因为要处理的 JSON 格式如下:

```
{"location": "中国","meat_consumption": [
    {"subject": "牛肉","measure": "每年人均消费千克数","time": "2000",
"value": "2.920215949"},
    {"subject": "牛肉","measure": "每年人均消费千克数","time": "2001",
"value": "2.989221358"},
    ...]
}
```

上述的 JSON 数据包括了两个顶层的关键字 location 和 meat_consumption。

meat_consumption 关键字映射的是一个 JSON 数组,而且 meat_consumption 的命名不符合 Kotlin 的命名规范,需要使用 @SerializedName 进行标注序列化和反序列化要使用的名称,因此定义的实体类,形式如下:

```
//模块 Ch08_05 定义实体类 MeatConsumption.kt
import com.google.gson.annotations.SerializedName
data class MeatConsumption(val location: String,@SerializedName("meat_consumption")
val meatConsumptionList: List<MeatConsumptionData>) {
    data class MeatConsumptionData(val subject: String,val measure: String,
        val time: Int,val value: Double)
}
```

定义 JSON 数据解析的实用类,代码如下:

```
//模块 Ch08_05 定义解析 JSON 数据的实用类 JSONUtils.kt
class JSONUtils {
    companion object{
        fun parseJSONData(jsonData: String): MeatConsumption{
            val gson=Gson()
            val meatConsumption=gson.fromJson<MeatConsumption>(jsonData,
                MeatConsumption::class.java)
            return meatConsumption
        }
    }
}
```

JSONUtils 采用了 GSON 库的 Gson 直接对 JSON 字符串进行解析生成对应的对象 MeatConsumption。与 Android 系统的 JSONObject API 相比较,GSON 库处理的方式更加简洁,代码如下:

```
//模块 Ch08_05 主活动测试 MainActivity.kt
class MainActivity: ComponentActivity() {
    override fun onCreate(savedInstanceState: Bundle?) {
        super.onCreate(savedInstanceState)
        enableEdgeToEdge()
        setContent {
            var meatConsumption: MutableState<MeatConsumption?>=remember{ mutableStateOf
                (null) }
            LaunchedEffect(Unit) {
                withContext(Dispatchers.IO){
```

```
            val jsonData=HttpUtils.getWebContent("http://10.0.2.2:5000/json/
            meat_consumption.json")                     //获得 JSON 数据字符串
            meatConsumption.value=JSONUtils.parseJSONData(jsonData)
                                                        //解析 jsonData
            Log.d("TAG","${meatConsumption.value?.location}消费情况")
            meatConsumption.value?.meatConsumptionList?.forEach {
                Log.d("TAG","${it.subject}在${it.time}消费：${it.value}${it.
                measure}")
            }
        }
    }
    Chapter08Theme {
        Scaffold(modifier=Modifier.fillMaxSize()) {
            innerPadding->Column(modifier=Modifier.padding(innerPadding)
                .fillMaxSize(),horizontalAlignment=Alignment.CenterHorizontally,
                verticalArrangement=Arrangement.Center){
                Text("2000-2010 年肉类消费情况",fontSize=24.sp)
            }
        }
    }
}
```

运行结果如图 8-7 所示。

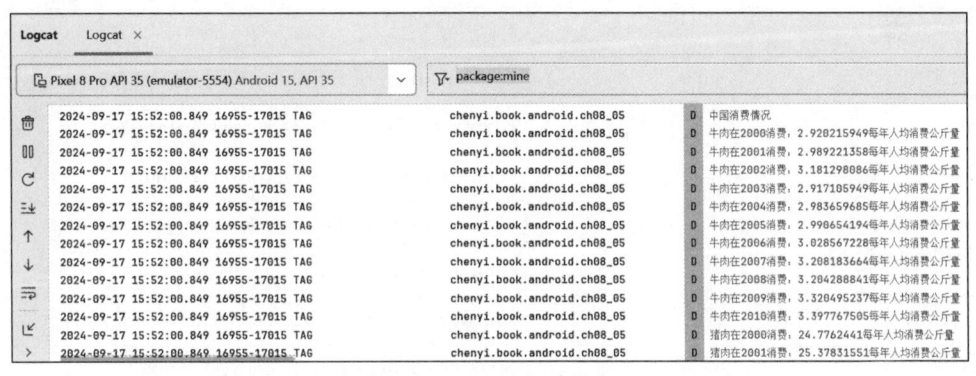

图 8-7 日志记录的结果

8.5 使用 Retrofit 库访问网络资源

在使用 Android 的 HttpURLConnection 访问网络资源时,往往需要处理过多的网络通信细节,例如创建网络连接、设置访问的方式,连接的时间,读取数据的实现,甚至输入输出流的处理、关闭网络连接等操作。只要一个环节存在问题,就会导致网络访问资源出现问题。

Square 公司推出的 Retrofit 库(https://square.github.io/retrofit/)改变了网络访问的方式。它实现了网络请求的封装。Retrofit 库采用了回调处理方式,使用该方式就可通过接口提交请求和相应的参数配置获得对应的响应,并将响应获得的数据解析成其他特定的格式。例如,将 JSON 数据解析成对象。与 HttpURLConnection API 一样,只要进行网络访问,就需要设置网络访问权限和明文访问许可。

具体操作参见 8.1 节。

要使用 Retrofit 库,需要在项目的 libs.versions.toml 文件中增加如下配置:

```
[versions]
retrofit="2.11.0"
[libraries]
retrofit={module="com.squareup.retrofit2:retrofit",version.ref="retrofit"}
retrofit-convert-gson={
    module="com.squareup.retrofit2:converter-gson",version.ref="retrofit"
}
```

然后在模块的构建文件 build.gradle.kts 中增加依赖:

```
dependencies {
    implementation(libs.retrofit)
    implementation(libs.retrofit.convert.gson)
    ...                                                                     //略
}
```

com.squareup.retrofit2:converter-gson 是 Retrofit 转换库,用于将 HTTP 请求获得的 JSON 数据借助 GSON 自动解析并转换成对象。Retrofit 还有一些其他的转换库,用于将其他的数据转换成对象,如表 8-3 所示。在开发过程中,可根据需要自行添加。

表 8-3 Retrofit 转换库列表

库 名	依 赖 库
Gson	com.squareup.retrofit2:converter-gson
Jackson	com.squareup.retrofit2:converter-jackson
Moshi	com.squareup.retrofit2:converter-moshi
Protobuf	com.squareup.retrofit2:converter-protobuf
Wire	com.squareup.retrofit2:converter-wire
Simple XML	com.squareup.retrofit2:converter-simplexml
JAXB	com.squareup.retrofit2:converter-jaxb
Scalars(基本类型、封装的、字符串)	com.squareup.retrofit2:converter-scalars

在依赖库的支持下,希望通过 HTTP 请求获得的响应,得到对应的对象数据。完成下列步骤。

(1) 定义实体,将请求的数据转换成对象的实体对象。例如,要访问的网络资源是 http://127.0.0.1:5000/json/students.json,则数据格式如下:

```
[{"id": "60001223","name": "张三","gender": "男","age": 20},
 {"id": "60001224","name": "李四","gender": "男","age": 19},
 {"id": "60001113","name": "王五","gender": "男","age": 21}
]
```

可以将对应 JSON 数据的关键字定义为 Student 实体类,代码如下:

```
//模块 Ch08_06 实体类 Student.kt
data class Student(val id: String,val name: String,val gender: String,
    val age: Int)
```

(2) 创建提供服务的接口类,定义业务服务的主要方法,代码如下:

```
//模块 Ch08_06 服务接口类 StudentService.kt
/** 定义 Service 接口 */
interface StudentService {
    @GET("students.json")
    fun getStudentList(): Call<List<Student>>
}
```

在上述的接口中,可以定义业务服务的各种形式的访问。StudentService 中通过@GET("students.json")设置请求访问服务器的 students.json 数据。getStudentList()的返回值 Call<List<Student>>,Call 表示向 Web 服务器发送请求获取响应的调用方法,要求从指定服务器响应数据需要转换成 List<Student> 的数据类型。

StudentService 接口只定义了 GET 的请求。访问网络还有多种形式,例如 POST、DELETE 等多种方式。这些不同的方式使得接口的访问地址会有所不同。表 8-4 对这些访问方式和接口访问地址做了总结归纳。表 8-4 中假设所访问的网络地址为 http://example.com/。

表 8-4　访问方式

方式	接口定义样例	访问地址和说明

GET	`interface StudentService{`
	`    @GET("students.json")`
	`    fun getStudent(): Call<Student>`
	`}`
		http://example.com/students.json 的接口地址是静态的

	`interface StudentService{`
	`    @GET("{page}/students.json")`
	`    fun getStudent(`
	`    @Path("page") page: Int): Call<Student>`
	`}`
		http://example.com/<page>/students.json 的接口地址是动态的。
		例如:
		http://example.com/1/students.json
		表示访问第一页的数据

	`interface StudentService{`
	`    @GET("{page}/students.json")`
	`    fun getStudent(`
	`    @Query("gender") gender: String,`
	`    @Query("age") age: Int: Call<Student>`
	`}`
		http://example.com/students.json?gender=<gender>&age=<age>的接口地址是动态的。
		例如:
		http://example.com/students.json?gender="男"&age=20
		表示检索性别男,年龄为 20 的所有记录

DELETE	`interface StudentService{`
	`    @DELETE("{id}")`
	`    fun delStudent(`
	`    @Path("id")id: String): Call<ResponseBody>`
	`}`
		执行格式:
		DELETE http://example.com/<id>
		例如:
		DELETE http://example.com/60001223
		表示请求删除 id=60001223 的记录。
		ResponseBody 表示服务器的响应

续表

方式	接口定义样例	访问地址和说明

POST	interface StudentService{
	　　@POST("new")
	　　fun newStudent(
	　　@Body student: Student): Call<
	　　ResponseBody>
	}
		执行对应的格式：
		POSThttp://example.com/create
		{"id":"29","name":"刘柳","gender":"女","age":20}
		表示当以 POST 方式提交数据给服务器，会将@Body 标注的 student 对象转换成 JSON 数据作为参数提交给服务器。服务器接收请求

	interface StudentService{
	　　@FormUrlEncoded
	　　@POST("update")
	　　fun updateStudent(@Field("name")
	　　name: String): Call<Student>
	}
		执行对应格式：
		POSThttp://example.com/update? name=<name>
		表示以 POST 提交表单数据给服务器

（3）定义创建服务对象，代码如下：

```kotlin
//模块 Ch08_06 创建 StudentServiceCreator.kt
object StudentServiceCreator {
    private val baseURL="http://10.0.2.2:5000/json/"      //访问网站的基址
    private val retrofit=Retrofit.Builder()               //创建 Retrofit 对象
        .baseUrl(baseURL)
        .addConverterFactory(GsonConverterFactory.create())
        .build()
    fun <T>createService(serviceClass: Class<T>)=retrofit.create(serviceClass)
}
```

因为在整个访问网络资源的过程中，只需要一个 Retrofit 对象来创建动态的服务代理。所以将 StudentServiceCreator 类定义为单例类，即该类只有一个对象就是它本身。通过 retrofit 对象来创建对应的业务服务实例对象。在 createService()函数中指定的是一个泛型类型参数，通过传递类型对象的实例，达到创建业务服务对象的目的。

（4）定义视图模型 StudentViewModel，处理访问网络获取数据的核心业务，代码如下：

```kotlin
//模块 Ch08_06 的 StudentViewModel.kt
//模块 Ch08_06 的 StudentViewModel.kt
class StudentViewModel: ViewModel() {
    private var _output:MutableStateFlow<List<Student>>=
        MutableStateFlow(mutableListOf<Student>())
    val output: StateFlow<List<Student>>=_output.asStateFlow()
    fun requestStudents(){
        val creator=StudentServiceCreator.createService(StudentService::class.java)
        creator.getStudentList().enqueue(
            object: Callback<List<Student>>{
                override fun onResponse(call: Call<List<Student>>,
                    response: Response<List<Student>>) {
                    //请求响应成功
                    response.body()?.also{
                        it:List<Student> -> _output.value=it
```

```
                }
            }
            override fun onFailure(call: Call<List<Student>>, t: Throwable) {
                                                                              //请求失败
                t.printStackTrace()
            }
        })
    }
}
```

(5) 为了让网络访问执行具有一定的可视性,定义对应 MainActivity 的主界面的可组合函数 MainScreen,代码如下:

```
//模块 Ch08_06 生成 MainActivity 的界面可组合函数 MainScreen
@Composable
fun MainScreen(modifier:Modifier,viewModel:StudentViewModel){
    val output=viewModel.output.collectAsState()
    Column(modifier=modifier.fillMaxSize(),
        horizontalAlignment=Alignment.CenterHorizontally,
        verticalArrangement=Arrangement.Center) {
        LazyColumn(modifier=Modifier.fillMaxWidth().wrapContentHeight()) {
            items(output.value) {
                Text("学号:${it.id}-姓名:${it.name}-性别:${it.gender}-年龄:${it.age}",
                    fontSize=20.sp)
            }
        }
        Button(onClick={viewModel.requestStudents()}) {Text("请求", fontSize=24.sp) }
    }
}
```

(6) 定义 MainActivity,代码如下:

```
//模块 Ch08_06 主活动 MainActivity.kt
class MainActivity : ComponentActivity() {
    override fun onCreate(savedInstanceState: Bundle?) {
        super.onCreate(savedInstanceState)
        enableEdgeToEdge()
        val viewModel:StudentViewModel=ViewModelProvider(this).get(StudentViewModel::
            class.java)
        setContent {
            Chapter08Theme {
                Scaffold(modifier=Modifier.fillMaxSize()) {
                    innerPadding ->MainScreen(modifier=Modifier.padding(innerPadding),
                        viewModel=viewModel)
                }
            }
        }
    }
}
```

运行结果如图 8-8 所示。

在 MainActivity 类的定义中可以发现 Button 中通过调用 StudentViewModel 对象 viewModel 的 requestStudents 函数实现了对网络的直接访问,并在主线程中直接修改了显示文本的内容。这与前面访问网络资源必须定义一个新线程,并结合消息处理机制更新 UI 形成反差。

实际运行时,当调用 creator.getStudentList 函数时,返回 Call＜List＜Student＞＞的对象,然后再调用这个对象的 enqueue()函数。在具体执行过程中,在 Retrofit 请求标注中设置网址,并通过 HTTP 请求网络资源。请求发生时,Retrofit 会自动开启内部的子线程完成网络请求,服务器做出的相应响应会被回调到 enqueue 函数传入的 Callback 中,对网络访问的响应做出 onResponse(响应成功)或 onFailure(响应失败)的回调处理,然后由 Retrofit 自动切换成主线程。由于 Retrofit 将线程的自动切换进行了封装,因此在上述代码中,好似主线程承担了所有的任务。实际情况是,Retrofit 自动对多线程进行管理,完成了移动应用发起的 HTTP 请求,从服务器获得响应以及将响应的数据自动转换成列表对象等一系列的操作。

例 8-6 以条状图的方式显示 2000—2010 年中国肉类消费量(原始数据来源于经济合作发展组织农业统计数据),并对数据加工处理。

图 8-8 访问的运行结果

JSON 数据可以从 http://127.0.0.1:5000/json/meat_consumption.json 本地服务器中获得,参见例 8-5。

因为本项目访问的是在线资源,因此必须在 AndroidManifest.xml 中配置网络访问权限和明文访问许可。本例使用了 Retrofit 库和 MPAndroidChart 库。其中,MPAndroidChart(https://github.com/PhilJay/MPAndroidChart)是图表库,用于 Android 应用绘制图表。在项目的 libs.versions.toml 配置 Retrofit 库和 MPAndroidChart 库的版本信息,形式如下:

```
[versions]
retrofit="2.11.0"
mpchart="v3.1.0"
[libraries]
retrofit={module="com.squareup.retrofit2:retrofit",version.ref="retrofit"}
retrofit-convert-gson={module="com.squareup.retrofit2:converter-gson",
    version.ref="retrofit"}
mpandroidchart={module="com.github.PhilJay:MPAndroidChart",version.ref="mpchart"}
```

在模块的配置构建文件 build.gradle.kts 增加相应的配置以及两个库的依赖,形式如下:

```
dependencies {
    implementation(libs.retrofit)
    implementation(libs.retrofit.convert.gson)
    implementation(libs.mpandroidchart)
    ...                                                                    //略
}
```

处理的数据对应的实体类,代码如下:

```kotlin
//模块Ch08_07定义实体类MeatConsumption.kt
data class MeatConsumption(
    val location: String,
    @SerializedName("meat_consumption")meatConsumptionList: List<MeatConsumptionData>) {
    data class MeatConsumptionData(val subject: String,val measure: String,val
        time: Int,val value: Double)
}
```

实体类属性定义与获取的JSON数据大体一致,只是对JSON关键字meat_consumption在类的属性中指定转换为meatConsumption以符合Kotlin的命名规范,代码如下:

```kotlin
//模块Ch08_07定义接口MeatConsumptionService.kt
interface MeatConsumptionService {
    @GET("meat_consumption.json")
    fun getMeatConsumptions(): Call<MeatConsumption>
}
```

定义网络访问服务构建器ConsumptionServiceCreator被定义成单一模式,是对象类,即该类只有一个对象。该类构建了Rotrofit对象retrofit,指定访问网站的基址http://10.0.2.2:5000/json,然后指定转换器为GSON转换器。ConsumptionServiceCreator定义的create方法获得一个指定泛型类型的服务对象,代码如下:

```kotlin
//模块Ch08_07定义服务器构建器ConsumptionServiceCreator.kt
object ConsumptionServiceCreator {
    private val retrofit=Retrofit.Builder()
        .baseUrl("http://10.0.2.2:5000/json/")
        .addConverterFactory(GsonConverterFactory.create())
        .build()
    fun <T>create(serviceClass:Class<T>):T=retrofit.create(serviceClass)
}
```

定义视图模型类MeatViewModel,利用它实现访问在线资源,获取生成柱状图的数据。因为要显示的柱状图的数据是固定的,所以MeatViewModel类中定义init代码段来访问在线资源。这样就实现了一次访问,后续通过肉的不同类别直接获取数据,代码如下:

```kotlin
class MeatViewModel: ViewModel() {
    val beefList=SnapshotStateList<BarEntry>()         //定义一组牛肉的条形数据
    val pigList=SnapshotStateList<BarEntry>()          //定义一组猪肉的条形数据
    val poultryList=SnapshotStateList<BarEntry>()      //定义一组家禽的条形数据
    val sheepList=SnapshotStateList<BarEntry>()        //定义一组羊肉的条形数据
    var _dataList:MutableStateFlow<List<BarEntry>?>=MutableStateFlow(null)
    val dataList:StateFlow<List<BarEntry>?>=_dataList.asStateFlow()
    init{
        requestFrmHttp()                               //初始化就访问在线资源
    }
    fun requestFrmHttp() {
        val creator=ConsumptionServiceCreator.create(MeatConsumptionService::
            class.java)
```

```
            creator.getMeatConsumptions().enqueue(object: Callback<MeatConsumption>{
                override fun onResponse(call: Call<MeatConsumption>,
                    response: Response<MeatConsumption>) {
                    val meatConsumption =response.body()
                    meatConsumption?.meatConsumptionList?.forEach {
                        when(it.subject){
                            "牛肉"->beefList.add(BarEntry(1+beefList.size.toFloat(),
                                it.value.toFloat(),it.time))
                            "猪肉"->pigList.add(BarEntry(1+pigList.size.toFloat(),
                                it.value.toFloat(),it.time))
                            "家禽"->poultryList.add(BarEntry(1+poultryList.size.toFloat(),
                                it.value.toFloat(),it.time))
                            "羊肉"->sheepList.add(BarEntry(1+sheepList.size.toFloat(),
                                it.value.toFloat(),it.time))
                        }
                    }
                }
                override fun onFailure(call: Call<MeatConsumption>, t: Throwable) {
                    t.printStackTrace()
                }
            })
        }
        fun requestBy(subject:String){ //按照不同类别获取柱状图需要的数据
            when(subject){
                "牛肉"->_dataList.value=beefList
                "羊肉"->_dataList.value=sheepList
                "家禽"->_dataList.value=poultryList
                "猪肉"->_dataList.value=pigList
            }
        }
    }
```

在这个例子中,需要使用 MPAndroidChart 库的 BarChart 控件。BarChart 控件是一个传统的 Android View 层次结构的控件,要使用 BarChart 绘制柱状图,需要执行绘制柱状图如下的主要的过程。

(1) 创建 BarChart 对象,并初始化,设置相应的参数,代码如下:

```
barChart.setDrawBorders(false)              //不显示边框
val description=Description()               //右下角的描述内容不显示
description.isEnabled=false
barChart.description=description
```

(2) 创建一系列的 BarEntry 对象,此处添加 BarEntry 对象的(X,Y)值,代码如下:

```
var bar1=BarEntry(1+beefList.size.toFloat(),it.value.toFloat(),it.time)
val list=mutableListOf<BarEntry>()
list.add(bar1)
```

(3) 创建 BarDataSet 对象,将一组 BarEntry 对象添加到 BarDataSet 对象中,代码如下:

```
val dataSet=BarDataSet(list,"Label")
dataSet.color=resources.getColor(android.R.color.holo_blue_dark,null)
```

(4) 创建 BarData 对象，添加 BarDaraSet 对象，代码如下：

```
val data=BarData(dataSet)
```

(5) 显示柱状图，代码如下：

```
barChart.setData(data)
```

如果修改 barChart 界面，可以调用 barChart.invalidate 函数来实现。可以通过 AndroidView 中的可组合项来渲染传统的 Android View 层次结构的控件。结合 AndroidView 显示 MPAndroidChart 库的 BarChart 控件可以处理成如下形式：

```
AndroidView(factory={
    BarChart(it).apply {       //初始化控件
        …                      //略
    }},
    update={                   //内部数据变更修改控件界面
        …                      //略
        invalidate()           //重新绘制
})
```

定义界面的 MainScreen 可组合函数。在 MainScreen 中通过 AndroidView 渲染 BarChart 控件，代码如下：

```
//模块 Ch08_07 模块可组合函数 MainScreen
@Composable
fun MainScreen(modifier:Modifier,viewModel:MeatViewModel) {
    val dataList=viewModel.dataList.collectAsState()
    var subject by remember { mutableStateOf("") }
    var dataSet: MutableState<BarDataSet?>=remember { mutableStateOf(null) }
    val context=LocalContext.current
    val colorMap: Map<String, Int>=mapOf(
        "牛肉" to android.R.color.holo_green_light,
        "羊肉" to android.R.color.holo_orange_light,
        "家禽" to android.R.color.holo_red_light,
        "猪肉" to android.R.color.holo_blue_light)              //类别映射颜色
    LaunchedEffect(subject) {                                    //根据类别不同处理
        if (subject.isNotBlank()) {
            dataList.value?.run {
                dataSet.value=BarDataSet(this, subject)          //生成柱状图的 BarDataSet
                dataSet.value?.color=context.getColor(colorMap.get(subject)!!)
                                                                 //设置颜色
            }
        }
    }
    Column(modifier=modifier.fillMaxSize(),
        horizontalAlignment=Alignment.CenterHorizontally,
```

```kotlin
            verticalArrangement=Arrangement.Center) {
            Text("2000-2010年中国人均肉类消费", fontSize=28.sp)
            AndroidView(modifier=Modifier.size(600.dp, 800.dp), factory={
                BarChart(it).apply {
                    setDrawBorders(false)                       //不显示边框
                    description=Description()                   //右下角的描述内容不显示
                    description.isEnabled=false
                    val barChart=this
                    dataSet.value?.apply{
                        barChart.data=BarData(dataSet.value)    //设置数据
                    }
                } },
                update={
                    val barChart=it
                     dataSet.value?.also {
                        barChart.data=BarData(dataSet.value)
                     }
                    barChart.notifyDataSetChanged()             //通知数据发生变更
                    barChart.invalidate()                       //重新绘制
                })
            Row(modifier=Modifier.fillMaxWidth()) {
                Button(onClick={
                    subject="牛肉"
                    viewModel.requestBy(subject)
                }) {Text("牛肉消费")}
                Button(onClick={
                    subject="猪肉"
                    viewModel.requestBy(subject)
                }) {Text("猪肉消费")}
                Button(onClick={
                    subject="家禽"
                    viewModel.requestBy(subject)
                }) {Text("家禽消费") }
                Button(onClick={
                    subject="羊肉"
                    viewModel.requestBy(subject)
                }){Text("羊肉消费")}
            }
        }
    }
}
```

在 MainActivity 中调用 MainScreen 可组合函数,生成主界面,代码如下:

```kotlin
//模块 Ch08_07 定义主活动 MainActivity.kt
class MainActivity: ComponentActivity() {
    override fun onCreate(savedInstanceState: Bundle?) {
        super.onCreate(savedInstanceState)
        val viewModel:MeatViewModel=ViewModelProvider(this).get(MeatViewModel::class
            .java)
        enableEdgeToEdge()
        setContent {
```

```
            Chapter08Theme {
                Scaffold(modifier=Modifier.fillMaxSize()) {
                    innerPadding->MainScreen(modifier=Modifier.padding(innerPadding),
                    viewModel=viewModel)
                }
            }
        }
    }
}
```

运行结果如图 8-9 所示。

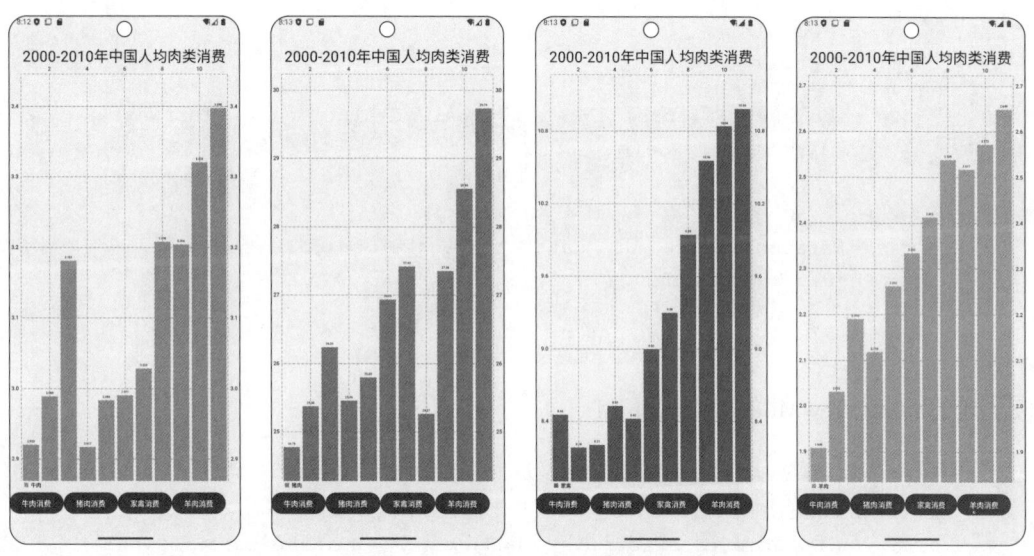

图 8-9　2000—2010 年中国人均肉类消费条状图显示结果

上述 MainActivity 可通过按钮实现不同肉类消费人均水平的柱状图切换。

在实际应用中,可以使用 WorkManager 组件在后台处理对于网络资源的访问。首先定义一个实体类 TableData,将访问的数据包含在 TableData 的 4 个列表中,代码如下:

```
data class TableData(val beefList:MutableList<BarEntry>=mutableListOf<BarEntry>(),
    val pigList:MutableList<BarEntry>=mutableListOf<BarEntry>(),
    val poultryList:MutableList<BarEntry>=mutableListOf<BarEntry>(),
    val sheepList:MutableList<BarEntry>=mutableListOf<BarEntry>())
```

定义在线访问的后台任务 LoadDataWork,代码如下:

```
class LoadDataWork (val context: Context, params: WorkerParameters): Worker(context,
    params){
    @SuppressLint("RestrictedApi")
    override fun doWork(): Result {
        val tableData=TableData()
        val creator=ConsumptionServiceCreator.create(MeatConsumptionService::class.java)
        creator.getMeatConsumptions().enqueue(object: Callback<MeatConsumption>{
            override fun onResponse(call: Call<MeatConsumption>,
                response: Response<MeatConsumption>) {
```

```
            val meatConsumption=response.body()
            meatConsumption?.meatConsumptionList?.forEach {
                when(it.subject){
                    "牛肉"->tableData.beefList.add(BarEntry(1+tableData.beefList
                        .size.toFloat(),it.value.toFloat(),it.time))
                    "猪肉"->tableData.pigList.add(BarEntry(1+tableData.pigList
                        .size.toFloat(),it.value.toFloat(),it.time))
                    "家禽"->tableData.poultryList.add(BarEntry(1+tableData.poultryList
                        .size.toFloat(),it.value.toFloat(),it.time))
                    "羊肉"->tableData.sheepList.add(BarEntry(1+tableData.sheepList
                        .size.toFloat(),it.value.toFloat(),it.time))
                }
            }
        }
        override fun onFailure(call: Call<MeatConsumption>, t: Throwable) {
            t.printStackTrace()
        }
    })
    val data=Data.Builder().put("data",tableData).build()
    return Result.Success(data)
  }
}
```

修改上述的 MeatViewModel,代码如下:

```
class MeatViewModel: ViewModel() {
    var data:TableData?=null
    var _dataList:MutableStateFlow<List<BarEntry>?>=MutableStateFlow(null)
    val dataList:StateFlow<List<BarEntry>?>=_dataList.asStateFlow()
    init{
        requestFrmHttp()                                    //初始化就访问在线资源
    }
    fun requestFrmHttp() {
        val constraints: Constraints=Constraints.Builder()
            .setRequiredNetworkType(NetworkType.CONNECTED)   //需要联网
            .build()
        val request=OneTimeWorkRequestBuilder<LoadDataWork>()
            .setConstraints(constraints)                     //设置约束
            .build()
        WorkManager.getInstance(context).enqueue(request)
        WorkManager.getInstance(context).getWorkInfoByIdLiveData(request.id)
            .observe(context as MainActivity) {
                if (it!=null&&it.state==WorkInfo.State.SUCCEEDED) {
                    data=it.outputData.keyValueMap.get("data") as TableData
                                                             //接收传递的数据
                }
            }
    }
}
```

```
fun requestBy(subject:String){          //按照不同类别获取柱状图需要的数据
    when(subject){
        "牛肉"->_dataList.value=data?.beefList
        "羊肉"->_dataList.value=data?.sheepList
        "家禽"->_dataList.value=data?.poultryList
        "猪肉"->_dataList.value=data?.pigList
    }
}
```

访问在线资源必须设置联网 NetworkType.CONNECTED 限制约束。在测试时，模拟器无法正常运行，必须使用真机模拟。读者可以自行修改上述代码进行测试。

8.6 智能聊天移动应用实例

8.6.1 功能需求分析和设计

智能聊天移动应用实例是结合在线聊天机器人的一个移动应用。它可以实现用户与机器人聊天并展示聊天的信息。当前，已有很多在线聊天机器人，"青云客网络"免费登录提供的人工智能聊天产品就是其中之一。"青云客网络"提供的接入人工智能聊天请求地址是 http://api.qingyunke.com/api.php，请求的方式是 GET 方式，需要提交的必选参数 key 设定为 free(表示免费)，appid 为对应的移动应用编号(是可选项，0 表示智能识别)，必选 msg 参数表示聊天的内容。

本智能聊天移动应用是一款结合"青云客网络"的人工智能聊天机器人实现的应用，具体要求的功能如下。

(1) 实现基本的聊天功能。
(2) 可以定义本移动应用的帮助信息，可查看聊天机器人的基本介绍信息。
(3) 可以定义应用的功能配置，修改移动应用字体的大小和背景图片。

这个智能聊天移动应用的主要功能包括"智能聊天""系统配置""查看帮助"。图 8-10 所示为用例图。

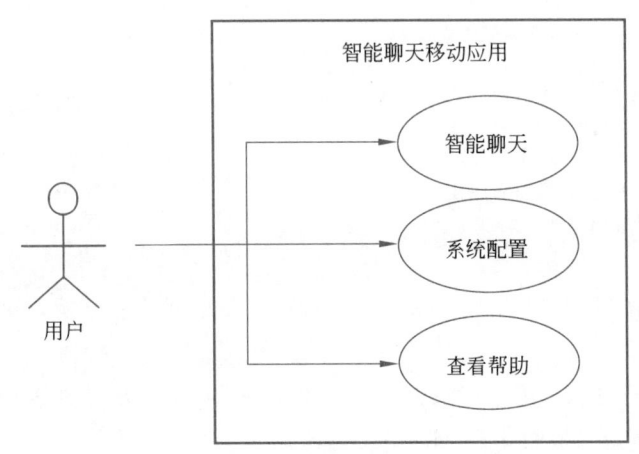

图 8-10 智能聊天移动应用的用例图

1. 智能聊天

智能聊天是本移动应用的核心功能。当启动智能聊天移动应用后，用户就可以通过智能聊天模块与智能聊天机器人对话。图 8-11 所示为智能聊天的时序图。

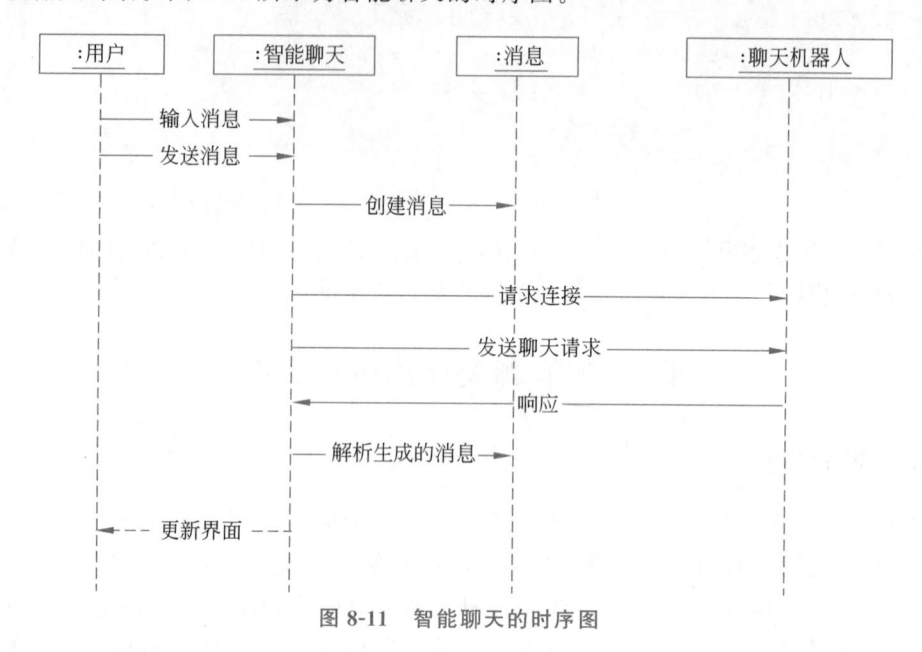

图 8-11 智能聊天的时序图

2. 系统配置

系统配置是对移动应用的界面进行配置，如图 8-12 所示。具体包括字体大小调整、背景图片选择等。

3. 查看帮助

查看帮助主要用于为用户提供移动应用的使用说明和功能介绍，如图 8-13 所示。

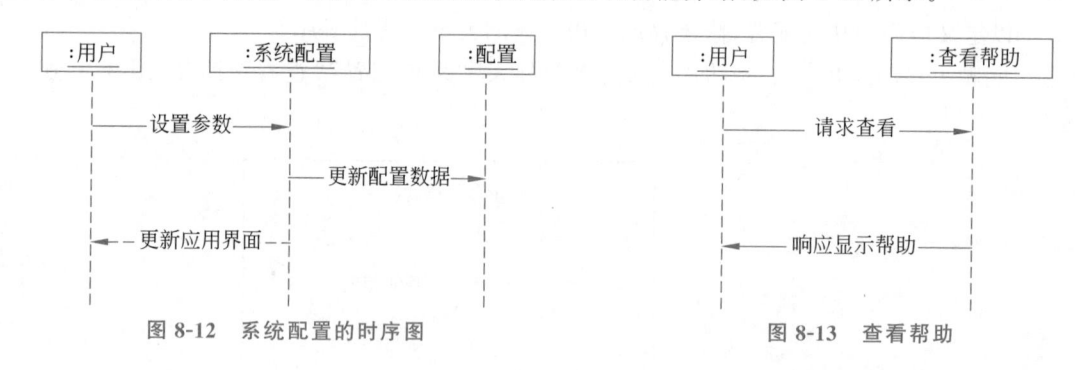

图 8-12 系统配置的时序图　　　　　图 8-13 查看帮助

4. 类设计

本移动应用开发采用 MVI(Model View Intent)模式，并根据功能分析，可以设计出如图 8-14 所示的类图。

(1) Message 是一个数据类，用于记录聊天的基本信息，包含聊天类别分为发送的聊天信息和接收的聊天信息。聊天的类别定义成枚举类 MessageType。

(2) ReceivedMsg 是一个数据类，表示从网络中获得的机器人聊天的数据。

(3) ConfigData 定义配置界面的相关状态数据，包括发送和接收的聊天消息，聊天界面字体的配

图 8-14 类图

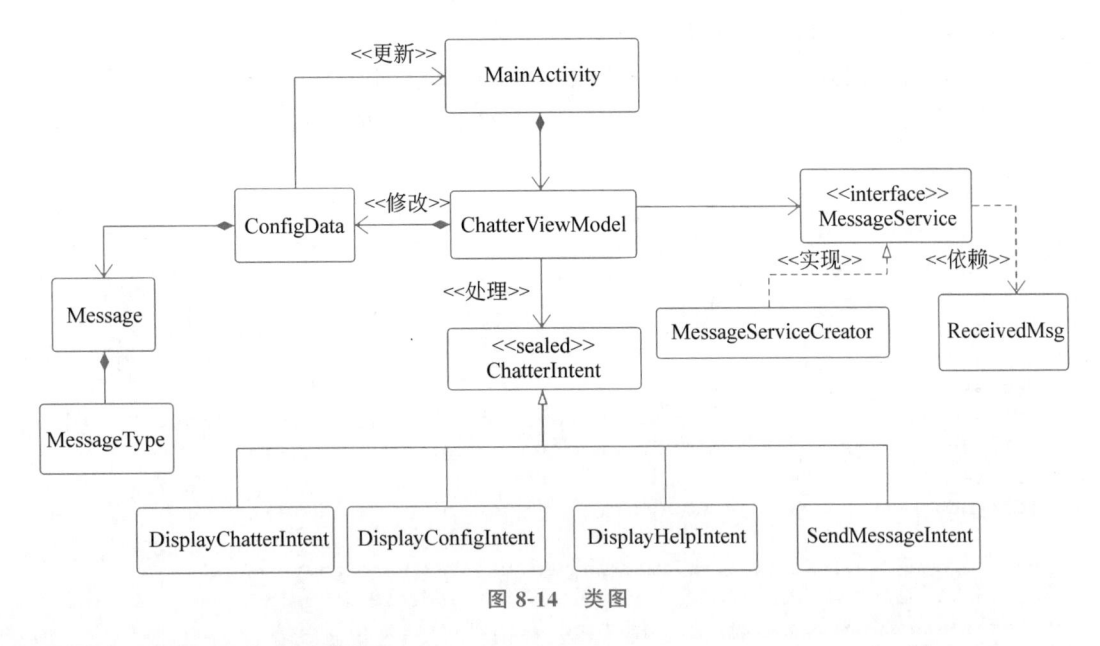

置,聊天背景图片的配置等相关数据。

(4) ChatterIntent 是一个密封类,定义要处理的意图。它具有 4 个子类:DisplayChatterIntent 用于表示进入聊天界面的意图;DisplayConfigIntent 用于表示进入配置界面的意图;DisplayHelpIntent 用于表示进入帮助界面的意图;SendMessageIntent 用于表示发送聊天信息给聊天服务器的意图。

(5) MessageService 是一个接口,定义了网络访问请求的接口。

(6) MessageServiceCreator 是一个利用 Retrofit2 库访问网络的单例类,用于网络访问。

(7) ChatterViewModel 是一个视图模型组件。一方面,它为视图提供状态数据,达到更新界面的目的;另一方面,它根据意图类别,处理不同意图的业务逻辑,修改状态数据。

(8) MainActivity 用于对移动应用的各个界面进行管理。主要处理界面的切换。在 MainActivity 中切换的可组合函数定义的界面如下。

- HomeScreen 可组合函数定义的是主界面,通过 Scaffold 脚手架结构定义整个界面的框架。在 HomeScreen 可组合函数中还调用 BottomMenu 可组合函数,用来定义底部的导航栏。
- ChatterScreen 可组合函数定义的是聊天的界面,定义聊天的文本输入框和显示已经发送和接收聊天信息的列表内容。在 ChatterScreen 可组合函数中调用 MessageCard 可组合函数实现聊天消息的列表显示。MessageCard 可组合项中根据消息类别的不同,调用 RightMessageCard 和 LeftMessageCard 可组合函数,将消息发送方和消息接收方在左右两部显示。
- HelpScreen 可组合函数定义的是展示帮助网页的界面。
- ConfigScreen 可组合函数定义的配置聊天界面的相关参数,包括背景图片、文字大小。ConfigScreen 还包含 FontConfigView、BgConfigView 和 ImageCard 可组合函数,它们分别表示配置字体、配置图片的局部可组合项。

8.6.2 系统的实现

本系统是基于互联网的移动应用,所有的智能聊天数据均来自网络。需要在配置相应的网络权限以及设置明文访问许可为"真"。在创建的 ChatterApp 模块中,需要对项目配置清单 AndroidManifest

.xml 进行系统配置网络权限和明文访问许可,参见 8.1 节。

本项目中在开发的过程中使用了 Retrofit 框架,可实现基于 HTTP 访问在线资源,并结合 RxJava 框架实现并发处理。为了导航处理方便,采用 JetPack Compose 的 Navigation 组件。因此在项目的 libs.versions.toml 文件中进行依赖库的版本管理,代码如下:

```toml
[versions]
navigationCompose="2.7.7"
retrofit="2.11.0"
rxjava="3.1.9"
rxandroid="3.0.2"
[libraries]
retrofit={module="com.squareup.retrofit2:retrofit",version.ref="retrofit"}
retrofit-convert-gson={module="com.squareup.retrofit2:converter-gson",
    version.ref="retrofit"}
retrofit-adpater-rxjava3={module="com.squareup.retrofit2:adapter-rxjava3",
    version.ref="retrofit"}
rxjava3={group="io.reactivex.rxjava3",name="rxjava",version.ref="rxjava"}
rxandroid={group="io.reactivex.rxjava3",name="rxandroid",version.ref="rxandroid"}
androidx-navigation-compose={ group ="androidx.navigation",
    name="navigation-compose", version.ref="navigationCompose" }
```

对模块的构建配置文件 build.gradle.kts 文件增加相应的依赖,形式如下:

```kotlin
dependencies {
    implementation(libs.retrofit)
    implementation(libs.retrofit.convert.gson)
    implementation(libs.retrofit.adpater.rxjava3)
    implementation(libs.rxjava3)
    implementation(libs.rxandroid)
    implementation(libs.androidx.navigation.compose)
    …                        //略
}
```

整个项目比较简单,主要有 3 个界面:"智能聊天""系统配置""系统帮助"。在这些界面中需要状态数据更新界面。

1. 定义数据类

定义聊天消息的枚举类 MessageType,表示聊天消息的类别,代码如下:

```kotlin
enum class MessageType{
    CLIENT_IN,                                                      //输入信息
    CLIENT_OUT                                                      //输出信息
}
```

定义聊天消息类 Message,表示聊天消息,代码如下:

```kotlin
data class Message(val name:String="",                              //消息发送者的名称
    val content:String="",                                          //消息发送的内容
    var date: LocalDateTime=LocalDateTime.now(),                    //消息发送的时间
    val type:MessageType=MessageType.CLIENT_OUT,                    //消息发送的类别
    var iconId:Int=R.mipmap.person)                                 //消息发送者图标
```

定义从"青云客"接收的聊天数据,用来映射对应 JSON 响应的数据,代码如下:

```
data class ReceivedMsg(val result:Int,                          //图标
    @SerializedName("content") val message:String)              //接收聊天信息
```

定义 ConfigData 类。它用来表示修改界面的状态的数据,代码如下:

```
data class ConfigData(var fontSize: TextUnit=20.sp,             //字体大小
    var fontType:String="中号",                                  //字体类型
    var bgImageId:Int=R.mipmap.bg5,                             //背景图片编号
    var bgImageName:String="背景 5",                             //背景图片的名称
    var messageContent:String="",                               //最后发送消息
    var route:String="homePage",                                //导航路径,默认为首页
    val messages: SnapshotStateList<Message>=mutableStateListOf()   //聊天的所有信息
)
```

2. 定义要处理业务的意图

在聊天移动应用中,主要的业务包括显示聊天界面、显示配置聊天的界面、显示帮助界面,以及处理聊天发送和接收的业务。根据这些业务需求,定义 ChatterIntent 密封类,并在密封类中定义了 4 个子类,分别表示不同的业务意图,代码如下:

```
sealed class ChatterIntent {
    data class SendMessageIntent(val content:String):ChatterIntent()    //聊天意图
    data object DisplayChatterIntent:ChatterIntent()                    //显示聊天界面意图
    data object DisplayConfigIntent:ChatterIntent()                     //显示配置界面意图
    data object DisplayHelpIntent:ChatterIntent()                       //显示帮助界面意图
}
```

3. 获取网络聊天数据

本移动应用利用"青云客网络"的"人工智能聊天"机器人,需要将聊天的信息基于 HTTP 发送到"青云客网络"所在的服务器。然后根据响应反馈的 JSON 数据解析和处理成移动终端可以接收的消息对象。

定义聊天消息服务访问接口 MessageService,MessageService 接口中返回的数据是 Observable 的 Flowable 对象,代码如下:

```
interface MessageService {
    @Headers("Accept: application/json")                        //定义请求头
    @GET("api.php")
    fun getMessage(@Query("key") key:String,                    //关键字
        @Query("appid") appid:Int,                              //应用编号
        @Query("msg") msg:String): Flowable<ReceivedMsg>        //返回消息
}
```

MessageServiceCreator 是创建 MessageService 对象的构建者,它结合 Retrofit2 和 RxJava3 处理实际的网络访问的消息服务创建者,代码如下:

```
object MessageServiceCreator {
    private const val QINGYUNKE_URL="http://api.qingyunke.com/"
    private val retrofit=Retrofit.Builder()
        .baseUrl(QINGYUNKE_URL)
        .addConverterFactory(GsonConverterFactory.create())
        .addCallAdapterFactory(RxJava3CallAdapterFactory.create())
        .build()
    fun <T>createService(serviceClass:Class<T>):T=retrofit.create(serviceClass)
}
```

在上述代码中，除了利用 GsonConverterFactory.create()通过 GSON 处理将 JSON 字符串转换成实体对象 ReceivedMsg。还增加了 RxJava3CallAdapterFactory.create()处理将 ReceivedMsg 封装到 Flowable，获得 Flowable 实例对象。

4. 定义视图模型

视图模型类 ChatterViewModel 是核心组件，它是实际处理业务意图的操作者，并为视图提供变更的状态数据，达到重组界面的目的，代码如下：

```
class ChatterViewModel: ViewModel() {
    var _configState:MutableStateFlow<ConfigData>=MutableStateFlow(ConfigData())
                                                                                //配置信息
    val configOutput:StateFlow<ConfigData>=_configState.asStateFlow()
    private val creator: MessageService=
        MessageServiceCreator.createService(MessageService::class.java)
    fun handleIntent(intent: ChatterIntent) {                                    //处理业务意图
        when (intent) {
            is ChatterIntent.DisplayChatterIntent->doChatter()                   //处理显示聊天界面
            is ChatterIntent.DisplayHelpIntent->doHelp()                         //处理显示帮助界面
            is ChatterIntent.DisplayConfigIntent->doConfig()                     //处理显示配置界面
            is ChatterIntent.SendMessageIntent->doSendMessage(intent.content)
        }
    }
    fun doChatter(){                                                             //显示聊天界面
        val currentState=_configState
        val newState=currentState.value.copy(route="chatterPage")
        _configState.value=newState                                              //变更状态
    }
    fun doHelp(){                                                                //设置帮助路由
        val currentState=_configState
        val newState=currentState.value.copy(route="helpPage")
        _configState.value=newState                                              //变更状态
    }
    fun doConfig(){                                                              //设置配置路由
        val currentState=_configState
        val newState=currentState.value.copy(route="configPage")
        _configState.value=newState
    }
    fun doSendMessage(messageContent:String){                                    //发送消息
        val currentState=_configState
        val newState=currentState.value.copy(messageContent=messageContent
```

```
            val sendMessage=Message("我",messageContent, LocalDateTime.now(),
                MessageType.CLIENT_OUT)
            newState.messages.add(sendMessage)
            var message:Message?=null
            val flowable=creator.getMessage("free",0,newState.messageContent)
                                                                    //访问网络资源
            flowable.subscribeOn(Schedulers.newThread())
                .observeOn(Schedulers.io())
                .doOnNext {
                    it: ReceivedMsg ->message=Message("青云客",it.message,LocalDateTime.now(),
                        MessageType.CLIENT_IN, R.mipmap.robot)
                }
                .observeOn(AndroidSchedulers.mainThread())
                 .subscribe {
                    if (message!=null) {
                        newState.messages.add(message!!)
                         _configState.value=newState                 //变更状态
                    }
                }
        }
}
```

上述定义中，doSendMessage 函数通过 HTTP 访问接口对网络资源进行访问，结合 Retrofit2 获得 Flowable 对象。对这个 Flowable 对象创建线程实现对数据的解析，将生成的 Message 加入消息队列中。然后再到 Android 的主线程中通知适配器的数据已发生变更，修改显示私有 MutableStateFlow 数据_configState，从而达到为界面提供的输出的 StateFlow 对象 ConfigOuput 也随之发生变更。这样的处理方式，使得网络处理和数据的访问以及界面的更新处理更加流畅、简单、方便。

5．定义智能聊天界面

在聊天界面中主要包括聊天信息文本输入框和显示已发送和已接收的聊天信息列表内容。根据消息类别的不同，设计成将发送消息在右边显示，而接收的消息在左边显示。

定义发送消息的 RightMessageCard 可组合函数，代码如下：

```
@Composable
fun RightMessageCard(message: Message, fontSize: TextUnit) {      //当前用户输入
    Row(modifier=Modifier.fillMaxWidth().background(Color.Transparent),
        horizontalArrangement=Arrangement.End) {
        Column{
            Text(text="${message.date}",fontSize=14.sp,color=Color.Black)
            Text(modifier=Modifier.align(Alignment.End),
                text="${message.content}",
                fontSize=fontSize,color=Color.Black)
        }
        Column(horizontalAlignment=Alignment.CenterHorizontally){
            Image(modifier=Modifier.size(60.dp, 60.dp).clip(CircleShape).background
                (Color.Blue),
                painter=painterResource(id=message.iconId),
                contentDescription=null)
            Text(text="${message.name}",
```

定义接收消息的 LeftMessageCard 可组合函数，代码如下：

```
@Composable
fun LeftMessageCard(message: Message, fontSize: TextUnit){              //接收其他客户的输入
    Row(modifier=Modifier
        .fillMaxWidth()
        .padding(start=2.dp)
        .background(Color.Transparent)){
            Column(horizontalAlignment=Alignment.CenterHorizontally){
                Image(modifier=Modifier.size(60.dp, 60.dp).clip(CircleShape).background
                    (Color.Blue),painter=painterResource(id=message.iconId),
                    contentDescription=null)
                Text(text="${message.name}",fontSize=fontSize,color=Color.Black)
            }
            Column{
                Text(text="${message.date}",fontSize=14.sp,color=Color.Black)
                Text(text="${message.content}",fontSize=fontSize,color=Color.Black)
            }
        }
}
```

将上述的两种不同类别的消息统一封装为 MessageCard，可以根据消息类别的不同调用不同的可组合函数生成聊天消息的列表单项，代码如下：

```
@Composable
fun MessageCard(message: Message, fontSize: TextUnit){
    Card(colors=CardDefaults.cardColors(containerColor=Color.Transparent,
        contentColor=Color.Green)){
        if (message.type==MessageType.CLIENT_OUT){
            RightMessageCard(message=message, fontSize=fontSize)
        } else {
            LeftMessageCard(message=message, fontSize=fontSize)
        }
    }
}
```

然后定义显示聊天界面的 ChatterScreen 可组合函数，代码如下：

```
@Composable
fun ChatterScreen(modifier:Modifier,viewModel:ChatterViewModel) {
    val configState=viewModel.configOutput.collectAsState()               //获取状态
    var content by remember{mutableStateOf("你好") }                       //输入文本
    val state=rememberLazyListState()
    val scope=rememberCoroutineScope()
```

```kotlin
Box(modifier=modifier.fillMaxSize()){
    Image(modifier=Modifier.fillMaxSize(),
        painter=painterResource(id=configState.value.bgImageId),
        contentScale=ContentScale.Crop,
        contentDescription=null)
    Column{
        LazyColumn(state=state,userScrollEnabled=true,      //显示消息列表
            modifier=Modifier.fillMaxWidth().height(750.dp)){
            items(configState.value.messages){
                it:Message->MessageCard(it,configState.value.fontSize)
                                                            //显示消息单项
            }
        }
        Row(modifier=Modifier.fillMaxWidth().wrapContentHeight(),
            horizontalArrangement=Arrangement.Start,
            verticalAlignment=Alignment.Bottom){
            TextField(colors=TextFieldDefaults.colors(      //设置输入框颜色
                unfocusedContainerColor=Color.Transparent,
                focusedContainerColor=Color.Transparent,
                unfocusedTextColor=Color.Black,
                focusedTextColor=Color.Black),
                modifier=Modifier.fillMaxWidth(),
                value=content,
                onValueChange={it:String->content=it},trailingIcon={
                    IconButton(onClick={
                        scope.launch {                      //处理发送聊天信息业务
                            viewModel.handleIntent(ChatterIntent.SendMessageIntent
                                (content))
                            state.animateScrollToItem(configState.value.messages.
                                size,1)
                        }
                    }){Icon(imageVector=Icons.Filled.Send,contentDescription=null) }
            }
        )
        }
    }
}
```

6. 定义系统配置界面

本系统中实现了简易的字体大小的变换和聊天背景的切换功能。利用 ConfigData 类的 bgImageId 和 fontSize 属性接受变更的背景图片编号和字体大小。系统配置界面由多个组合函数构成。

定义字体配置的可组合函数 FontConfigView，代码如下：

```kotlin
@Composable
fun FontConfigView(configData: ConfigData){
    var fontType by remember { mutableStateOf(configData.fontType) }
    Text(text=stringResource(id=R.string.title_font_size),fontSize=36.sp)
```

```
Row{RadioButton(selected=fontType=="小号",
    onClick={
        fontType="小号"
        configData.fontType="小号"
        configData.fontSize=16.sp
        }
    )
    Text("小号字体",fontSize=16.sp)
    RadioButton(selected=fontType=="中号",
        onClick={
            fontType="中号"
            configData.fontType="中号"
            configData.fontSize=24.sp
            }
    )
    Text("中号字体",fontSize=24.sp)
    RadioButton(selected=fontType=="大号",
        onClick={
            fontType="大号"
            configData.fontType="大号"
            configData.fontSize=28.sp
            }
    )
    Text("大号字体",fontSize=28.sp)
    }
}
```

定义显示可供选择背景图片的网格单项可组合函数 ImageCard,代码如下:

```
@Composable
fun ImageCard(imageId:Int){                                    //imageId 图片资源编号
    Card(
        colors=CardDefaults.cardColors(containerColor=Color.White,
            contentColor=Color.Unspecified),
            elevation=CardDefaults.cardElevation(defaultElevation=5.dp),
            border=BorderStroke(1.dp, Color.Green)
    ){
        Image(modifier=Modifier.size(100.dp,120.dp),
            alignment=Alignment.Center,
            painter=painterResource(id=imageId),
            contentDescription=null)
    }
}
```

定义背景图片配置的可组合函数 BgImageView,代码如下:

```
@Composable
fun BgConfigView(configData: ConfigData){
    var currentImage by remember{ mutableStateOf(configData.bgImageId) }
    val bgImages=listOf(R.mipmap.bg1, R.mipmap.bg2, R.mipmap.bg3,
        R.mipmap.bg4, R.mipmap.bg5)                            //要配置背景图片数据
    Text(text=stringResource(id=R.string.title_bg_images),fontSize=36.sp)
```

```
        LazyHorizontalGrid(rows=GridCells.Fixed(2), userScrollEnabled=true){
            items(bgImages){
                imageId:Int ->Row{
                    Column(modifier=Modifier.width(140.dp)){
                        ImageCard(imageId)
                        Row(verticalAlignment=Alignment.CenterVertically,
                            horizontalArrangement=Arrangement.Center){
                            RadioButton(selected=currentImage==imageId,onClick={
                                currentImage=imageId
                                configData.bgImageId=imageId
                                configData.bgImageName="背景${bgImages.indexOf(imageId)+1}"
                            })
                            Text("背景${bgImages.indexOf(imageId)+1}",fontSize=18.sp)
                        }
                    }
                }
            }
        }
    }
```

定义配置界面可组合函数 ConfigScreen,代码如下：

```
@Composable
fun ConfigScreen(modifier:Modifier,viewModel: ChatterViewModel){
    val configState=viewModel.configOutput.collectAsState()
    Box(modifier=modifier.fillMaxSize().padding(5.dp),
        contentAlignment=Alignment.Center){
        Column{
            Spacer(modifier=Modifier.padding(20.dp))
            FontConfigView(configState.value)          //定义字体配置单选按钮
            Spacer(modifier=Modifier.padding(20.dp))
            BgConfigView(configState.value)            //定义背景图片选择网格
        }
    }
}
```

7. 定义查看帮助界面

帮助界面是结合 AndroidView 和 WebView 调用 res/raw/help.html 文件来显示帮助信息,代码如下：

```
@Composable
fun HelpScreen(modifier: Modifier){
    Box(modifier=modifier.fillMaxSize(),contentAlignment=Alignment.Center){
        AndroidView(factory={
            ctx: Context ->WebView(ctx).apply{
                settings.javaScriptEnabled=true
                webViewClient=WebViewClient()
                loadUrl("file:///android_res/raw/help.html")
            }
```

```
        })
    }
}
```

8. 定义导航图 NavigationGraphScreen,实现在不同界面的切换

定义界面类,代码如下:

```
/** route:String 导航路径
  * title:String 导航名称
  * icon:Int 导航图标
**/
sealed class Screen(val route:String,val title:String,val icon:Int){
    data object ChatterPage:Screen(route="chatterPage",title="聊天",
        icon=R.mipmap.chatter)
    data object ConfigPage:Screen(route="configPage",title="配置",
        icon=R.mipmap.config)
    data object HelpPage:Screen(route="helpPage", title="帮助",icon =R.mipmap.help)
}
```

定义子类,代码如下:

```
@Composable
fun NavigationGraphScreen(modifier: Modifier, navController: NavHostController,
    viewModel:ChatterViewModel) {
    NavHost(navController=navController, startDestination="chatterPage") {
        composable(route="chatterPage") {
            viewModel.handleIntent(ChatterIntent.DisplayChatterIntent)
            ChatterScreen(modifier,viewModel)
        }
        composable(route="configPage") {
            viewModel.handleIntent(ChatterIntent.DisplayConfigIntent)
            ConfigScreen(modifier,viewModel)
        }
        composable(route="helpPage") {
            viewModel.handleIntent(ChatterIntent.DisplayHelpIntent)
            HelpScreen(modifier)
        }
    }
}
```

9. 定义主界面,组合脚手架结构,组装主界面

为了处理脚手架结构的底部导航栏方便,定义界面列表数据,代码如下:

```
val screens=listOf(Screen.ChatterPage, Screen.ConfigPage, Screen.HelpPage)
```

定义底部导航栏的可组合函数 BottomMenu,代码如下:

```
@Composable
fun BottomMenu(navController: NavController, viewModel: ChatterViewModel){
    val configState=viewModel.configOutput.collectAsState()
```

```
BottomAppBar{
    screens.forEach{
        screen: Screen->NavigationBarItem(selected=screen.route==configState
            .value.route,onClick={
                configState.value.route=screen.route
                navController.navigate(screen.route)
            },
            label={Text("${screen.title}")},
            icon={
                Icon(modifier=Modifier.size(60.dp,30.dp),
                    painter=painterResource(screen.icon),
                    tint=Color.Green,
                    contentDescription=null)
            }
        )
    }
}
```

组装脚手架结构的可组合函数 HomeScreen,代码如下:

```
@OptIn(ExperimentalMaterial3Api::class)
@Composable
fun HomeScreen(navController:NavHostController,viewModel:ChatterViewModel){
    val configData=viewModel.configOutput.collectAsState()
    val scope=rememberCoroutineScope()
    Scaffold(modifier=Modifier.fillMaxSize(),
        bottomBar={BottomMenu(navController=navController,viewModel=viewModel)},
        topBar={
            TopAppBar(title={Text("iChatter",fontSize=20.sp)},
                navigationIcon={
                    IconButton(onClick={
                        viewModel.handleIntent(ChatterIntent.DisplayChatterIntent)
                        scope.launch {
                            navController.navigate(configData.value.route)
                        }
                    }){
                        Icon(painterResource(id=R.mipmap.robot),tint=Color.Unspecified,
                            contentDescription="logo")
                    }
                }
            )
        }){
        innerPadding->NavigationGraphScreen(modifier=Modifier.padding(innerPadding),
            navController=navController,viewModel) //调用导航图
    }
}
```

HomeScreen 的中心区调用的是导航图界面 NavigationGraphScreen。底部调用 BottomMenu 生成底部导航栏。在顶部栏定义导航图标实现对返回聊天界面的切换。

10. MainActivity 调用 HomeScreen,生成移动应用的主界面

代码如下:

```
class MainActivity: ComponentActivity() {
    override fun onCreate(savedInstanceState: Bundle?) {
        super.onCreate(savedInstanceState)
        enableEdgeToEdge()
        setContent {
            val navController=rememberNavController()
            val viewModel=ViewModelProvider(this).get(ChatterViewModel::class.java)
            Chapter08Theme {
                HomeScreen(navController=navController, viewModel=viewModel)
            }
        }
    }
}
```

运行结果如图 8-15 所示。

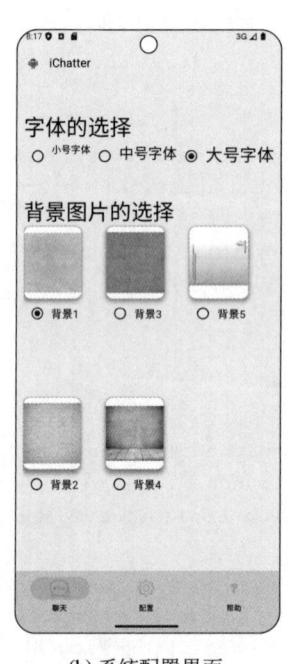

(a) 智能聊天界面　　(b) 系统配置界面　　(c) 帮助界面

图 8-15　智能聊天的运行结果

习　题　8

一、选择题

1. 在访问在线资源时,移动应用的 AndroidManifest.xml 必须配置＿＿＿＿权限。

　　A. android.permission.INTERNET

　　B. android.permission.NETWORK

C. android.permission.WIFI

D. android.permission.WIRELESS

2. 可以通过在 AndroidManifest.xml 文件 application 元素增加_____属性，并设置为 true，实现在线资源的明文访问。

  A. android：networkSecuriryConfig   B. android：usesCleartextTraffic

  C. cleartextTrafficPermitted   D. android：cleartextTrafficPermitted

3. 可以使用_____组件显示在线的 Web 页面。

  A. VideoView   B. WebView

  C. TextView   D. ImageView

4. 使用 HttpURLConnection 访问在线资源时，假设已知 HttpURLConnection 对象用 connection 方法访问网易网站，若要求请求方式为 GET，设置连接的时间为 5s，则下列代码正确的是_____。

  A.

  connection.requestMethod="GET"
  connection.connectTimeOut=5

  B.

  connection.requestMethod="GET"
  connection.connectTimeOut=5000

  C.

  connection.requestMethod="GET"
  connection.readTimeOut=5

  D.

  connection.requestMethod="GET"
  connection.readTimeOut=5000

5. 已知字符串 jsonData = {"name"："张三"，"gender"："男"，"age"：20}，则利用 jsonData 生成 JSONObject 对象的正确表达是_____。

  A. JSONobject(jsonData)

  B. JsonObject(jsonData)

  C. JSONObject(jsonData)

  D. JSONOBJECT(jsonData)

6. 已知字符串 jsonData = {"name"："张三"，"gender"："男"，"age"：20}，则利用 GSON 解析成实体类 Student 中的对象，可表示为_____。其中，Student 类的定义如下：

```
data class Student(val name: String, val gender: String, val age: Int)
```

  A.

  val gson=Gson()
  val student=gson.toJson(jsonData,Student::class.java)

B.

　　val gson＝Gson(JSONObject(jsonData))
　　val student＝gson.toJson(Student∷class.java)

C.

　　val gson＝Gson()
　　val student＝gson.fromJson(jsonData,Student∷class.java)

D.

　　val gson＝Gson(JSONObject(jsonData))
　　val student＝gson.fromJson(Student∷class.java)

二、问答题

1. 叙述使用 HttpURLConnection 以 POST 方式访问在线资源的步骤。
2. 叙述使用 Retrofit 库访问网络资源的步骤。

三、上机实践题

1. 结合 WebView 组件，编程访问百度首页。
2. 使用 HttpURLConnection 方法访问在线资源 https://cdn.mdeer.com/data/yqstaticdata.js，将获取的数据转换为 JSON 格式在日志中显示。
3. 结合 Retrofit 库和 RxJava 库，访问资源 https://cdn.mdeer.com/data/yqstaticdata.js，并将获取的数据转换为解析 JSON 格式，将其在合适的界面中显示。
4. 结合 Retrofit 库和 RxJava 库，利用"青云客"网址创建一个在线与机器人聊天的移动应用。

第 9 章　数据的持久化处理和 ContentProvider 组件

现实生活中，移动应用需要对大量数据进行处理，有时是读取外部存储器的文件、在线资源或者在移动应用的运行过程中产生的大量数据，并对这些数据进行本地存储和读取操作，可为移动应用功能的实现提供数据保证。在移动应用的运行过程中，从本地获取历史数据比从互联网获取更加快捷、方便，因此在移动应用中需要对数据进行持久化处理。数据持久化处理的本质是将内存中的瞬时数据保存到本地存储设备，以保证在关机的情况下数据不会丢失。这些被保存的数据所处的状态称为永久状态，如图 9-1 所示。

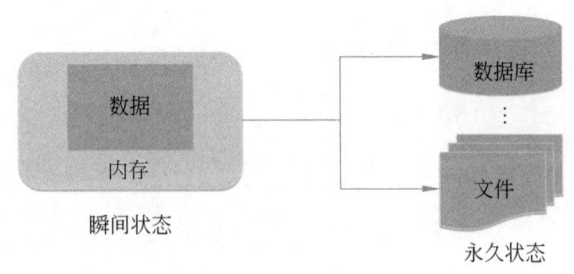

图 9-1　数据的持久化处理

本章介绍 4 种数据持久化的处理方式：文件处理、SQLite3 数据库的处理、DataStore 存储处理和 Room 组件实现数据库处理。此外，ContentProvider 是 Android 四大基本组件之一，用于数据的共享。本章也会对 ContentProvider 基本组件进行介绍。

9.1　DataStore 存储处理

DataStore 是 Android JetPack 开发套件的成员。DataStore 是一种适用于小型数据库的轻量级存储方案，只在移动应用内部使用。DataStore 基于 Kotlin 协程和 Flow 构建而成，以异步、一致的事务方式存储数据。它用于取代 Android 的 SharePreferences 处理方案。DataStore 有两种不同的实现方式。

(1) Preferences DataStore 方式。它以键值对的方式存储数据。这种方式不需要预定架构，也不确保类型安全。

(2) Proto DataStore 方式。它通过协议缓冲区存储类型化对象。这种方式能够确保类型安全。

DataStore 采用异步的方式存储数据，支持 RxJava2 和 RxJava3 进行异步处理。需要配置相应的依赖。首先，在项目的 libs.versions.toml 中增加配置，代码如下：

```
[versions]
datastore="1.1.1"
[libraries]
androidx-datastore-core-android={ group="androidx.datastore",
    name="datastore-core-android", version.ref="datastore" }
androidx-datastore-preferences={group="androidx.datastore",
```

```
        name="datastore-preferences",version.ref="datastore"}
androidx-datastore-preferences-rxjava2={group="androidx.datastore",
        name="datastore-preferences-rxjava2",version.ref="datastore"}
androidx-datastore-preferences-rxjava3={group="androidx.datastore",
        name="datastore-preferences-rxjava3",version.ref="datastore"}
androidx-datastore={group="androidx.datastore",name="datastore",
        version.ref="datastore"}
androidx-datastore-rxjava2={group="androidx.datastore",name="datastore-rxjava2",
        version.ref="datastore"}
androidx-datastore-rxjava3={group="androidx.datastore",name="datastore-rxjava3",
        version.ref="datastore"}
```

在项目对应的模块的 build.gradle.kts 构建文件中增加依赖,代码如下:

```
dependencies {
    implementation(libs.androidx.datastore.core.android)
    //支持 DataStore Preference
    implementation(libs.androidx.datastore.preferences)
    //可选项
    implementation(libs.androidx.datastore.preferences.rxjava2)
    //可选项
    implementation(libs.androidx.datastore.preferences.rxjava3)
    //支持 DataStore Proto
    implementation(libs.androidx.datastore)
    //可选项
    implementation(libs.androidx.datastore.rxjava2)
    //可选项
    implementation(libs.androidx.datastore.rxjava3)
    …          //略
}
```

1. Preferences DataStore 方式

利用 preferenceDataStore 创建委托来创建顶部的 Preferences DataStore 实例,代码如下:

```
val Context.dataStore: DataStore<Preferences> by preferencesDataStore(name="settings")
```

然后从创建的 DataStore＜Preferences＞实例中读取键值对。在表 9-1 中展示了创建 Preferences 的关键字的所有函数。

表 9-1 创建偏好关键字函数

函　　数	说　　明
intPreferencesKey(name：String)	创建 Int 类型关键字,name 为关键字名称
doublePreferencesKey(name：String)	创建 Double 类型关键字,name 为关键字名称
stringPreferencesKey(name：String)	创建 String 类型关键字,name 为关键字名称
booleanPreferencesKey(name：String)	创建 Boolean 类型关键字,name 为关键字名称
floatPreferencesKey(name：String)	创建 Float 类型关键字,name 为关键字名称
longPreferencesKey(name：String)	创建 Long 类型关键字,name 为关键字名称

续表

函　　数	说　　明
stringSetPreferencesKey(name：String)	创建 StringSet 类型关键字，name 为关键字名称
byteArrayPreferencesKey(name：String)	创建 ByteArray 类型关键字，name 为关键字名称

创建 Preference 偏好的关键字必须是唯一的。对要处理的键值对，通过 DataStore 对象的 data 属性，通过 Flow 流提供存储的值，代码如下：

```
val EMAIL=stringPreferencesKey("email")                    //字符串类型关键值
val emailFlow: Flow<String>=context.dataStore.data.map {
    preferences ->preferences[EMAIL]?:""
}
```

创建好偏好键值对后，需要将偏好键值对写入 Preferences DataStore 中。Preferences 的 DataStore 的 edit()函数实现以事务的方式是写入具体的数据，更新 DataStore 中的数据，代码如下：

```
context.dataStore.edit {
    preferences:MutablePreferences->val storedEmailValue=settings[EMAIL]?:""
    preferences[EMAIL]=email
}
```

例 9-1 创建一个移动应用，可以按照指定账号登录，为了方便下次登录，可以进行"记住密码"的配置。而且配置自动登录，下次登录可以自动加载上次录入的登录信息。登录成功后进入显示用户账号的页面。

定义一个通过的 DataStoreUtils 实用类，实现对偏好键值对的存取处理，代码如下：

```
//模块 Ch09_01 的 DataStoreUtils.kt
class DataStoreUtils(val context: Context) {
    private val Context.dataStore: DataStore<Preferences>by preferencesDataStore(name="
        settings")
    private val ACCOUNT=stringPreferencesKey("account")          //创建 account 关键字
    private val PASSWORD=stringPreferencesKey("password")        //密码关键字
    private val REMEMBERED=booleanPreferencesKey("remembered")   //记住密码
    private val AUTOLOGIN=booleanPreferencesKey("autoLogin")     //自动登录
    val accountFlow:Flow<String>=context.dataStore.data.map{
        preferences:Preferences->preferences[ACCOUNT]?:""
    }
    val passwordFlow: Flow<String>=context.dataStore.data.map {
        preferences:Preferences->preferences[PASSWORD]?:""
    }
    val rememberedFlow:Flow<Boolean>=context.dataStore.data.map{
        preferences:Preferences->preferences[REMEMBERED]?:false
    }
    val autoLoginFlow:Flow<Boolean>=context.dataStore.data.map{
        preferences:Preferences->preferences[AUTOLOGIN]?:false
```

```kotlin
    }
    suspend fun autoLogin(autoLogined:Boolean){                                    //自动登录
        context.dataStore.edit{
            preferences:MutablePreferences->preferences[AUTOLOGIN]=autoLogined
        }
    }
    suspend fun storeAccountInfo(account:String,password:String,remembered:Boolean,
        autoLogined: Boolean){
        context.dataStore.edit{
            preferences:MutablePreferences->preferences[ACCOUNT]=account          //账号
            preferences[PASSWORD]=password                                         //密码
            preferences[REMEMBERED]=remembered                                     //记住账号
            preferences[AUTOLOGIN]=autoLogined                                     //自动登录
        }
    }
}
```

定义一个登录界面可组合函数 LoginScreen，它提供配置账号和密码输入，以及配置记住密码和自动登录的界面。密码输入是不能明文显示的，因此配置密码框对应的 TextField 增加 visualTransformation 属性，并配置密码输入可见过滤器 PasswordVisualTransformation，使得输入密码增加回文遮挡符，代码如下：

```kotlin
//模块 Ch09_01 的可组合函数 LoginScreen
@Composable
fun LoginScreen(modifier:Modifier=Modifier,dataStoreUtils: DataStoreUtils,
    navController: NavHostController){
    var accountInput by remember{mutableStateOf("guest@example.com")}
    var passwordInput by remember{ mutableStateOf("") }
    var rememberInput by remember{ mutableStateOf(false) }
    var autoLoginInput by remember{ mutableStateOf(false) }
    val scope=rememberCoroutineScope()                                             //创建协程
    LaunchedEffect(Unit,autoLoginInput){                                           //读取数据
        dataStoreUtils.autoLoginFlow.collect{
            autoLogin:Boolean->autoLoginInput=autoLogin
            if (autoLogin){                                                        //判断是否自动登录
                dataStoreUtils.accountFlow.collect{
                    account:String->dataStoreUtils.passwordFlow.collect{
                        password:String->
                        accountInput=account                                        //读取保存的账号数据
                        passwordInput=password                                      //读取保存的密码数据
                    }
                }
            }
        }
    }
    Column(modifier=modifier.fillMaxSize()
        .background(colorResource(id=android.R.color.holo_orange_light)),
        verticalArrangement=Arrangement.Center,
        horizontalAlignment=Alignment.CenterHorizontally) {
        Text("Preferences DataStore 应用示例",fontSize=28.sp)
```

```
            TextField(modifier=Modifier.fillMaxWidth(),          //输入账号的文本输入框
                value=accountInput,
                onValueChange={
                    it:String->accountInput=it
                },
                label={Text("输入账号")},
                leadingIcon={
                    Icon(imageVector=Icons.Filled.AccountBox,contentDescription ="账号")
                },
                colors=TextFieldDefaults.colors(unfocusedContainerColor=Color.White,
                    focusedContainerColor=Color.White)          //设置背景颜色
            )
            TextField(modifier=Modifier.fillMaxWidth(),          //输入密码文本输入框
                value=passwordInput,
                onValueChange={
                    it:String->passwordInput=it
                },
                visualTransformation=PasswordVisualTransformation(),//创建密码输入可见过滤器
                leadingIcon={
                    Icon(imageVector=Icons.Filled.Lock, contentDescription="密码")
                },
                colors=TextFieldDefaults.colors(unfocusedContainerColor=Color.White,
                    focusedContainerColor=Color.White)
            )
            Row{
                Checkbox(checked=rememberInput,
                    onCheckedChange={
                    rememberInput=it
                    if (rememberInput)
                        scope.launch {                           //写入数据
                            dataStoreUtils.storeAccountInfo(account=accountInput,
                                password=passwordInput, remembered=rememberInput,
                                autoLogined=autoLoginInput)
                        }
                    }
                )
                Text("记住密码")
                Checkbox(checked=autoLoginInput,
                    onCheckedChange={                            //复选框
                    autoLoginInput=it
                    if (autoLoginInput){
                        scope.launch {
                            dataStoreUtils.autoLogin(autoLoginInput) //保存账号和密码数据
                        }
                    }
                    }
                )
                Text("自动登录")
            }
            Button(onClick={navController.navigate("home")}){    //导航home路径指向的HomeScreen
                Text("登录",fontSize=24.sp)
            }
        }
    }
}
```

定义登录成功后进入用户账号的用户界面 HomeScreen,代码如下:

```
//模块 Ch09_01 的可组合函数 HomeScreen
@Composable
fun HomeScreen(modifier:Modifier,dataStoreUtils: DataStoreUtils,
    navController: NavController){
    var account by remember{mutableStateOf("")}
    LaunchedEffect (Unit){
        dataStoreUtils.accountFlow.collect{
            it:String->account=it                    //读取 Preferences 的 DataStore 存储数据
        }
    }
    Column(modifier=modifier.fillMaxSize()
        .background(colorResource(id=android.R.color.holo_orange_light)),
            horizontalAlignment=Alignment.CenterHorizontally){
        Text("欢迎 $account 进入 JetPack 世界",fontSize =20.sp)    //显示登录后的账号
        Button(onClick={navController.navigate("login")          //返回登录界面
        }){
            Text("重新登录",fontSize=20.sp)
        }
        Text("首页",fontSize=32.sp)
    }
}
```

为了实现在不同界面的切换,结合 Navigation 组件(配置参考第 4 章)定义导航图界面 NavigationGraphScreen,配置显示的各个界面。并设置 LoginScreen 可组合函数对应的登录界面为开始路径。NavigationGraphScree 为根据导航路径切换界面提供方便,代码如下:

```
//模块 Ch09_01 的可组合函数 NavigationGraphScreen
@Composable
fun NavigationGraphScreen(modifier:Modifier){
    val navController: NavHostController=rememberNavController()
    val context=LocalContext.current
    val dataStoreUtils=DataStoreUtils(context)
    NavHost(navController=navController, startDestination="login") {
        composable(route="login"){
            LoginScreen(modifier=modifier,
                dataStoreUtils=dataStoreUtils,navController=navController)
        }
        composable(route="home"){
            HomeScreen(modifier=modifier,
                dataStoreUtils=dataStoreUtils,navController=navController)
        }
    }
}
```

因为可组合函数不能独立上下文存在,所以在 MainActivity 中调用 NavigationGraphScreen,实现主界面的显示,代码如下:

```
//模块 Ch09_01 的 MainActivity.kt
class MainActivity: ComponentActivity() {
    override fun onCreate(savedInstanceState: Bundle?) {
        super.onCreate(savedInstanceState)
        enableEdgeToEdge()
        setContent {
            Chapter09Theme {
                Scaffold(modifier=Modifier.fillMaxSize()) {
                    innerPadding->NavigationGraphScreen(modifier=
                        Modifier.padding(innerPadding))           //调用导航图
                }
            }
        }
    }
}
```

运行结果如图 9-2 所示。

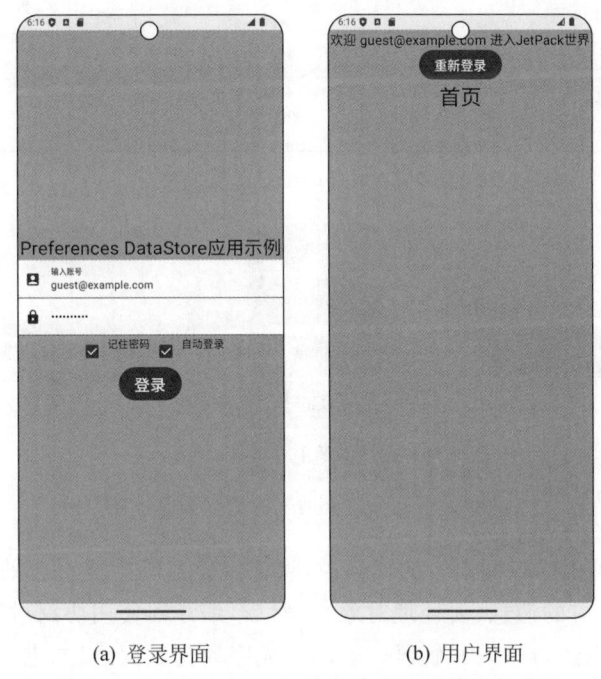

(a) 登录界面　　　　(b) 用户界面

图 9-2　Preferences DataStore 应用示例的运行结果

通过 Preferences DataStore 可以将数据保存在移动终端中。浏览 Device Explorer 工具，可以在对应移动应用的目录 data/data/files/datastore 中发现保存偏好配置 settings.preferences_pb 文件，文件名与代码中创建的 preferencesDataStore(name="settings") 的名称一致，如图 9-3 所示。

2. Proto DataStore 方式

Proto DataStore 采用 Protocol Buffers(协议缓存区)定义架构，按键存取值获取正确的数据，实现小批量数据的处理。要了解 Proto DataStore 方式，就要首先了解 Protocol Buffers。

Protocol Buffers 是 Google 公司推出的一种语言中立、平台中立的数据序列化的方法，比处理 XML 数据更加灵活、更有效率。在图 9-4 中展示了 Protocol Buffer 数据序列化的方式。

首先需要通过在 .proto 文件定义架构，在文件定义消息类型来指定数据的结构以及要序列化的服

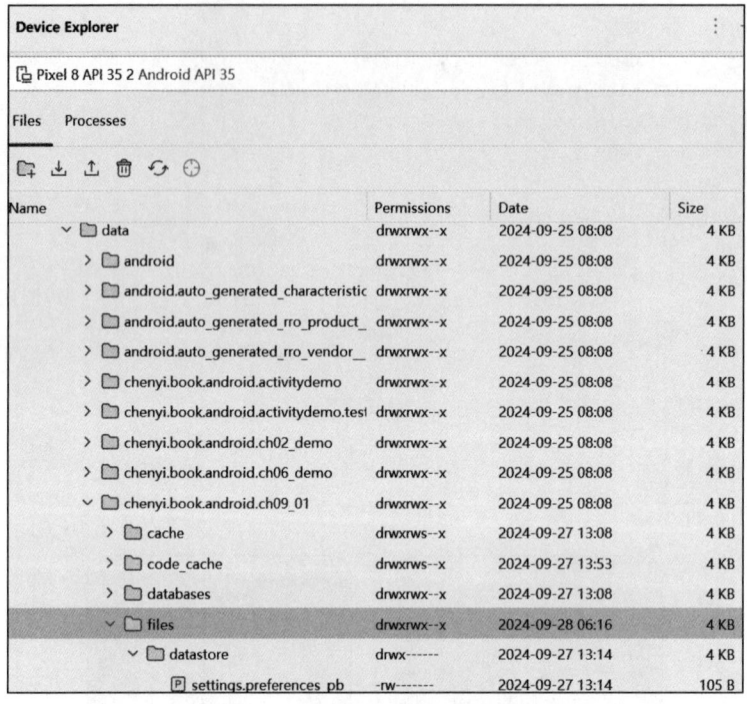

图 9-3 保存的偏好配置文件

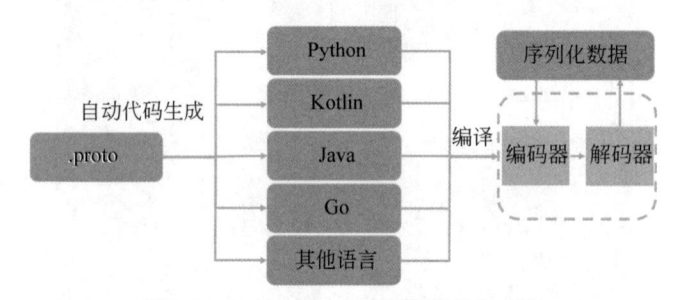

图 9-4 Protocol Buffer 数据序列化流程

务。接着根据 proto 文件可以在不同语言的代码中,将该文件自动生成对应的结构类型(消息的逻辑结构)。根据结构类型创建对应的消息(要处理对象)。然后,编译器使用协议对其进行编译。按照预先定义的架构用于对序列化数据进行编码和解码。解码生成序列化数据,以二进制数据的形式保存起来;也可以对可序列化数据进行编码,在代码中将这些二进制数据转换成代码需要的对象。

在具体使用 Proto DataStore 存储类型化数据,必须执行如下步骤。

Proto DataStore 实现使用 DataStore 和协议缓冲区将类型化对象保留在磁盘上。要使用 Proto DataStore 还需要额外配置协议缓存区插件以及增加协议缓存区和 Proto DataStore 的依赖。在项目的 libs.versions.toml 文件中增加配置关于协议缓存区和 Proto DataStore 的相关内容,代码如下:

```
[versions]
datastore="1.1.1"
protobuf="0.9.4"
protobufJavalite="3.19.4"
[libraries]
androidx-datastore-core-android={
```

```
        group="androidx.datastore",
        name="datastore-core-android", version.ref="datastore"
}
google-protobuf={
        group="com.google.protobuf",name="protobuf-javalite",
        version.ref="protobufJavalite"
}
[plugins]
google-protobuf={
        id="com.google.protobuf",version.ref="protobuf"
}
```

在项目的模块 build.gradle.kts 中配置 protobuf 插件和相应的依赖，在 android 元素内部需要指定 pb 转换文件的源目录，以及配置 protobuf 参数，代码如下：

```
import com.google.protobuf.gradle.proto
plugins {
    alias(libs.plugins.google.protobuf)
}

android{
    sourceSets.named("main") {
        proto{
            srcDir("src/main/proto")
        }
    }              //配置转换文件的源目录
    ...            //略
}
dependencies {
    implementation(libs.androidx.datastore.core.android)
    implementation(libs.google.protobuf)
    ...            //略
}
protobuf {
    protoc {
        artifact="com.google.protobuf:protoc:3.23.4"
    }
    generateProtoTasks {
        all().forEach {
            task->task.builtins {
                create("java"){
                    option("lite")
                }
            }
        }
    }
}
```

在 app/src/main/proto 目录下定义 ProtoBuf 架构文件，指定对象的要处理数据类型，代码如下：

```
syntax="proto3";                                          //语法 proto3
option java_package="com.example.application";            //设置字节码生成类的包
option java_multiple_files=true;                          //是否是 java 多文件
message SomePreferences {
    int32 value1=1;                                       //第一个关键字的类型
    float value2=2;                                       //第二个关键字的类型
}
```

表 9-2 中展示了标量值的类型对应 Java 或 Kotlin 语言的数据类型的情况。

表 9-2 proto3 常见标量值类型对应的 Java/Kotlin 类型

proto3 类型	Java/Kotlin	proto3 类型	Java/Kotlin
double	double	fixed32	int
float	float	fixed64	long
int32	int	sfixed32	int
int64	long	sfixed64	long
uint32	int	bool	boolean
uint64	long	string	String
sint32	int	bytes	ByteString
sint64	long		

创建 Proto DataStore 用于存储类型化对象。必须定义一个实现 Serializer<T> 的类，其中 T 是 proto 文件中定义的类型。此序列化器类会告知 DataStore 如何读取和写入数据类型，代码如下：

```
object SomePreferencesSerializer:Serializer<SomePreferences >{
    override val defaultValue: Settings=SomePreferences.getDefaultInstance()
                                                                              //默认值
    override suspend fun readFrom(input: InputStream): SomePreferences {      //读取数据
        try {
            return SomePreferences.parseFrom(input)
        } catch (exception: InvalidProtocolBufferException) {
            throw CorruptionException("Cannot read proto.", exception)
        }
    }
    override suspend fun writeTo(t: SomePreferences,
        output: OutputStream)=t.writeTo(output)                               //写入数据
}
```

注意：当创建 ProtoBuf 架构文件后，一定要构建项目模块，根据架构文件生成对应的字节码类型，如 SomePreferences，才可以在此基础上创建类型化对象。

使用 dataStore 所创建的属性委托来创建 DataStore<T> 实例，其中 T 是在 proto 文件中定义的类型。如果在 Kotlin 文件顶层调用该实例一次，便可在应用的所有其余部分通过此属性委托访问该实例，代码如下：

```
val Context.settingsDataStore: DataStore<SomePreferences >by dataStore(
    fileName=" SomePreferences.pb",
    serializer=SomePreferencesSerializer()
)
```

使用 DataStore.data 获取相应属性的 Flow，代码如下：

```
val value1Flow: Flow<Int>=context.settingsDataStore.data.map {
    preferences->preferences.value1
}
```

```
val value2Flow: Flow<Float>=context.settingsDataStore.data.map {
    preferences->preferences.value2
}
```

调用 Proto DataStore 的 updateData()函数,实现以事务的方式更新存储的对象,代码如下:

```
context.settingsDataStore.updateData {
    preferences->preferences.toBuilder()
        .setValue1(preferences.value1 +1)
        .setValue2(preferences.value2 * 2)
        .build()
}
```

例 9-2 创建一个简单移动应用,设置背景音乐的开或关,当设置背景音乐微开时,播放背景音乐;设置背景音乐为关时,关闭背景音乐。同时,还可以设置背景音乐左右声道的音量。

定义 MusicApp 为 Application 的子类,方便获取整个移动应用的上下文,代码如下:

```
class MusicApp: Application() {
    companion object{
        var instance:MusicApp?=null
            private set
    }
    override fun onCreate() {
        super.onCreate()
        instance=this
    }
}
```

在 AndroidManifest.xml 中配置应用名称,代码如下:

```
<application android:name=".MusicApp"
    ...                                              //略
>
    ...                                              //略
</application>
```

在 Android Studio 中切换到 Project 开发视图,在 src/main 目录下创建 proto 目录,再在 src/main/proto 目录中创建架构文件 MusicPreferences.proto,代码如下:

```
syntax="proto3";                                     //语法 proto3
option java_package="chenyi.book.android.ch09_01";   //设置 java 生成类所在的包
message MusicPreferences{
    bool opened=1;                                   //第一个关键字表示打开音乐
    float leftVolume=2;                              //第二个关键字取值 int32
    float rightVolume=3;                             //第三个关键字取值 int32
}
```

文件一旦创建好后,一定要构建项目,使之生成对应 Kotlin 的类型,可以在模块的 build/generated/source/proto/debug 目录下查看到生成的代码。在这个模块中会生成一个 MusicPreferencesOuterClass,在这个类内部会生成对应的内部类 MusicPreferences,如图 9-5 所示。

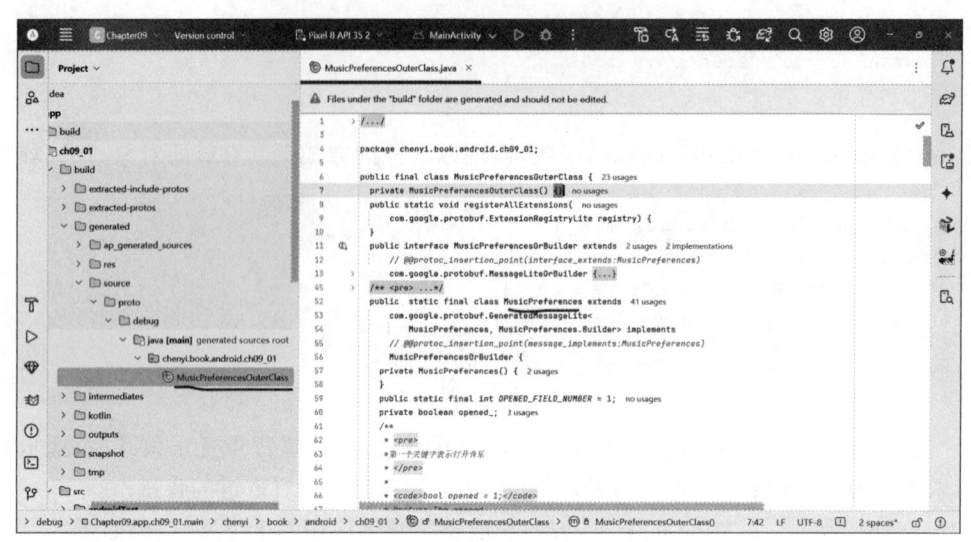

图 9-5 构建后自动生成类文件

处理序列化，创建一个实现 Serizer 的类表示序列化器，并指定要需要的类型，与 proto 文件指定的类型一致。定义的序列化器的类会告知 Proto DataStore 如何实现读写操作，代码如下：

```kotlin
class MusicPreferencesSerializer: Serializer<MusicPreferences>{
    override val defaultValue:MusicPreferences
        get()=MusicPreferences.getDefaultInstance()
    override suspend fun readFrom(input: InputStream): MusicPreferences {
        try{
            return MusicPreferences.parseFrom(input)
        }catch(e:InvalidProtocolBufferException){
            throw CorruptionException("无法读取 MusicPreferences.proto 文件")
        }
    }
    override suspend fun writeTo(t:MusicPreferences,
        output: OutputStream)=t.writeTo(output)
}
```

创建顶层的 Proto Dastore 实例，在文件顶层创建的 Proto DataStroe 对象实例一次，就可以在整个移动应用中通过属性委托访问该对象实例，代码如下：

```kotlin
val Context.musicPreferencesDataStore: DataStore<MusicPreferences>by dataStore(
    fileName="MusicPreferences.pb",serializer=MusicPreferencesSerializer()
)
```

定义数据类 MusicConfig，保存背景音乐的相关配置信息，代码如下：

```kotlin
data class MusicConfig(var leftVolume:Float,var rightVolume:Float,var opened:Boolean)
```

为了方便对 Proto DataStore 数据的读取，定义一个 ProtoDataStoreUtils 类，代码如下：

```kotlin
class ProtoDataStoreUtils(val context: Context) {
    val musicConfigFlow:Flow<MusicConfig>=context.musicPreferencesDataStore.data.map {
        it:MusicPreferences->MusicConfig(it.leftVolume,it.rightVolume,it.opened)
```

```
                                                              //读取数据
    }
    suspend fun updateLeftVolume(leftVolume:Float){
        context.musicPreferencesDataStore.updateData {
            it:MusicPreferences->it.toBuilder()
                .setLeftVolume(leftVolume)
                .build()                              //写入左声道数据
        }
    }
    suspend fun updateRightVolume(rightVolume:Float){
        context.musicPreferencesDataStore.updateData {
            it:MusicPreferences->it.toBuilder()
                .setRightVolume(rightVolume)
                .build()                              //写入右声道数据
        }
    }
    suspend fun updateOpenState(opened:Boolean=false){
        context.musicPreferencesDataStore.updateData {
            it: MusicPreferences ->it.toBuilder()
                .setOpened(opened)
                .build()                              //写入是否播放音乐
        }
    }
    suspend fun updateData(leftVolume:Float,rightVolume:Float,opened:Boolean=false){
        context.musicPreferencesDataStore.updateData {
            it:MusicPreferences->it.toBuilder()
                .setLeftVolume(leftVolume)
                .setRightVolume(rightVolume)
                .setOpened(opened)
                .build()                              //写入左右声道和播放音乐状态
        }
    }
}
```

本移动应用采用 WorkManager 组件来处理后台音乐的播放，WorkManager 组件的配置参考第 7 章，此处不再介绍。定义 MusicWorker 控制音乐播放，并从 Proto Data 中读取左右声道数据设置音频播放的左右声道，代码如下：

```
class MusicWorker(val context: Context,params: WorkerParameters):Worker(context,params){
    override fun doWork(): Result {
        val mediaPlayer=MediaPlayer.create(context,R.raw.song)
        val job=Job()
        val scope=CoroutineScope(job)
        scope.launch {                              //在协程范围读取数据，并设置声道
            protoDataStoreUtils.musicConfigFlow?.collect{
                it: MusicConfig->mediaPlayer.setVolume(it.leftVolume,it.rightVolume)
            }
        }
        mediaPlayer.start()                         //播放音乐
        return Result.success()
    }
}
```

定义 MusicScreen 展示背景音乐配置的界面。为了控制背景音乐的播放，另外定义了 playMusic 外部函数，代码如下：

```kotlin
val protoDataStoreUtils=ProtoDataStoreUtils(MusicApp.instance!!.applicationContext)
@Composable
fun MusicScreen(modifier: Modifier){
    var opened by remember{ mutableStateOf(false) }                        //启动音乐
    var leftVolume by remember{ mutableFloatStateOf(0.5f) }                //左声道
    var rightVolume by remember{ mutableFloatStateOf(0.5f) }               //右声道
    val scope=rememberCoroutineScope()                                     //协程范围
    val context=LocalContext.current                                       //上下文
    LaunchedEffect (Unit){
        protoDataStoreUtils.musicConfigFlow.collect{
            it: MusicConfig ->opened=it.opened
            leftVolume=it.leftVolume
            rightVolume=it.rightVolume
            if (opened)playMusic(context)
        }
    }
    Column(modifier=modifier
        .fillMaxSize()
        .background(colorResource(id=android.R.color.holo_orange_light)),
        horizontalAlignment=Alignment.CenterHorizontally){
        Text("音乐配置界面",fontSize=32.sp)
        Row {
            Text("背景音乐", fontSize=24.sp)
            RadioButton(
                selected=opened,
                onClick={
                    opened=!opened
                    if (opened) {
                        playMusic(context)
                    }
                    scope.launch {
                        protoDataStoreUtils.updateOpenState(opened)          //写入音乐播放开关
                    }
                }
            )
        }
        Text("音量", fontSize=24.sp)
        Slider(value=leftVolume,
            onValueChange={
                it: Float->leftVolume=it
                scope.launch {
                    protoDataStoreUtils.updateLeftVolume(leftVolume)         //写入左声道
                }
            }
        )
        Slider(value=rightVolume,
            onValueChange={
                it:Float->rightVolume=it
                scope.launch {
                    protoDataStoreUtils.updateRightVolume(rightVolume)       //写入右声道
                }
            }
        )
```

```kotlin
        }
    }

fun playMusic(context: Context){                                        //播放音乐
    val constraints: Constraints=Constraints.Builder()
        .setRequiresBatteryNotLow(true)
        .build()
    val request=OneTimeWorkRequestBuilder<MusicWorker>()
        .setConstraints(constraints).build()
    WorkManager.getInstance(context).enqueue(request)                   //加入队列
}
```

在 MainActivity 中调用 MusicScreen，实现主界面，代码如下：

```kotlin
class MainActivity: ComponentActivity() {
    override fun onCreate(savedInstanceState: Bundle?) {
        super.onCreate(savedInstanceState)
        enableEdgeToEdge()
        setContent {
            Chapter09Theme {
                Scaffold(modifier=Modifier.fillMaxSize()) {
                    innerPadding->MusicScreen(modifier=Modifier.padding(innerPadding))
                }
            }
        }
    }
}
```

运行结果如图 9-6 所示。

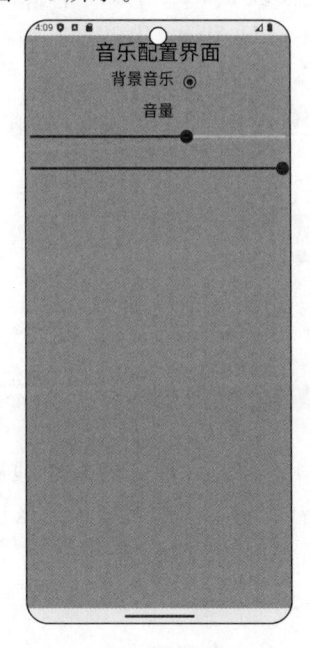

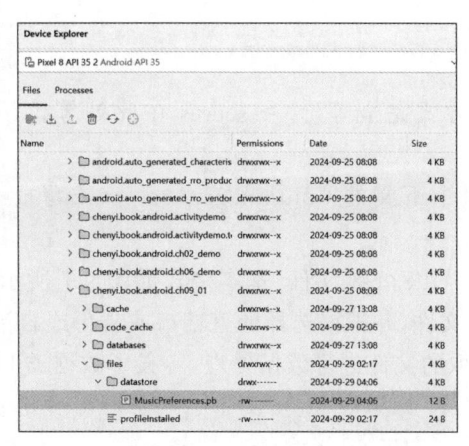

(a) 运行界面　　　　　　　　　(b) 生成DataStore文件

图 9-6　配置背景音乐的运行结果

正确使用 DataStore 必须注意以下几点。

（1）在一个进程中只能有一个 DataStore 实例。无论 DataStore 实例是 Preference DataStore 实例还是 Proto DataStore 实例都有且仅有一个，不能同时存在。否则会抛出 IllegalStateException。

（2）DataStore 的通用类型是不可变的。因此，采用 Proto Buffer 协议缓存区处理 DataStore，可以保证数据的不可变性以及执行高效。

（3）同一个文件不能混用 SingleProcessDataStore（单进程的 DataStore）和 MultiProcessDataStore（多进程的 DataStore）。

9.2 文件处理

Android 提供了文件的读取和存储功能。在前几章的应用示例中，有对文件的简单应用。例如，使用了 HTML 文件作为系统应用的帮助文件，以及播放 raw 目录下的 MP4 音频文件。因此有必要了解如何进行文件处理。文件处理涉及输入输出流对文件的读取和存储，主要分成以下 3 种情况。

第一种情况是文件放在应用程序的 res/raw 目录下，这些文件在编译的时候和其他文件一起被打包。可以通过输入流直接读取 res/raw 目录下的资源文件，也可以通过资源的编号例如 R.raw.file 来读取资源文件，代码如下：

```
val in: InputStream =getResources().openRawResource(R.raw.file)
```

或

```
val in: InputStream =resources.openRawResource(R.raw.file)
```

第二种情况是文件放在应用程序的 assets 目录下。在系统编译时，assets 目录下的文件不会被编译，而是与移动应用直接打包在一起。可以通过输入流直接读取 assets 目录中的资源文件。值得注意的是，来自 res/raw 和 assets 中的文件只可以读取而不能进行写入操作，代码如下：

```
val in: InputStream =getResources().getAssets().open(fileName)
```

或

```
val in=resources.assets.open(fileName)
```

上述两种表达方式完全等价。resouces 在此处等价于 getResources 函数，assets 等价于 getAssets 函数。

此外，可以通过上下文的 Context.getAssets（或写成 Context.assets）函数来获得 assets 目录的访问。

例 9-3 已知有 cities.csv 文件（基本格式如图 9-7 所示）记录了大部分的省市地区，保存在模块的 assets 目录，读取该文件，并根据选择的直辖市或省份或自治区，将下级的城市或地区读取出来。

为了方便对读取的文件进行数据处理，定义了数据类 District，代码如下：

```
//模块 Ch09_02 District.kt
/** 定义实体类 District 表示地方
 * state: 表示直辖市或省
```

```
 * city: 表示城市或地区或自治州
 * county: 表示地区或县
 */
data class District(val state: String, val city: String, val county: String)
```

```
北京市,北京市,东城区
北京市,北京市,西城区
北京市,北京市,崇文区
北京市,北京市,宣武区
北京市,北京市,朝阳区
北京市,北京市,丰台区
北京市,北京市,石景山区
北京市,北京市,海淀区
北京市,北京市,门头沟区
北京市,北京市,房山区
北京市,北京市,通州区
北京市,北京市,顺义区
北京市,北京市,昌平区
北京市,北京市,大兴区
北京市,北京市,平谷区
北京市,北京市,怀柔区
北京市,北京市,密云县
北京市,北京市,延庆县
天津市,天津市,和平区
天津市,天津市,河东区
```

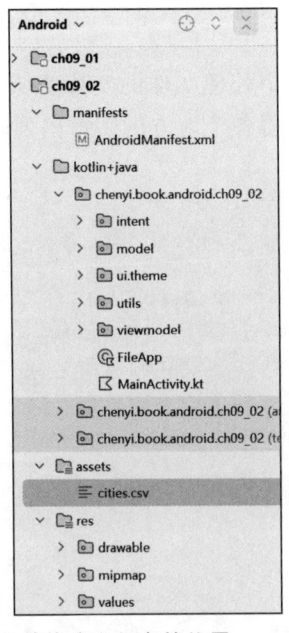

图 9-7　cities.csv 文件格式和保存的位置

为了读取 assets 目录下的文件，往往需要通过上下文来实现，为了更方便地获得移动应用的上下文，定义一个 Application 的子类 FileApp，并在 AndroidManifest.xml 设置为移动应用。当移动应用启动时，系统会自动将这个类初始化。通过此方式，可使得获得移动应用的上下文更加方便。在下列的代码中定义了一个 Context 对象，它实际对应的是 applicationContex 的上下文，全局都可以使用。因为 Application 对象在整个运行期间只有唯一的一个对象实例，这使得 Context 对象也具有唯一性，不会产生内存泄漏，代码如下：

```
//模块 Ch09_02 FileApp.kt
class FileApp: Application() {
    companion object{
        @SuppressLint("StaticFieldLeak")
        lateinit var context: Context          //定义移动应用的上下文
    }
    override fun onCreate() {
        super.onCreate()
        context = applicationContext           //将整个移动应用的上下文赋值给 context 变量
    }
}
```

为了让 FileApp 应用类发挥作用，需要在 application 元素种设置 android:name=".FileApp"属性，使得 AndroidManifest.xml 指定 FileApp 类为系统的应用，配置如下：

```
<!--模块 Ch09_02 AndroidManifest.xml -->
<?xml version="1.0" encoding="utf-8"?>
```

```xml
<manifest xmlns:android="http://schemas.android.com/apk/res/android"
    package="chenyi.book.android.ch09_2">
    <application …
        android:name=".FileApp">            …
    </application>
</manifest>
```

在此前提下，可以定义一个处理文件的实用类 FileUtils，通过 FileUtils 类可以获得 assets 目录保存的文件数据，以获取达到获得各个地方的字符串列表，代码如下：

```kotlin
//模块 Ch09_02 FileUtils.kt
class FileUtils {
    companion object{
        val districtList=getDistricts()
        fun getDistricts(): List<District>{                    //读取文件获得地点列表
            val list=mutableListOf<District>()                 //定义保存地区的列表
            val context=FileApp.context                        //获得移动应用的上下文
            val inputStream=context.assets.open("cities.csv")
                                                               //读 assets 目录 cities.csv 文件获得输入流
            val inputReader=inputStream.bufferedReader(Charsets.UTF_8)
                                                               //打包成 UTF-8 缓冲字符输入流
            inputReader.use{
                it.lines().forEach {
                    line: String ->                            //对输入流逐行进行处理
                        val content=line.split(",")            //将每行字符按照","进行分隔成字符
                                                               //串数组
                        list.add(District(content[0],content[1],content[2]))
                                                               //将创建 District 对象加入列表中
                }
            }
            return list
        }
        fun getStates(): List<String>{                         //获得所有的直辖市和省的名称
            val stateList=mutableListOf<String>()
            districtList.forEach {
                districit: District->if(districit.state !in stateList)
                    stateList.add(districit.state)
            }
            return stateList
        }
        fun getCities(state: String): List<String>{    //获得特定直辖市或省的下级自治州或城市或地区
            val citieList=mutableListOf<String>()
            districtList.forEach {
                district: District->if(state==district.state && district.city !in
                    citieList){
                        citieList.add(district.city)
                }
            }
```

```
            return citieList
        }
        fun getCounties(state: String,city: String): List<String>{        //获得县级地区列表
            val countyList=mutableListOf<String>()
            districtList.forEach {
                district: District ->if(state==district.state&&city=district.city&&district.
                    county !in countyList)
                    countyList.add(district.county)
            }
            return countyList
        }
    }
}
```

本例采用 MVI 模式实现对地区的选择。将该移动应用的状态定义在 DistrictState 类中，封装了所有省（直辖市）级名称、市级名称和县级名称的列表，并记录到当前选择的省（直辖市）市县的变量，代码如下：

```
//模块 Ch09_02 定义状态类 DistrictState.kt
data class DistrictState(
    val provinces:MutableList<String>=mutableListOf(),        //省直辖市列表
    var cities:MutableList<String>=mutableListOf(),           //城市列表
    var counties:MutableList<String>=mutableListOf(),         //县级地区列表
    var province:String="",                                    //当前省（直辖市）
    var city:String="",                                        //当前城市
    var county:String=""                                       //当前区
)
```

定义要处理的业务意图类 DistrictIntent，在 DistrictIntent 密封类中封装了 4 个内部类 ChooseProvinceIntent、ChooseCityIntent、ChooseCountyIntent 和 DisplayDistrictIntent。这些内部类分别表示选择当前省（直辖市）名意图、选择当前城市名意图、选择当前县级地区名意图和显示当前地区的意图，代码如下：

```
//模块 Ch09_02 定义 DistrictIntent.kt
sealed class DistrictIntent {
    data object ChooseProvinceIntent:DistrictIntent()
    data class ChooseCityIntent(val province:String):DistrictIntent()
    data class ChooseCountyIntent(val province:String,val city:String):DistrictIntent()
    data class DisplayDistrictIntent(val province: String,val city:String,
        val county:String):DistrictIntent()
}
```

在 MVI 模式中业务处理的实质处理是由视图模型来承担的。在本例中定义 DistrictViewModel 类，根据意图对象的不同做出不同的业务处理。在 DistrictViewModel 中定义的可变私有 MutableStateFlow<DistrictState> 对象，在 DistrictViewModel 中修改可变的状态流，然后调用 asStateFlow 函数获得 StateFlow<DistrictState>对象发送给视图界面，从而达到更新界面的目的，代码如下：

```
//模块 Ch09_02 定义视图模型 DistrictViewModel.kt
class DistrictViewModel: ViewModel() {
    var _districtState=MutableStateFlow(DistrictState())
```

```kotlin
        val output=_districtState.asStateFlow()                    //定义保存DistrictState对象的可变状态流
        fun handleIntent(intent:DistrictIntent){                    //定义状态流
            when (intent){                                          //处理意图
                is DistrictIntent.ChooseProvinceIntent->chooseProvince()    //处理选择省/直辖市
                is DistrictIntent.ChooseCityIntent->chooseCity(intent.province)  //处理选择城市名称
                is DistrictIntent.ChooseCountyIntent->chooseCounty(province=intent.province,
                    city=intent.city)                               //处理选择县级地区的名称
                is DistrictIntent.DisplayDistrictIntent->           //处理显示选择的地区完整名称
                    displayDistrict(province=intent.province,
                        city=intent.city,county=intent.county)
            }
        }
        fun chooseProvince(){                                       //获取省/直辖市列表
            val currentState=_districtState.value                   //获得当前状态值
            val newState=currentState.copy()                        //产生DistrictState对象的副本
            if (newState.provinces.isEmpty())
                newState.provinces.addAll(FileUtils.getStates())
                                                                    //修改DistrictState省/直辖市名列表
            newState.cities.clear()                                 //清除DistrictState城市名列表
            newState.counties.clear()                               //清除DistricttState县级地区名列表
            _districtState.value=newState                           //修改状态值
        }
        fun chooseCity(province:String){                            //获取城市列表
            val currentState=_districtState.value
            val newState=currentState.copy(province=province)
                                    //产生DistrictState对象的副本,并修改province属性值
            newState.cities.clear()
            newState.counties.clear()
            newState.cities.addAll(FileUtils.getCities(province))
                                                                    //修改DistrictState的城市名列表
            _districtState.value=newState
        }
        fun chooseCounty(province:String,city:String){              //获取县级列表
            val currentState=_districtState.value
            val newState=currentState.copy(province=province,city=city)
            newState.counties.clear()
            newState.counties.addAll(FileUtils.getCounties(province,city))
            _districtState.value=newState
        }
        fun displayDistrict(province: String,city:String,county:String){    //修改地区名称
            val currentState=_districtState.value
            val newState=currentState.copy(province=province,city=city,county=county)
            _districtState.value =newState
        }
    }
```

为了实现三级地区名称选择,需要通过列表选择框来完成。JetPack Compose 组件中并没有提供常见 UI 界面的列表选择框 Spinner 组件,因此需要自定义列表选择框。自定义 MySpinner 可组合函数实现列表选择框 UI 控件。MySpinner 自定义列表选择框中将 TextField 可组合项和 ExposedDropdownMenuBox 可组合项组装成,代码如下:

```kotlin
//模块 Ch09_02 的可组合函数 MySpinner
@OptIn(ExperimentalMaterial3Api::class)
@Composable
fun MySpinner(items:List<String>,action:(String)->Unit){
    var expanded by remember {mutableStateOf(true) }            //下拉状态
    var content by remember{mutableStateOf("") }                //显示选择文本内容
    ExposedDropdownMenuBox(modifier=Modifier.padding(bottom=20.dp),expanded=expanded,
        onExpandedChange={expanded=!expanded }) {
        TextField(readOnly=true,                                //只读
            value="$content",                                   //设置显示文本
            onValueChange={                                     //输入发送变化处理
                content=it
            },
            trailingIcon={                                      //输入框尾部的图标
                ExposedDropdownMenuDefaults.TrailingIcon(expanded=expanded)
            },
            colors=ExposedDropdownMenuDefaults.textFieldColors( //设置颜色
                unfocusedContainerColor=Color.White,
                unfocusedTextColor=Color.Black,
                focusedContainerColor=Color.White,
                focusedTextColor=Color.Black),
                modifier=Modifier.menuAnchor()
        )
        ExposedDropdownMenu(                                    //定义下拉菜单
            expanded=expanded,
            onDismissRequest={ expanded=false }) {
            items.forEach {                                     //遍历列表
                DropdownMenuItem(                               //定义下拉菜单项
                    text={Text(it,fontSize=24.sp) },
                    onClick={                                   //点击动作处理
                        content=it                              //修改输入框的显示内容
                        expanded=false
                        action.invoke(it)                       //交互动作调用
                    }
                )
            }
        }
    }
}
```

定义可组合函数 HomeScreen 实现三级地区名称的选择。在下列的 HomeScreen 定义了 3 个自定义列表选择框 MySpinner，通过三级联动的处理，实现选择"直辖市/省/自治区"，再选择"城市或自治州或地区"，最后选择"县或地区或自治县"，代码如下：

```kotlin
//模块 Ch09_02 定义可组合函数 HomeScreen
@Composable
fun HomeScreen(modifier:Modifier,viewModel:DistrictViewModel){
    val output=viewModel.output.collectAsState()
    val fontColor=colorResource(id=R.color.teal_200)
    LaunchedEffect(Unit) {
        viewModel.handleIntent(DistrictIntent.ChooseProvinceIntent)
                                                                //处理省/直辖市业务
    }
```

```
            Column(modifier=modifier.fillMaxSize().padding(top=10.dp),
                verticalArrangement=Arrangement.Center,
                horizontalAlignment=Alignment.CenterHorizontally){
                Text("省市列表",fontSize=36.sp,color=fontColor)
                MySpinner(items=output.value.provinces) {                    //处理选择城市业务
                    viewModel.handleIntent(DistrictIntent.ChooseCityIntent(it))
                }
                MySpinner(items=output.value.cities) {                       //处理选择县级地区业务
                    viewModel.handleIntent(DistrictIntent.ChooseCountyIntent(output.value.
                        province,it))
                }
                MySpinner(items=output.value.counties) {
                    viewModel.handleIntent(DistrictIntent.DisplayDistrictIntent(output
                        .value.province,output.value.city,it))               //处理显示地区名
                }
                if (output.value.province.isNotEmpty())
                    Text("${output.value.province}-${output.value.city}-${output.value.county}",
                        fontSize=24.sp,color =fontColor)                     //显示选择结果
            }
        }
```

MainActivity 调用 HomeScreen 可组合函数生成主界面，实现三级地区选择的动作处理，并显示最终选中的地区，代码如下：

```
//模块 Ch09_02 的 MainActivity.kt
class MainActivity: ComponentActivity() {
    override fun onCreate(savedInstanceState: Bundle?) {
        super.onCreate(savedInstanceState)
        enableEdgeToEdge()
        val viewModel=ViewModelProvider(this).get(DistrictViewModel::class.java)
                                                                            //生成视图模型
        setContent {
            Chapter09Theme {
                Scaffold(modifier=Modifier.fillMaxSize()) {
                    innerPadding->HomeScreen(modifier=Modifier.padding(innerPadding),
                        viewModel=viewModel)
                }
            }
        }
    }
}
```

运行结果如图 9-8 所示。

第三种情况是，可以结合 FileInputStream 文件，按输入字节流和 FileOutputStream 文件输出字节流，对移动应用私有文件夹中的 data（数据）进行读写操作。可以通过 FileOutputStream 类的 openFileOutput(name：String,mode：Int)函数打开相应的输出流，通过 FileInputStream 类的 openFileInput (name：String)函数打开相应的输入流，形式如下：

图 9-8　地区选择运行结果

```
val outputStream: FileOutputStream=Context.openFileOutput(fileName,mode)
                                                    //创建文件输出字节流
val inputStream: FileInputStream=Context.openFileInput(fileName)
                                                    //创建文件输出字节流
```

其中,参数 mode 表示打开文件的模式,可以取值为 MODE_PRIVATE 和 MODE_APPEND:MODE_PRIVATE 只能被当前程序读写,MODE_APPEND 可以用追加方式打开文件。

默认情况下,使用输入输出流保存文件仅对当前应用程序可见,对于其他应用程序不可见,不能访问其中的数据。如果用户卸载了该移动应用,则保存数据的文件也会被一起删除。

例 9-4　访问百度的在线 LOGO 图片(网址:https://www.baidu.com/img/flexible/logo/pc/result.png),将图片保存到本地移动终端中,然后生成并显示副本图片。

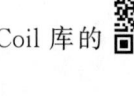

因为本例需要在后台访问在线资源,需要使用 WorkManager 库、Retroift2,又因为使用 Coil 库的 AsyncImage 可组合函数显示在线图片,因此在项目的 libs.versions.toml 文件配置如下内容:

```
[versions]
coil="2.7.0"
retrofit="2.11.0"
work="2.9.1"
rxandroid="3.0.2"
[libraries]
coil-compose={group="io.coil-kt", name="coil-compose",version.ref="coil"}
retrofit={module="com.squareup.retrofit2:retrofit",version.ref="retrofit"}
retrofit-convert-gson={
    module="com.squareup.retrofit2:converter-gson",version.ref="retrofit"
}
retrofit-adpater-rxjava3={
    module="com.squareup.retrofit2:adapter-rxjava3",version.ref="retrofit"
}
androidx-work={group="androidx.work",name="work-runtime-ktx",version.ref="work"}
```

在项目模块的 build.gradle.kts 构建文件,配置上述依赖库,代码如下:

```kotlin
//模块 Ch09_03 的 build.gradle.kts
dependencies {
    ...
    implementation(libs.coil.compose)
    implementation(libs.androidx.work)
    implementation(libs.retrofit)
    implementation(libs.retrofit.convert.gson)
    implementation(libs.retrofit.adpater.rxjava3)
}
```

本移动应用的用途是从指定网址下载百度 LOGO 图片,结合 Retrofit 库实现对网络资源的访问,因此需要定义一个获得访问资源的接口 DownloadService,代码如下:

```kotlin
//模块 Ch09_03 下载的接口 DownloadService.kt
interface DownloadService {
    @GET("result.png")
    fun getImage(): Call<ResponseBody>
}
```

将下载的文件写入本地移动应用的私有目录 files 中。为了方便后续加载图片,将私有目录 files 的文件读取称为 Bitmap 对象。定义文件处理实用类 FileUtils 的代码如下:

```kotlin
//模块 Ch09_03 文件实用类 FileUtils.kt
object FileUtils {
    private fun isFileExist(context:Context,fileName:String):Boolean{    //判断文件是否存在
        val files=context.fileList()
        files.forEach {
            if (it.contains(fileName)) {
                return true
            }
        }
        return false
    }
    fun readImageFile(context:Context,fileName:String):Bitmap?{          //读取图片文件
        var bitmap:Bitmap?=null
        if (isFileExist(context,fileName)) {
            try {
                val inputStream=context.openFileInput(fileName)          //创建文件输入流
                inputStream.use {
                    bitmap=BitmapFactory.decodeStream(inputStream)
                                                                          //将文件输入流写入 Bitmap 对象中
                }
            } catch (e: IOException) {
                e.printStackTrace()
            }
        }
        return bitmap
    }
    fun copyFile(context: Context, responseBody: ResponseBody, fileName:String){
```

```kotlin
                                                                //将响应写入文件
        var outputStream=context.openFileOutput(fileName, Context.MODE_PRIVATE)
        val fileReader=ByteArray(1024)
        var inputStream=responseBody.byteStream()
        try{
            outputStream.use{
                while (true){                    //从响应中获取输入流读取字节数组到 fileReader 中
                    val read:Int=inputStream.read(fileReader)
                    if (read==-1){
                        break
                    }
                    outputStream.write(fileReader,0,read)
                }
                outputStream.flush()
            }
        }catch(e: IOException){
            e.printStackTrace()
        }finally {                                               //关闭输入输出流
            inputStream.close()
            outputStream.close()
        }
    }
}
```

因为下载在线图片和复制到本地都是在后台运行的,所以在本移动应用中定义了后台服务 DownloadWork 实现下载和复制图片的业务,代码如下:

```kotlin
//模块 Ch09_03 的 DownloadWork.kt
class DownloadWork(val context:Context,params:WorkerParameters):Worker(context,params){
    private val baseUrl="https://www.baidu.com/img/flexible/logo/pc/"
                                                                //基址
    override fun doWork(): Result {
        downloadImageFile("image.png")                          //将文件下载并保存本地
        val data=Data.Builder().putString("imageFile","image.png").build()
                                                                //生成的文件名发送出去
        return Result.success(data)
    }
    private fun downloadImageFile(fileName:String){
        val retrofit=Retrofit.Builder()
            .baseUrl(baseUrl)
            .build()                                            //创建 Retrofit 对象
        val service=retrofit.create(DownloadService::class.java) //创建请求网络服务对象
        service.getImage().enqueue(object:Callback<ResponseBody>{ //请求服务插入队列
            override fun onResponse(call: Call<ResponseBody>,
                response: Response<ResponseBody>) {
                response.body()?.run{                           //响应成功
                    FileUtils.copyFile(context,this,fileName)   //将下载的文件保存到本地
```

```
            }
        }
        override fun onFailure(call: Call<ResponseBody>, t: Throwable) {
            t.printStackTrace()
        }
    })
}
```

显示主界面的可组合函数 HomeScreen 定义了两个显示图片的可组合项,一个是显示在线图片的 AsyncImage 可组合项,另一个用于显示复制的图片的 Image 可组合项;另外,定义 Button 可组合项实现下载业务的调用,将图片下载到本地并复制,代码如下:

```
//模块 Ch09_03 的 HomeScreen 可组合函数
@Composable
fun HomeScreen(modifier:Modifier,action:()->Unit){
    val imageUrl="https://www.baidu.com/img/flexible/logo/pc/result.png"
    val context=LocalContext.current
    val emptyImage=ImageBitmap.imageResource(id=R.mipmap.empty)
    var image by remember{mutableStateOf(emptyImage) }
    var download by remember{ mutableStateOf(false) }
    LaunchedEffect(download) {
        val bitmap=FileUtils.readImageFile(context,"image.png")     //读取副本图片文件
        bitmap?.let{
            image=it.asImageBitmap()                                //将 Bitmap 转换成 ImageBitmap 类型
        }
    }
    Column(modifier=modifier.fillMaxSize(),
        horizontalAlignment=Alignment.CenterHorizontally,
        verticalArrangement=Arrangement.Center){
        AsyncImage(modifier=Modifier.size(400.dp,200.dp),
            model=imageUrl,
            contentDescription="在线百度 logo 图片")                 //加载在线图片
        Button(onClick={
            action.invoke()                                          //执行下载业务
            download=true                                            //修改状态
        }){
            Text("生成本地副本")
        }
        if (download)
            Image(bitmap=image, modifier=Modifier.size(400.dp,200.dp),
                contentDescription="显示副本")                       //显示副本图片
    }
}
```

在 MainActivity 中调用 HomeScreen,构成显示的主界面内容。在 MainActivity 中定义的 downloadFile 函数可以作为参数传递给 HomeScreen,使得 HomeScreen 具备了下载并复制图片文件的能力,代码如下:

```kotlin
//模块 Ch09_03 的 MainActivity.kt
class MainActivity: ComponentActivity() {
    override fun onCreate(savedInstanceState: Bundle?) {
        super.onCreate(savedInstanceState)
        enableEdgeToEdge()
        setContent {
            Chapter09Theme {
                Scaffold(modifier=Modifier.fillMaxSize()) {
                    innerPadding ->HomeScreen(modifier=Modifier.padding(innerPadding)){
                        downloadFile()
                    }
                }
            }
        }
    }
    private fun downloadFile(){
        val constraints=androidx.work.Constraints.Builder()
            .setRequiredNetworkType(NetworkType.CONNECTED)
            .setRequiresStorageNotLow(true)
            .build()                                                //创建下载的约束条件
        val request=OneTimeWorkRequestBuilder<DownloadWork>()
            .setConstraints(constraints)                            //配置约束条件
            .build()                                                //创建任务请求
        WorkManager.getInstance(this).enqueue(request)              //将请求加入请求队列
    }
}
```

运行结果如图 9-9 所示。

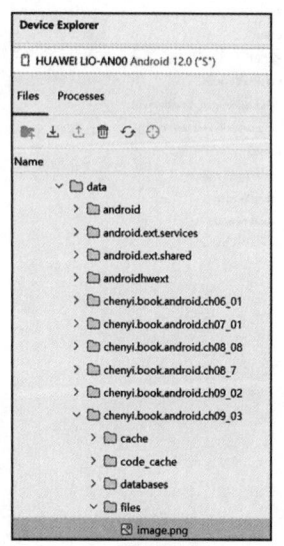

图 9-9　运行结果

本移动应用因为使用 WorkManager 组件，设置了联网访问的约束，模拟器的 ConnectivityManager.activeNetwork 不具有 NetworkCapabilities.NET_CAPABILITY_VALIDATED 功能，所以只能在真机中测试运行。观察运行结果可以发现，下载的图片经过复制后保存到了 data/data/ chenyi.book.android.

ch09_03/files 目录下。本移动应用通过文件输入流可以将一个图片文件转换成 Bitmap 对象，并调用 asImageBitmap()函数转换成能在 Image 可组合项中显示的 ImageBitmap 对象，最后在 Image 可组合项显示。在实际的应用中，往往需要将在线的资源缓存到本地，通过 JAVA IO 流库可以方便地实现这样的应用需求。

9.3 Room 组件

在实际的移动应用中，往往需要处理大量结构复杂多样的数据。针对此问题，Android 提供了内置的 SQLite 数据库。SQLite 是一个基于 C 语言的轻量级的数据库引擎，用于资源有限的设备进行适量的数据存取。从本质上来说，保存数据的 SQLite 只是一个文件，其内部只支持 NULL、INTEGER、REAL、TEXT 和 BLOB 这 5 种数据类型，可以接收 varchar(n)，char(n)，decimal(p,s)等数据类型，SQLite 在运算或保存数据时会转换成上述 5 种类型。如果要存储布尔类型的数据，可以存储为 Integer，0 值为 false，1 为 true。SQLite 使用 SQL 命令的完整关系型数据库。每个使用 SQLite 的移动应用都有一个该数据库的实例，并且在默认情况下仅限当前应用使用。

可在 Android Studio 中可以使用 Database Navigator 插件查看 SQLite 数据库中的内容。如果 Android Studio 没有安装 Database Navigator，可以通过 File|Settings 菜单进入配置界面，在 Settings 对话框中选中 Plugins 选项，然后在搜索栏中输入 Database Navigator 进行检索。如果已经安装，则会显示如图 9-10 的 installed 效果。如果没有安装，则选择 install 进行安装。成功后，菜单选项会显示 DB Navigator 的菜单项。可以选择 Database Browser 查看数据库。

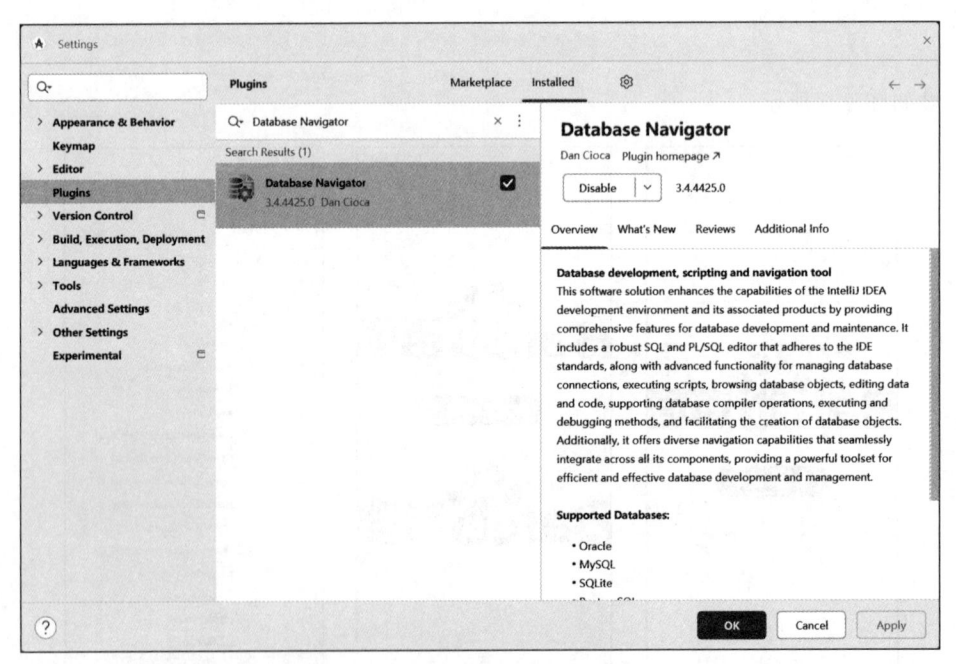

图 9-10　Database Navigator 插件

2018 年，Android JetPack 推出的 Room 组件可以采用基于 ORM(Object Relational Mapping，对象关系映射)库进行数据库的持久化处理，即在 SQLite 的基础上提供了一个抽象层，让对象和数据库之间映射的元数据，将实体类对象自动持久化到关系数据库中，如图 9-11 所示。这种方式使得处理数据库操作更加方便。

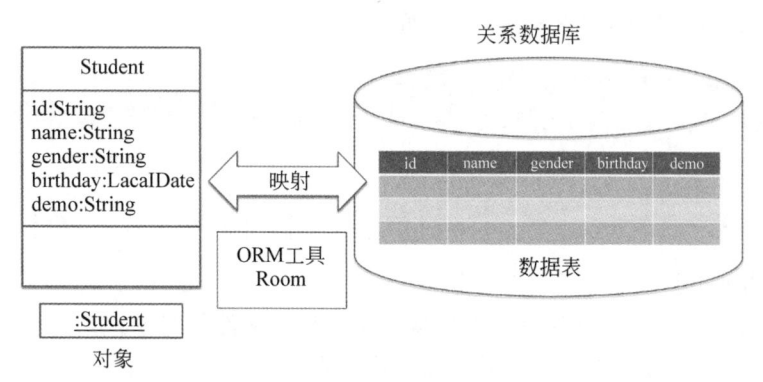

图 9-11 ORM 映射示意

9.3.1 用 Room 实现数据库的基本操作

要使用 Room 库,必须对应用的项目模块进行配置。又因为 Room 组件使用注解配置代码,必须对代码中出现的注解进行解析。本章采用主流注解解析工具 KSP(Kotlin Symbol Processing,Kotlin 符号处理)处理注解。首先需要在项目的 libs.versions.toml 文件中增加配置 Room 和 KSP 的依赖、插件的相应的版本信息,代码如下:

```
[versions]
room="2.6.1"
ksp=" 1.9.20-1.0.14"
[libraries]
androidx-room-runtime={group="androidx.room",name="room-runtime",version.ref="room"}
androidx-room-compiler={
    group="androidx.room",name="room-compiler",version.ref="room"
}

androidx-room-ktx={group="androidx.room",name="room-ktx",version.ref="room"}
androidx-room-rxjava3={
    group="androidx.room",name="room-rxjava3",ver    sion.ref="room"}
    [plugins]
    google-devtools-ksp={id="com.google.devtools.ksp",version.ref="ksp"}
}
```

在对应应用模块所在项目的顶层构建配置文件 build.gradle.kts 中增加 KSP 插件,代码如下:

```
plugins {
    ...                                                              //略
    alias(libs.plugins.google.devtools.ksp) apply false
}
```

在应用模块的模块构建配置文件 build.gradle.kts 中,需要做 3 处配置。
增加用于处理注解的 KSP 插件,代码如下:

```
plugins {
    ...                                                              //略
    alias(libs.plugins.google.devtools.ksp)
}
```

配置编译器,增加如下内容,确保实现,代码如下:

```
ksp {
    arg("room.schemaLocation", "$projectDir/schemas")
    arg("room.incremental", "true")
    arg("room.expandProjection","true")
}
```

上述配置的说明如下。

(1) room.schemalLocation:启用并配置模块数据库导出文件的路径。

(2) room.incremental:启用 Gradle 增量注释处理器。

(3) room.expandProjection:配置 Room,以重写查询。

完成上述配置后增加依赖,代码如下:

```
dependencies {
    implementation(libs.androidx.room.runtime)
    //使用 Kotlin 标注处理工具 ksp
    ksp(libs.androidx.room.compiler)
    implementation(libs.androidx.room.rxjava3)
    //可选项,使用 Kotlin 扩展和协程
    implementation(libs.androidx.room.ktx)
    //可选项,使用 RxJava3 支持 Room
    implementation(libs.androidx.room.rxjava3)
    ...                                                                    //略
}
```

Room 中有 3 个非常重要的组成部分:数据库(Database)、数据访问对象(Data Access Object)和实体类(Entity)。图 9-12 展示了三者之间的关系。

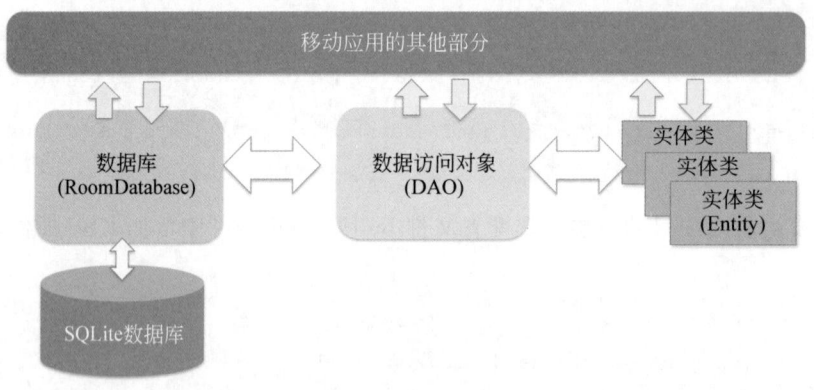

图 9-12　Room 组件 3 个成员的相互关系

(1) 实体类(Entity):映射并封装了数据库对应的数据表中对应的结构化数据。实体定义了数据库中的数据表。实体类中的数据域与表的列一一对应。

(2) 数据访问对象(Data Access Object,DAO):在 DAO 中定义了访问数据库的常见的操作(例如插入、删除、修改和检索等),以达到创建映射数据表的实体类对象,以及对该实体类对象实例的属性值进行设置和获取的目的。

(3) 数据库(RoomDatabase):表示对数据库基本信息的描述,包括数据库的版本、名称、包含的实

体类和提供的 DAO 实例。Room 组件中所有的数据库必须扩展为 RoomDatabase 抽象类,从而实现对实际 SQLite 数据库的封装。

下面,通过一个简单学生成绩登记移动应用实例来说明 Room 组件对数据库的访问。

1. 定义实体类

定义 Student 实体类,增加@Entity 注解,使之成为 Room 的实体类,并于具体的数据表的相应字段进行映射,代码如下:

```
//模块 Ch09_04 定义 Student 实体类
@Entity(tableName="students")
@Parcelize
data class Student(@PrimaryKey(autoGenerate=true)
    @ColumnInfo(name="studentId",typeAffinity=ColumnInfo.INTEGER) val id:Long,
    @ColumnInfo(name="studentNo",typeAffinity=ColumnInfo.TEXT) val no:String?,
    @ColumnInfo(name="studentName",typeAffinity=ColumnInfo.TEXT) val name:String,
    @ColumnInfo(name="studentScore",typeAffinity=ColumnInfo.INTEGER)
val score:Int,
    @ColumnInfo(name="studentGrade",typeAffinity=ColumnInfo.TEXT) val grade:String?):
        Parcelable{
    @Ignore
    constructor(no:String,name:String,score:Int,grade:String):this(0,no,name,score,
        grade)
}
```

Student 实体类用@Entity 注解设置 tableName="students"指定映射的数据表名为 students。使之成为 Room 的实体类。如果希望采用实体类名作为数据表名,去掉 tableName 参数的设置。

可以通过@ColumnInfo 注解增加了实体类 name 属性与数据表对应字段的映射,也可以通过 typeAffinity 属性指定对应 SQLite 数据库中数据表中各字段的数据类型。当然,在 Room 映射的实体类对象和数据表中的记录,并不一定需要设置 typeAffinity 属性。如果未指定,则默认值为 UNDEFINED,Room 会根据字段的类型和可用的类型转换器 TypeConverters 对其进行解析。@ColumnInfo 注解中,Student 类映射的 students 数据表的内容如图 9-13 所示。

studentId	studentNo	studentName	studentScore	stuentGrade

图 9-13 数据表的示意图

此时,Student 类的 id 属性与数据表 students 的 studentId 字段对应,Student 类的 no 属性与 students 的 studentNo 字段对应,Student 类的 name 属性与 students 的 studentName 字段对应,Student 类的 score 属性与 students 的 studentScore 字段对应,Student 类的 grade 属性与 students 的 studentGrade 字段对应。

为了创建实体类的对象实例更加灵活,往往需要多个构造函数进行不同情况下对象实例的创建。由于 Room 只能识别和使用一个构造函数,往往是主构造函数创建对象。如果在实体类中定义了多个其他的构造函数,可以使用@Ignore 让 Room 忽略这些构造函数。

为了创建实体类的对象实例更加灵活,往往需要多个构造函数,用于不同情况下对象实例的创建。由于 Room 只能识别和使用一个构造函数,往往使用主构造函数创建对象。如果在实体类中定义了多个其他的构造函数,可以使用@Ignore 让 Room 忽略这些构造函数。

如果希望数据表中的字段与类的属性名称保持一致,则可以将@ColumnInfo注解进行删除。因此一个最为简单的实体类定义可以写成如下形式:

```
@Entity
data class Student(@PrimaryKey(autoGenerate=true) val id: Long,val no: String,
    val name: String, val score: Int, val grade: String)
```

因为@Entity没有指定数据表,而是将类名作为数据表的名称,这时生成的数据表名为Student,并且这个数据表的字段名称分别为id、no、name、score和grade。在本例中,仍采用较为完整的定义方式,模块Ch09_04定义实体类Student。另外,在Student类定义实现了Parcelable是定义为可序列化的处理,为应用的数据传递提供支持。

2. 定义数据访问对象

数据访问对象(DAO)用于封装数据访问的业务处理,用于访问存储在SQLite数据库中的数据,可以根据应用的需求定义相关的增加、删除、修改和检索操作的方法。需要通过@Dao注解将DAO定义为接口,在接口中,使用注解定义实现SQL操作的方法。这些注解可以由@Insert、@Update、@Delete或@Query构成,代码如下:

```
//模块Ch09_04 定义DAO接口 StudentDAO.kt
@Dao
interface StudentDAO {
    @Insert
    fun insertStudent(student:Student):Long                    //插入记录
    @Update
    fun updateStudent(student:Student)                         //修改记录
    @Delete
    fun deleteStudent(student:Student)                         //删除记录
    @Query("select * from students")
    fun queryAllStudents():List<Student>                       //检索所有记录
    @Query("select * from students where studentNo=:no")
    fun queryStudentByNo(no:String):Student                    //检索指定学号的学生记录
}
```

在上述代码中,只有在@Dao注解给接口StuentDAO的前提下,Room才会将StudentDAO处理成DAO。StudentDAO根据业务要求,封装了对数据库的增加、删除、修改和检索等基本操作的方法。@Insert注解标注了insertStudent函数,将参数student的对象插入数据库中,并返回在数据表存储主键studentId的值。@Update注解标注了updateStudent函数,将一个student对象修改对应数据库中的对应记录。@Delete注解标注了deleteStudent函数,可以实现将一个student对象的记录从数据库中删除。@Query注解标注了queryAllStudents函数和queryStudentByNo函数,通过特定标注参数设定的操作分别执行检索所有记录和按照学号检索返回指定的学生的记录。

3. 定义数据库类

RoomDatabase是一个抽象类。应用定义的数据库类必须扩展为RoomDatabase类,用于封装Android系统的SQLite数据库,是嵌入Android中SQLite数据库上一层的数据库封装。RoomDatabase的实例用于创建和返回数据库对象实例,每个移动应用只有一个RoomDatabase对象。

Room使用@Database注解标注数据库,并通过参数的设置,指定数据库的数据表对应的实体类。如果数据库中有多个数据表,则需要将它们都加入entities属性值对应的数组中,并用"["和"]"括起来,然后

用","进行分隔,用 version 属性指定数据库的当前版本,代码如下:

```kotlin
//模块 Ch09_04 定义数据库类 StudentDatabase.kt
//创建数据库 1.0 版,定义数据表与 Student 实体类对应
@Database(entities=[Student::class], version=1)
abstract class StudentDatabase: RoomDatabase() {
    abstract fun studentDao(): StudentDAO                    //数据访问对象
    companion object{
        private var instance: StudentDatabase?=null
        @Synchronized
        fun getInstance(context: Context): StudentDatabase {
                                            //单例模式创建为一个 StudentDatabase 对象实例
            instance?.let{
                return it
            }
            return Room.databaseBuilder(context, StudentDatabase::class.java,
                "studentDB.db").build()
        }
    }
}
```

StudentDatabase 是一个 Room Database 类。studentDao()函数用于获取 StudentDAO 数据访问对象。因为数据库具有唯一性,因此采用了单例设计模式,将 StudentDatabase 类维护自身的唯一对象实例,并通过 getInstance 函数来获得这个唯一的对象实例。另外,通过 Room.databaseBuilder 构建器中构建了 StudentDatabase 的唯一实例,在构建的过程中指定的 Context 的对象(即 context)表示必须是当前移动应用的上下文。它具有唯一性,并不对应不同 Activity 的上下文或 Service 的上下文。可通过类 StudentDatabase::class.java 指定 Class 类型;并在最后一个参数中指定数据库 studentDB.db。

4. 为应用增加数据库的操作

使用 Room 执行数据库的操作不能通过主线程完成。这是因为执行数据库的操作比较耗时,会导致影响主线程中的 UI 无法正常的显示和渲染。因此利用 Room 对数据库进行操作,可以使用协程或者 RxJava 来处理异步任务。RxJava 3 框架在第 5 章已经介绍,操作十分简单。在处理 Room 数据库时,仍需要采用 RxJava 3 来处理异步任务,需要在项目的 libs.versions.toml 中增加配置,代码如下:

```
[versions]
rxjava="3.1.9"
rxandroid="3.0.2"
[libraries]
rxjava3={group="io.reactivex.rxjava3",name="rxjava",version.ref="rxjava"}
rxandroid={group="io.reactivex.rxjava3",name="rxandroid",version.ref ="rxandroid"}
```

在应用模块的构建配置文件 build.gradle.kts 中增加 RxJava 3 库的依赖,代码如下:

```
//增加 RxJava 库的依赖
implementation(libs.rxjava3)
implementation(libs.rxandroid)
```

定义视图模型 StudentViewModel 类,实现对数据库的插入和检索操作,代码如下:

```kotlin
//模块Ch09_04的StudentViewModel.kt
class StudentViewModel: ViewModel() {
    var _output:MutableStateFlow<List<Student>?>=MutableStateFlow(null)
    val output: StateFlow<List<Student>?>=_output.asStateFlow()
    private fun judgeGrade(score:Int)=when {                    //判断成绩等级
        score>=90 ->"优秀"
        score>=80 ->"良好"
        score>=70 ->"中等"
        score>=60 ->"及格"
        else ->"不及格"
    }
    fun handleQuery() {
        Flowable.create<List<Student>>( {
            emitter ->val dao=StudentDatabase.getInstance(StudentApp.context).studentDao()
                                                                //获得Dao对象
            val students=dao.queryAllStudents()                 //检索记录
            emitter.onNext(students)                            //发射数据
        }, BackpressureStrategy.DROP)                           //设置背压策略
            .subscribeOn(Schedulers.io())                       //指定被观察者的线程处理I/O操作
            .observeOn(AndroidSchedulers.mainThread())          //指定观察者的线程为主线程
            .subscribe({
                it: List<Student> ->_output.value=it            //修改数据
            }) {
                e: Throwable? ->e?.printStackTrace()
            }
    }
    fun handleInsert(no:String,name:String,score:Int,){
        var student=Student(no,name,score,judgeGrade(score))
        Flowable.create<List<Student>>({
            emitter ->val dao =StudentDatabase.getInstance(StudentApp.context).studentDao()
                                                                //获得Dao对象
            var id=dao.insertStudent(student)                   //插入记录
            val students=dao.queryAllStudents()                 //检索记录
            emitter.onNext(students)                            //发射数据
        }, BackpressureStrategy.DROP)                           //设置背压策略
            .subscribeOn(Schedulers.io())                       //指定被观察者的线程处理I/O操作
            .observeOn(AndroidSchedulers.mainThread())          //指定观察者的线程为主线程
            .subscribe({
                it: List<Student> ->_output.value=it            //更新数据
            }) {
                e: Throwable? ->e?.printStackTrace()
            }
    }
}
```

在上述代码定义的 handleQuery 和 handleInsert 中采用 RxJava 3 库的 Flowable 来处理 StudentDatabase 的数据访问操作，接收 Student 的学号、姓名和成绩数据，根据成绩判断成绩登记，再生成 Student 对象插入数据库中，并检索 StudentDatabase 中所有的 Student 记录，返回一个包含所有 Student 数据的列表，一旦检索成功，利用 Flowable 将包含所有学生记录的列表进行发射，并在主线程

中修改可变状态流中的数据。

定义登记学生成绩的可组合函数 RegisterScreen，代码如下：

```
//模块 Ch09_04 RegisterScreen
@Composable
fun RegisterScreen(viewModel: StudentViewModel,navController: NavHostController){
    var no by remember{ mutableStateOf("") }
    var name by remember{ mutableStateOf("") }
    var score by remember{ mutableStateOf("") }
    Column(modifier=Modifier.fillMaxSize(),
        horizontalAlignment=Alignment.CenterHorizontally,
        verticalArrangement=Arrangement.Center){
            Text(modifier=Modifier.padding(bottom=100.dp),text="学生成绩等级界面",
                fontSize=36.sp)
         Row(verticalAlignment=Alignment.CenterVertically){
             Text("学生学号：",fontSize=24.sp)
             TextField(value=no, onValueChange ={it:String->no =it})
         }
         Row(modifier=Modifier.padding(top=30.dp),
             verticalAlignment=Alignment.CenterVertically){
             Text("学生姓名：",fontSize=24.sp)
             TextField(value=name, onValueChange={it:String->name=it})
         }
         Row(modifier=Modifier.padding(top=30.dp),
             verticalAlignment=Alignment.CenterVertically){
             Text("学生分数：",fontSize=24.sp)
             TextField(value="$score", onValueChange={it:String->score=it})
         }
         Button(modifier=Modifier.padding(top=100.dp), onClick={
             viewModel.handleInsert(no,name,score.toInt())
             navController.navigate("display")
         }){Text("登记成绩",fontSize =32.sp) }
    }
}
```

定义显示已登记成绩的学生成绩列表的可组合函数 DisplayScreen，因为采用了 LazyColumn 可组合项，定义的 StudentCard 表示单个学生记录的显示，代码如下：

```
//模块 Ch09_04 StudentCard 和 DisplayScreen
@Composable
fun StudentCard(student: Student){
    Row(modifier=Modifier.fillMaxWidth(),
        verticalAlignment=Alignment.CenterVertically){
        Text(modifier=Modifier.padding(start=20.dp,top=10.dp).size(120.dp,50.dp),
            textAlign=TextAlign.Justify,text=student.no?:"",fontSize=24.sp)
        Text(modifier=Modifier.padding(top=10.dp).size(120.dp,50.dp),text=student.name,
            fontSize=24.sp)
```

```
            Text(modifier=Modifier.padding(top=10.dp).size(60.dp,50.dp),
                text="${student.score}",fontSize=24.sp)
            Text(modifier=Modifier.size(80.dp,50.dp),text=student.grade?:"",fontSize=24.sp)
    }
}
@Composable
fun DisplayScreen(viewModel: StudentViewModel){
    val students=viewModel.output.collectAsState()                    //获取状态
    LaunchedEffect (students){                                        //显示学生列表
        viewModel.handleQuery()
    }
    Column(modifier=Modifier.fillMaxSize(),
        verticalArrangement=Arrangement.Center,
        horizontalAlignment=Alignment.CenterHorizontally){
            Text(modifier=Modifier.padding(bottom=50.dp),text="学生成绩列表界面",
                fontSize=36.sp)
            LazyColumn {
                students.value?.let{
                    items(it) {
                        StudentCard(it)
                    }
                }
            }
    }
}
```

为了实现从登记成绩界面 RegisterScreen 跳转到显示学生成绩列表的界面 DisplayScreen, 定义一个导航图可组合函数 NavigationGraphScreen, 代码如下：

```
//模块 Ch09_04 NavigationGraphScreen
@Composable
fun NavigationGraphScreen(modifier:
Modifier,viewModel:StudentViewModel){
    val navController: NavHostController=rememberNavController()
    NavHost(navController=navController, startDestination="register" ) {
        composable(route="register"){                                 //登记学生成绩界面
            RegisterScreen(viewModel=viewModel,navController=navController)
        }
        composable(route="display"){                                  //学生成绩列表界面
            DisplayScreen(viewModel=viewModel)
        }
    }
}
```

最后, 在 MainActivity 创建视图模型 StudentViewModel 对象, 调用 NavigationGraphScreen, 生成主界面, 代码如下：

```kotlin
//模块 Ch09_04 MainActivity.kt
class MainActivity : ComponentActivity() {
    override fun onCreate(savedInstanceState: Bundle?) {
        super.onCreate(savedInstanceState)
        enableEdgeToEdge()
        val viewModel=ViewModelProvider(this).get(StudentViewModel::class.java)
        setContent {
            Chapter09Theme {
                Scaffold(modifier=Modifier.fillMaxSize()) {
                    innerPadding->NavigationGraphScreen(modifier=
                        Modifier.padding(innerPadding), viewModel=viewModel )
                }
            }
        }
    }
}
```

运行结果如图 9-14 所示。

(a) 成绩登记界面　　(b) 学生成绩列表界面

图 9-14　运行结果

9.3.2　用 Room 实现迁移数据库

伴随着移动应用需求的变化和版本的更新，数据库也需要不断地升级。在数据库升级时，会希望保留原有的数据。因此，Room 提供了数据库迁移的方式来解决数据库的升级。

Room 库提供了 Migration 类来实现数据库的增量迁移。每个 Migration 子类都可用 Migration.migrate 函数实现新旧版本数据库之间的迁移。当移动应用需要升级数据库时，Room 库会利用一个或多个 Migration 子类运行 migrate 函数，在运行时将数据库迁移到最新版本，代码如下：

```kotlin
val MIGRATION_1_2=object: Migration(1, 2) {                    //数据库从版本1迁移到版本2
    override fun migrate(database: SupportSQLiteDatabase) {    //迁移方法定义
        database.execSQL("CREATE TABLE courses(courseId TEXT primary key not null, " +
            "courseName TEXT not null, " +"courseDemo TEXT)")   //创建一个新的数据表 course
    }
}
val MIGRATION_2_3=object: Migration(2, 3) {                    //数据库从版本2迁移到版本3
    override fun migrate(database: SupportSQLiteDatabase) {    //迁移方法定义
        database.execSQL("ALTER TABLE students ADD COLUMN studentAddress TEXT")
                        //修改数据表 students,增加一个新的字段 address,数据类型为 TEXT 字符串
    }
}
Room.databaseBuilder(applicationContext, StudentDatabase::class.java, "studentDB.db")
    .addMigrations(MIGRATION_1_2, MIGRATION_2_3).build()       //执行数据库迁移
```

例 9-5　在 9.3.1 节操作的数据库的基础上迁移 StudentDB.db 数据库,修改 StudentDB.db 中的数据表 students,增加一个新的字段表示学生的地址信息,数据类型为字符串。增加新的数据表 courses 表示课程,字段包括:

课程编号:字符串
课程名称:字符串
课程说明:字符串

其中,"课程编号"为主键。

在本例中,首先建立 StudentDB.db 数据库以及创建该数据库下的数据表 students 成功后才可以执行下列操作来实现数据库的迁移。

(1) 为数据库 StudentDB.db 的数据表 students 新增加字段。为数据库 studentDB.db 的数据表 students 增加一个 studentAddress 字段,代码如下:

```kotlin
//模块 Ch09_05 面向 studentDB.db 第2版数据库实体类 Student 对应的文件
Student.kt
//数据库 StudentDatabase 第2版 定义的学生实体类
@Entity(tableName="students")
@Parcelize
data class Student(@PrimaryKey(autoGenerate=true)
    @ColumnInfo(name="studentId") val id:Long,
    @ColumnInfo(name="studentNo") val no:String?,
    @ColumnInfo(name="studentName") val name:String,
    @ColumnInfo(name="studentScore") val score:Int,
    @ColumnInfo(name="studentGrade") val grade:String?,
    @ColumnInfo(name="studentAddress") val address:String?):Parcelable{
    @Ignore
    constructor(no:String,name:String,score:Int,grade:String,address:String):
        this(0,no,name,score,grade,address)
}
```

在 Student 实体类中增加属性 address 映射数据表的 studentAddress 字段并修改相应的构造函数。由于 Room 只能识别和使用一个构造器,所以如果存在多个构造器,就可以使用@Ignore 让 Room 忽略

那些与数据库处理无关的构造器。在Student类中使用constructor定义的副构造器使用@Ignore注解表示让Room对其忽略,代码如下:

```kotlin
//模块Ch09_05第一次修改数据库StudentDatabase.kt第2版的StudentDatabase
@Database(entities=[Student::class], version=2)
abstract class StudentDatabase: RoomDatabase() {
    abstract fun studentDao(): StudentDAO
    companion object{
        private var instance: StudentDatabase?=null
        val MIGRATION_1_2=object : Migration(1, 2) {             //数据库从版本1迁移到版本2
            override fun migrate(database: SupportSQLiteDatabase) {      //迁移方法定义
                database.execSQL("ALTER TABLE students ADD COLUMN studentAddressTEXT")
                        //修改数据表students,增加一个新的字段address,数据类型为TEXT字符串
            }
        }
        @Synchronized
        fun getInstance(context: Context): StudentDatabase{
                            //单例模式创建为一个StudentDatabase对象实例
            instance?.let{
                return it
            }
            return Room.databaseBuilder(context, StudentDatabase::class.java,
                "studentDB.db").addMigrations(MIGRATION_1_2).build().apply{
                instance=this
            }
        }
    }
}
```

在MainActivity中测试上述的数据库,代码如下:

```kotlin
//模块Ch09_05的MainActivity的测试核心代码:
class MainActivity: AppCompatActivity() {
    override fun onCreate(savedInstanceState: Bundle?) {
        super.onCreate(savedInstanceState)
        setContent{
            testVersion2()
        }
    }
    fun testVersion2(){                                    //数据库版本2的测试函数
        Flowable.create<Student>({
            emitter ->val dao=StudentDatabase.getInstance(StudentApp.context).studentDao()
                                                            //获得Dao对象
            dao.insertStudent(Student("6001013","李四",87,"良好",
                "江西省南昌红谷大道999号"))
                                                            //插入记录
            val students=dao.queryAllStudents()             //检索记录
```

```
            for(student in students)
                emitter.onNext(student)
        }, BackpressureStrategy.DROP)             //设置背压策略
            .subscribeOn(Schedulers.io())         //指定被观察者的线程处理I/O操作
            .observeOn(AndroidSchedulers.mainThread())  //指定观察者的线程为主线程
            .subscribe({
                it:Student ->Log.d("CH09_05","${it}")
            }) {
                e: Throwable? ->e?.printStackTrace()
            }
    }
}
```

利用 Device File Explorer，将模拟器中的 data/data/chenyi.book.android.ch09_05/database 目录下的数据库文件 StudentDB.db 导出到指定目录，例如 C:/database 目录中。然后，将使用 DB Navigator 查看数据库 StudentDB.db，可以发现数据库中的数据表已经发生变更，新增了字段，数据库的结果如图 9-15 所示。

图 9-15 增加新字段的数据表 students

（2）再次升级数据库 StudentDB.db 的新建数据表 courses。数据库中需要新建数据表 courses 来存储课程信息，可以通过下列的操作来实现，代码如下：

```
//模块 Ch09_05 定义课程实体类 Course 对应的 Course.kt
@Entity(tableName="courses")
@Parcelize
data class Course(@PrimaryKey @ColumnInfo(name="courseId") val id:String,
    @ColumnInfo(name="courseName") val name:String,
    @ColumnInfo(name="courseDemo") val demo:String?):Parcelable
```

在 Course 类中，通过@Entity 将 Course 类映射成数据表 courses。使用@ColumnInfo 注解将类的 id 属性映射为数据表的 courseId 字段。同理，类的 name 属性映射对应数据表的 courseName 字段，类的 demo 属性映射对应数据表的 courseDemo 字段，如图 9-16 所示。

Course		couseId	courseName	courseDemo
id:String	⇔			
name:String				
demo:String				

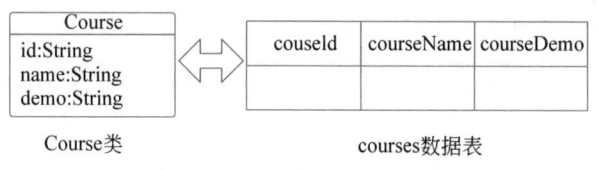

Course类　　　　　　　　　　　courses数据表

图 9-16　实体类与数据表映射

访问 Course 对象的代码如下：

```kotlin
//模块 Ch09_05 定义访问 Course 对象的 CourseDAO.kt
@Dao
interface CourseDAO {
    @Insert
    fun insertCourse(course: Course)                              //插入记录
    @Update
    fun updateCourse(course: Course)                              //修改记录
    @Delete
    fun deleteCourse(course: Course)                              //删除记录
    @Query("delete from courses where courseId=: id")             //按照课程编号删除记录
    fun deleteByCourseId(id: String)
    @Query("select * from courses")
    fun queryAllCourses(): List<Course>                           //检索所有的课程记录
}
```

可以在原有的 StudentDatabase 基础上增加第二次数据库迁移的操作，代码如下：

```kotlin
//模块 Ch09_05 studentDB.db 第 3 版数据库,第二次修改数据库
StudentDatabase.kt
@Database(entities=[Student::class,Course::class], version=3)
abstract class StudentDatabase:RoomDatabase() {
    abstract fun studentDao():StudentDAO
    abstract fun courseDao():CourseDAO
    companion object{
        private var instance:StudentDatabase?=null
        private val MIGRATION_1_2=object : Migration(1, 2) {
            override fun migrate(database: SupportSQLiteDatabase) {
                                                                  //迁移方法定义
                //修改数据表 students,增加一个新的字段 studentAddress,数据类型为 TEXT 字符串
                database.execSQL("ALTER TABLE students ADD COLUMN studentAddress TEXT")
            }
        }
        private val MIGRATION_2_3=object: Migration(2, 3) {
                                                                  //数据库从版本 2 迁移到版本 3
            override fun migrate(database: SupportSQLiteDatabase) {    //迁移方法定义
                database.execSQL("CREATE TABLE courses(courseId TEXT primary key not
                    null," +"courseName TEXT not null," +"courseDemo TEXT)")
                                                                  //创建一个新的数据表 course
            }
        }
        @Synchronized
        fun getInstance(context: Context): StudentDatabase{
                                                //单例模式创建为一个 StudentDatabase 对象实例
```

```
            instance?.let{return it}
            return Room.databaseBuilder(context, StudentDatabase::class.java,
                "studentDB.db").addMigrations(MIGRATION_1_2,MIGRATION_2_3).build().apply{
                instance=this
            }
        }
    }
}
```

需要在 MainActivity 中执行 StudentDatabase 实现数据库迁移,代码如下:

```
//模块 Ch09_05 在主活动定义 MainActivity.kt 测试核心代码
class MainActivity: AppCompatActivity() {
    override fun onCreate(savedInstanceState: Bundle?) {
        super.onCreate(savedInstanceState)
        setContent{
            testVersion3()                                      //测试升级到第三版数据库
        }
    }

    fun testVersion2(){                                         //数据库版本 2 的测试函数
        Flowable.create<Student>({
            emitter->val dao=StudentDatabase.getInstance(StudentApp.
            context).studentDao()                               //获得 Dao 对象
            dao.insertStudent(Student("6001013","李四",87,"良好",
                "江西省南昌红谷大道 999 号"))
                                                                //插入记录
            val students=dao.queryAllStudents()                 //检索记录
            for (student in students)
                emitter.onNext(student)
        }, BackpressureStrategy.DROP)                           //设置背压策略
            .subscribeOn(Schedulers.io())                       //指定被观察者的线程处理 I/O 操作
            .observeOn(AndroidSchedulers.mainThread())          //指定观察者的线程为主线程
            .subscribe({
                it:Student ->Log.d("CH09_05","${it}")
            }) {
                e: Throwable? ->e?.printStackTrace()
            }
    }
    fun testVersion3(){                                         //数据库版本 3 的测试函数
        Flowable.create<Course>({
            emitter->val dao=StudentDatabase.getInstance(StudentApp.context).courseDao()
                                                                //获得 Dao 对象
            dao.insertCourse(Course("CNO690232","Android 移动应用开发","秋级学期开课"))
                                                                //插入记录
            val courses=dao.queryAllCourses()                   //检索记录
```

```
            for (course in courses)
                emitter.onNext(course)
        }, BackpressureStrategy.DROP)              //设置背压策略
            .subscribeOn(Schedulers.io())          //指定被观察者的线程处理I/O操作
            .observeOn(AndroidSchedulers.mainThread())  //指定观察者的线程为主线程
            .subscribe({
                Log.d("CH09_05","课程记录：${it}")
            }) {
                e: Throwable? ->e?.printStackTrace()
            }
        }
    }
```

经过测试，利用 Device File Explorer，将模拟器中的 data/data/chenyi.book.android.ch09_05/database 目录下升级后的数据库文件 StudentDB.db 导出到指定目录，例如 C:/database 目录中。然后，将使用 DB Navigator 查看生成的数据库 StudentDB.db，可以发现数据库中的数据表已经发生变更，创建了新的数据表 courses，数据库的结果如图 9-17 所示。

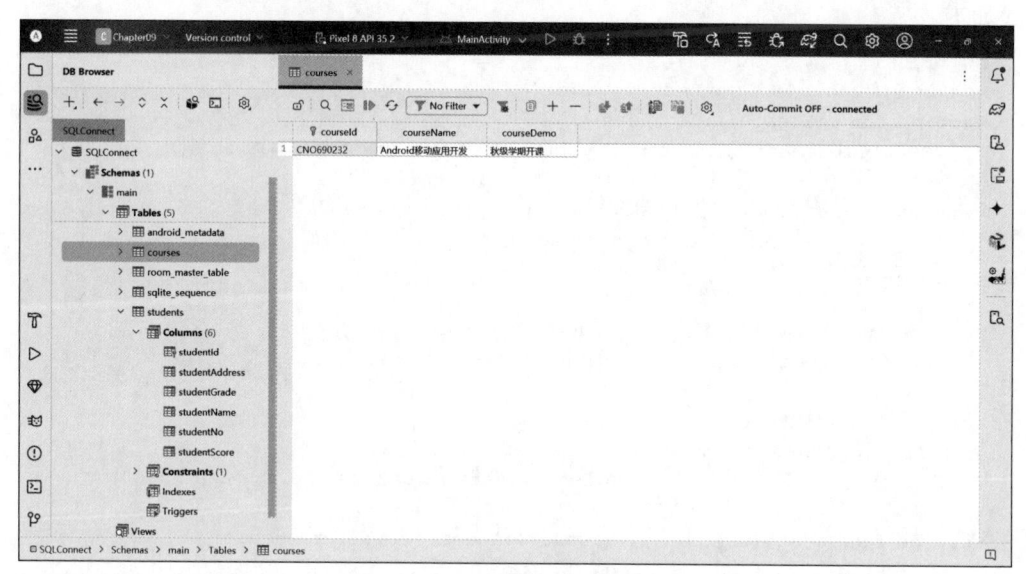

图 9-17 创建新的数据表 courses

9.4 ContentProvider 组件

移动应用之间常常需要共享数据，例如通讯录、媒体库、短信、地理位置等数据都可以在第三方移动应用之间共享。Android 提供的 ContentProvider 组件可以实现数据的共享。ContentProvider 组件的主要功能是为移动应用提供内容数据。这些内容数据可以从文件、数据库或云端的网络资源中获得，ContentProvider 组件将数据封装，并提供访问数据的标准方式，然后通过不同应用的 ContentResolver 对象根据特定的内容 URI 访问 ContentProvider 组件封装的数据，将获得的数据提供给移动应用使用。通过 ContentProvider 组件，实现了数据与应用界面的代码分离，在不同的移动应用共享数据的目的。

实际上，ContentProvider 在不同的应用程序共享数据才能发挥更大的作用。在图 9-18 展示了 ContentProvider 组件共享数据的情况。

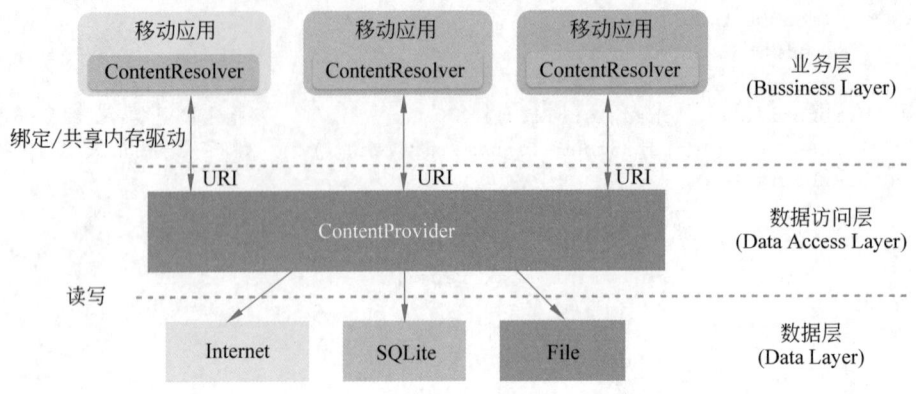

图 9-18 ContentProvider 组件共享数据

9.4.1 创建 ContentProvider 组件

要创建一个 ContentProvider 组件实现跨应用共享数据,必须先将其定义为 ContentProvider 类的子类,形式如下:

```kotlin
class MyContentProvider: ContentProvider() {
    override fun onCreate(): Boolean {
        TODO("Implement this to initialize your content provider on startup.")
    }
    override fun getType(uri: Uri): String? {
        TODO("Implement this to handle requests for the MIME type of the data at the given
            URI")
    }
    override fun insert(uri: Uri, values: ContentValues?): Uri? {
        TODO("Implement this to handle requests to insert a new row.")
    }
    override fun delete(uri: Uri, selection: String?,
        selectionArgs: Array<String>?): Int {
        TODO("Implement this to handle requests to delete one or more rows")
    }
    override fun update(uri: Uri, values: ContentValues?, selection: String?,
        selectionArgs: Array<String>?): Int {
        TODO("Implement this to handle requests to update one or more rows.")
    }
    override fun query(uri: Uri, projection: Array<String>?, selection:String?,
        selectionArgs: Array<String>?, sortOrder: String?): Cursor? {
        TODO("Implement this to handle query requests from clients.")
    }
}
```

自定义的 ContentProvider 组件需要用到 ContentProvider 类的 6 个函数。

(1) onCreate 函数。该函数用于 ContentProvider 组件的初始化,如果初始化成功,则返回 true,如果初始化失败,则返回 false。

(2) getType(uri:Uri)函数。该函数根据传入的内容 URI 返回相应的 MIME 类型,给定内容 URI 的数据类型。

(3) insert(uri:Uri, values:ContentValues?)函数。该函数用于向 ContentProvider 组件插入数

据。数据的内容被封装在 ContentValues 对象中,会根据 URI 指定的访问协议和权限插入指定的表中。

(4) delete(uri: Uri, selection: String?, selectionArgs: Array<String>?)函数。该函数用于按照设置的选择条件,删除 URI 指定的表或记录。

(5) update(uri: Uri, values: ContentValues?, selection: String?, selectionArgs: Array<String>?)函数。该函数用于按照设置的条件和提供数据的 ContentValues 对象,更新 ContentProvider 组件的数据。

(6) query(uri: Uri, projection: Array<String>?, selection: String?, selectionArgs: Array<String>?, sortOrder: String?)函数。该函数用于按照设置的条件,检索相应的记录。

在 ContentProvider 组件的 6 个函数定义中都涉及了 Uri 变量。Uri 变量是内容 URI 统一资源标识符,是 ContentProvider 组件提供的标准访问方式。ContentProvider 组件提供的内容 URI 具有如下形式:

```
schema://authority/path/id
```

其中参数说明如下。

(1) schema:表示访问协议,使用 ContentProvidere 组件时,访问协议为 content,表示提供内容。

(2) authority:表示访问权限,用于指定 ContentProvider 组件的唯一的名称,与配置清单文件 AndroidManifest.xml 指定 provider 元素的 android:authorities 配置一致,用来区分不同的 ContentProvider 组件,通常结尾会使用 Provider 表示。

(3) path:指定 ContentProvider 组件提供的数据类型。

(4) id:表示特定记录时才会使用编号。值得注意的是,id 是一个可选项,一旦内容 URI 出现 id 的设置,表示它是基于 ID 的内容 URI,用于访问特定记录的数据;如果没有 id 值,则表示它基于目录的 URI,访问特定目录/表(访问数据表针对 SQLite 数据库)下的数据。

内容 URI 与 ContentProvider 组件的 getType()函数有着千丝万缕的联系。getType 函数返回的是内容 URI 对应的 MIME 类型。MIME 类型即接收数据的格式类型,通过 MIME 类型,可以适当地处理数据。常见的 MIME 类型有表示纯文本的 text/plain、表示 JSON 数据的 application/json。在 Android 移动应用中定义的 MIME 类型所用的格式由两部分构成:

```
类型.子类型/有 ContentProvider 权限的特定内容
```

Android 的移动应用中规定 MIME 类型为 vnd,表示 MIME 不是官方的 IETF MIME 类型,而是"供应商前缀",数据格式是由供应商自定义的。其子类型主要有两种形式。

形式 1:

```
android.cursor.item
```

表示 URI 模式用于一行。

形式 2:

```
android.cursor.dir
```

如果将自定义的 ContentProvider 内容提供者的权限设置为 chenyi.book.android.ch09.provider,且这个自定义 ContentProvider 公开的是 city 表,则访问 city 表的 MIME 类型如下。

(1) vnd.android.cursor.item/chenyi.book.android.ch09.provider.city:表示访问一行。

• 349 •

（2）vnd.android.cursor.dir/chenyi.book.android.ch09.provider.city：表示访问多行。

在 getType() 函数中需要对内容 URI 对象进行匹配，获得指定的 MIME 类型。可以通过调用 android.content.UriMatcher 对象的 match() 函数实现 uri 的匹配。

例 9-6 定义如图 9-19 所示的 ContentProvider 应用架构。利用数据库 district.db 实现共享国内主要城市的信息。district.db 中定义了数据表 city 保存了各省市地区的基本信息。

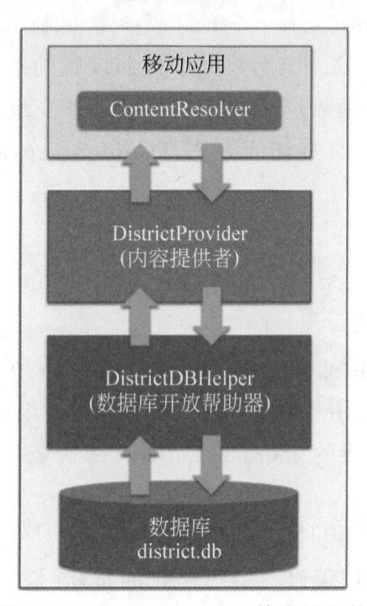

图 9-19 ContentProvider 的应用架构

在这个例子中应用的是本地的 district.db 数据库，该数据库有数据表 city，结构如图 9-20 所示。

图 9-20 数据表 city 的存储结构

使用 ContentProvider 组件的具体做法是定义 ContentProvider 组件，在本例中定义为 DistrictProvider。移动应用通过 Context 对象的 getContentResolver 函数获得 ContentResolver 对象，再通过 ContentResolver 对象来获得 DistrictProvider；DistrictProvider 通过 Room 组件的 Database 对象来访问数据库。

定义 Room 数据库类 DistrictDatabase，它是 RoomDatabase 的子类，用于获取数据库实例对象，代

码如下:

```kotlin
//模块 Ch09_06 数据库 DistrictDatabase.kt
@Database(entities=[District::class], version=1)
abstract class DistrictDatabase: RoomDatabase() {
    companion object{
        private var instance: DistrictDatabase?=null
        @Synchronized
        fun getInstance(context: Context): DistrictDatabase{
                                        //单例模式创建 DistrictDatabase 对象实例
            instance?.let{
                return it
            }
            return Room.databaseBuilder(context, DistrictDatabase::class.java, "
                districtDB.db").build()
        }
    }
}
```

定义一个 ContentProvider 组件(本例中定义为 DistrictProvider),实现对数据的封装,并提供 uri 的接口,方便按照指定的访问格式,对数据进行访问读取和写入,代码如下:

```kotlin
//模块 Ch09_06 内容提供者组件的定义 DistrictProvider.kt
class DistrictContentProvider: ContentProvider() {
    lateinit var database:DistrictDatabase              //获得数据库对象
    val TABLENAME="district"
    private val districtDir=0                           //定义 district 表的多行
    private val districtItem=1                          //定义 district 表的单行
    private val authority="chenyi.book.android.ch09.provider"
                                                        //定义 ContentProvider 的权限
    private val uriMatcher: UriMatcher=UriMatcher(UriMatcher.NO_MATCH)
    init{
        uriMatcher.addURI(authority,"district",districtDir)
        uriMatcher.addURI(authority,"district/# ",districtItem)
    }
    override fun onCreate(): Boolean=context?.let{context: Context->
                                                        //初始化 MyContentProvider
        database=DistrictDatabase.getInstance(context)  //创建数据库
        true
    }?:false
    override fun getType(uri: Uri): String? {           //返回 MIME 类型
        when(uriMatcher.match(uri)){                    //匹配内容 URI,返回 MIMI 对象
            districtDir->return "vnd.android.cursor.dir/chenyi.book.android
                .ch09.provider.district"
            districtItem->return "vnd.android.cursor.item/chenyi.book.android
                .ch09.provider.district"
        }
```

```kotlin
        return null
    }
    override fun insert(uri: Uri, values: ContentValues?): Uri? {      //插入记录
        val db=getWritableDatabase()                                    //获取可写数据库对象
        var returnUri:Uri?=null
        try {
            db.beginTransaction()                                       //开始事务
            values?.let{
                it:ContentValues->returnUri=ContentUris.withAppendedId(uri,
                    db.insert(TABLENAME, SQLiteDatabase.CONFLICT_REPLACE,it))
                                                                        //插入数据
            }
            db.setTransactionSuccessful()                               //设置事务处理成功标记
        } catch(e:Exception) {
            e.printStackTrace()
        } finally {
            db.endTransaction()                                         //结束事务
        }
        return returnUri
    }

    override fun delete(uri: Uri, selection: String?,
        selectionArgs: Array<String>?): Int {
        var removedNumber:Int=-1                                        //删除的行数
        val db=getWritableDatabase()
        try {
            db.beginTransaction()
            removedNumber=db.delete(TABLENAME,selection,selectionArgs)
                                                                        //删除记录返回删除行数
            db.setTransactionSuccessful()                               //成功处理事务
        } catch(e:Exception){
            e.printStackTrace()
        } finally {
            db.endTransaction()                                         //结束事务
        }
        return removedNumber
    }
    override fun update(uri: Uri, values: ContentValues?, selection: String?,
        selectionArgs: Array<String>?): Int {                           //修改记录
        var updatedNumber:Int=-1                                        //修改记录的行数
        val db=getWritableDatabase()
        try {
            db.beginTransaction()
            values?.let{
                it:ContentValues->updatedNumber=db.update(TABLENAME,
                    SQLiteDatabase.CONFLICT_NONE,it,selection,selectionArgs)
                                                                        //要修改的记录没有冲突就修改
            }
        } catch(e:Exception) {
            e.printStackTrace()
        } finally {
            db.endTransaction()
        }
```

```kotlin
            return updatedNumber
    }
    override fun query(uri: Uri, projection: Array<String>?, selection: String?,
        selectionArgs: Array<String>?, sortOrder: String? ): Cursor? {     //检索记录
        val sql:String=parseSelectSql(projection,selection,selectionArgs)
        val db=getReadableDatabase()                                       //获取可读的数据库
        var cursor:Cursor?=null
        try {
            db.beginTransaction()
            cursor=db.query("$sql")
        }catch(e:Exception){
            e.printStackTrace()
        }finally{
            db.endTransaction()
        }
        return cursor
    }
    private fun parseSelectSql(projection: Array<String>?, selection: String?,
        selectionArgs: Array<String>?):String{
                                    //解析出入的投影、选择条件和参数组装成检索 SQL
        if (projection==null)
            throw RuntimeException("投影字段为空!")
        var querySQl="select distinct ${projection?.get(0)} from $TABLENAME"
        selection?.let {
            it: String ->if (it?.contains("and")!!) {
            /*包含两个参数 select distinct countyName from district where provinceName=?
            and cityName=?*/
                var selectionOption=
                    selection?.replace("? and", "\"${selectionArgs?.get(0)!!}\" and")
                selectionOption=
                    selectionOption?.replace("?", "\"${selectionArgs?.get(1)!!}\"")
                querySQl="select distinct ${projection?.get(0)} from
                    $TABLENAME where $selectionOption"
            } else if (it.contains("?")!!) {
                /*包含一个参数 select distinct cityName from district where provinceName=?*/
                val selectionOption=selection?.replace("?", "\"${selectionArgs?.get
                    (0)!!}\"")
                querySQl="select distinct ${projection?.get(0)} from
                    $TABLENAME where $selectionOption"
             }
        }
        return querySQl
    }
    private fun getReadableDatabase():SupportSQLiteDatabase=
        database.openHelper.readableDatabase                               //可读数据库对象
    private fun getWritableDatabase():SupportSQLiteDatabase=
        database.openHelper.writableDatabase                               //可读写数据库对象
}
```

在模块的应用配置清单文件 AndroidManifest.xml 中会出现 DistrictProvider 类的配置。若将配置文件 android：enabled 设置为 true，表示可以使用这个组件；若将 android：exported 设置为 true，表示支持其他应用调用这个组件，可以让其他的移动应用共享其中封装的数据，代码如下：

```xml
<!--模块 Ch09_06 配置清单文件 AndroidManifest.xml -->
<?xml version="1.0" encoding="utf-8"?><manifest xmlns:android="http://schemas.android
    .com/apk/res/android">
    <application …>
        <provider
            android:name=".DistrictContentProvider"
            android:authorities="chenyi.book.android.ch09.provider"
            android:enabled="true"
            android:exported="true" />
        …                                                                       //略
    </application>
</manifest>
```

要让自定义的 DistrictProvider 可用,必须要将 DistrictProvider 加载到移动终端。因此,需要运行模块 ch09_06,使 DistrictProvider 加载并保存到移动终端中。

9.4.2 使用 ContentProvider 组件

要使用 ContentProvider 组件,需要通过调用 Context(上下文)对象的 getContentResolver 函数获得 ContentResolver 对象。通过 ContentResolver 调用对应的函数,对指定的内容提供者封装的数据共享和修改。ContentResolver 对象提供的对 ContentProvider 组件共享数据和修改数据的常用函数,如表 9-3 所示。

表 9-3 ContentResolver 访问内容提供者常见的函数

函 数	说 明
insert(Uri, ContentValues)	在给定的 URI 中插入一行,返回新插入行的 Uri
update(Uri, ContentValues, String, String[])	修改内容 URI 中的行
delete(Uri, String, String[])	删除由内容 URI 指定的行
query(Uri, String[], String, String[], String)	查询给定的 URI,返回带有结果集 Cursor 对象
query(Uri, String[], Bundle, CancellationSignal)	查询给定的 URI,返回一个支持取消的结果集的 Cursor 对象
getType(Uri)	返回给定内容 URL 的 MIME 类型

例 9-7 使用例 9-6 定义的 DistrictProvider 实现选择地区的操作。

在例 9-6 中,虽然运行模块 ch09_06,加载了 DistrictProvider 到移动终端中,但是并没有对数据库做任何处理。新建一个模块 ch09_07,与例 9-6 的模块不同。本例通过模块 ch09_07 实现对数据库的数据的变更。提供一个 cities.csv(数据格式同例 9-3)放置在模块 ch09_07 的 assets 目录中,将文件中的数据保存到数据库中。并通过检索,生成地区联动可选择的控件。

(1) 定义数据库处理的实体类 DBUtils,代码如下:

```
//模块 Ch09_07 访问内容提供者的实用类 DBUtils.kt
class DBUtils(val context: Context){
```

```kotlin
fun writeToDB(){
    val inputStream=context.assets.open("cities.csv")/
                                        //将assets文件的记录写入数据库district.db
    val inputReader=inputStream.bufferedReader(Charsets.UTF_8)
    val contentResolver=context.contentResolver
    val uri=Uri.parse("content://chenyi.book.android.ch09.provider/district")
    inputReader.use{
        it.lines().forEach {
            line:String->val content=line.split(",")
            val values=contentValuesOf(
                "state" to content[0],
                "city" to content[1],
                "county" to content[2])
            contentResolver.insert(uri,values)
        }
    }
}
fun getProvinces():List<String>{
    val provinces=mutableListOf<String>()
    val uri=Uri.parse("content://chenyi.book.android.ch09.provider/district")
    val contentResolver=context.contentResolver
    val cursor=contentResolver.query(uri,arrayOf("provinceName"),null,null,null)
                                //获得city数据表所有provinceName字段的值
    cursor?.apply{
        while(moveToNext()){
            val province=getString(0)
            if (province !in provinces) {
                provinces.add(province)
            }
        }
        close()                          //关闭游标
    }
    return provinces
}
fun getCities(province:String):List<String>{    //获得市级地区列表
    val cities=mutableListOf<String>()
    val uri=Uri.parse("content://chenyi.book.android.ch09.provider/district")
    val contentResolver=context.contentResolver
    val cursor=contentResolver.query(uri,arrayOf("cityName"),"provinceName =?",
        arrayOf(province),null)       //输入省或直辖市获得市级地区
    cursor?.apply{
        while (moveToNext()){
            val cityValue=getString(0)
            if (cityValue !in cities)
                cities.add(cityValue)
        }
        close()
    }
    return cities
}
```

```kotlin
fun getCounties(province:String,city:String):List<String>{     //获得县级地区列表
    val counties=mutableListOf<String>()
    val uri=Uri.parse("content://chenyi.book.android.ch09.provider/district")
    val contentResolver=context.contentResolver
    val cursor=contentResolver.query(uri,arrayOf("countyName"),
        "provinceName=? and cityName=?",arrayOf(province,city),null)
                                    //获得指定省或直辖市以及城市名的所有 countyName 字段
    cursor?.apply{
        while (moveToNext()){
            val countyValue=getString(0)
            if (countyValue !in counties )
                counties.add(countyValue)
        }
        close()
    }
    return counties
}
```

（2）定义视图模型 DistrictViewModel，DistrictViewModel 利用 DBUtils 实现对数据库的访问和操作，代码如下：

```kotlin
//模块 Ch09_07 的 DistrictViewModel.kt
class DistrictViewModel: ViewModel() {
    var _districtState=MutableStateFlow(DistrictState())
                                    //定义保存 DistrictState 对象的可变状态流
    val output=_districtState.asStateFlow()          //定义状态流

    fun handleIntent(context: Context, intent:DistrictIntent){ //处理意图
        val dbUtils=DBUtils(context)
        when (intent){
            is DistrictIntent.ChooseProvinceIntent->          //处理选择省/直辖市
                chooseProvince(dbUtils)
            is DistrictIntent.ChooseCityIntent->              //处理选择城市名称
                chooseCity(dbUtils, intent.province)
            is DistrictIntent.ChooseCountyIntent->            //处理选择县级地区的名称
                chooseCounty(dbUtils,province=intent.province,city=intent.city)
            is DistrictIntent.DisplayDistrictIntent->         //处理显示选择的地区完整名称
                displayDistrict(province=intent.province,city =intent.city,
                    county=intent.county)
        }
    }
    fun chooseProvince(dbUtils:DBUtils){                      //获取省/直辖市列表
        val currentState=_districtState.value                 //获得当前状态值
        val newState=currentState.copy()                      //产生 DistrictState 对象的副本
        if (newState.provinces.isEmpty())
            newState.provinces.addAll(dbUtils.getProvinces())
                                    //修改 DistrictState 的省/直辖市名列表
```

```
        newState.cities.clear()                    //清除DistrictState的城市名列表
        newState.counties.clear()                  //清除DistricttState的县级地区名列表
        _districtState.value=newState              //修改状态值
    }
    fun chooseCity(dbUtils:DBUtils,province:String){    //获取城市列表
        val currentState=_districtState.value
        val newState=currentState.copy(province=province)
                                //产生DistrictState对象的副本修改province属性值
        newState.cities.clear()
        newState.counties.clear()
        newState.cities.addAll(dbUtils.getCities(province))
                                //修改DistrictState的城市名列表
        _districtState.value=newState
    }
    fun chooseCounty(dbUtils:DBUtils,province:String,city:String){   //获取县级列表
        val currentState=_districtState.value
        val newState=currentState.copy(province=province,city=city)
        newState.counties.clear()
        newState.counties.addAll(dbUtils.getCounties(province,city))
        _districtState.value=newState
    }
    fun displayDistrict(province: String,city:String,county:String){   //修改地区名称
        val currentState=_districtState.value
        val newState=currentState.copy(province=province,city=city,county=county)
        _districtState.value=newState
    }
}
```

(3) 定义可组合函数 HomeScreen，调用三次 MySpriner 可组合函数（同例 9-3 的定义）分别生成对应省直辖市地区列表选项框、对应市级地区列表选项框和对应县级地区列表选项框，然后根据下拉不同列表框，处理不同意图（意图定义在 DistrictIntent 类及其子类，与例 9-3 的同定义），然后分别进行处理，代码如下：

```
//模块Ch09_07的可组合函数HomeScreen
@Composable
fun HomeScreen(modifier:Modifier,viewModel:DistrictViewModel) {
    val output=viewModel.output.collectAsState()
    val fontColor=colorResource(id=R.color.teal_200)
    val context=LocalContext.current
    LaunchedEffect(Unit) {
        viewModel.handleIntent(context,DistrictIntent.ChooseProvinceIntent)
                                                    //处理选择省直辖市列表
    }
    Column(modifier=modifier.fillMaxSize().padding(top=10.dp),
        verticalArrangement=Arrangement.Center,
        horizontalAlignment=Alignment.CenterHorizontally) {
        Text("省市列表",fontSize=36.sp,color=fontColor)
        MySpinner(items=output.value.provinces) {      //处理选择省直辖市
            viewModel.handleIntent(context,DistrictIntent.ChooseCityIntent(it))
        }
```

```
        MySpinner(items=output.value.cities) {                //处理选择城市地区意图
            viewModel.handleIntent(context,DistrictIntent.ChooseCountyIntent(output.
                value.province,it))
        }
        MySpinner(items=output.value.counties) {              //处理选择县级地区意图
            viewModel.handleIntent(context,DistrictIntent.DisplayDistrictIntent(output.
                value.province,output.value.city,it))
        }
        Spacer(modifier=Modifier.fillMaxWidth().height(10.dp))
        if (output.value.province.isNotEmpty())
            Text("${output.value.province}-${output.value.city}-${output.value.
                county}",fontSize=24.sp,color=fontColor)
    }
}
```

(4) 在 MainActivity 中调用 HomeScreen 可组合函数，渲染界面，代码如下：

```
//模块 Ch09_07 的 MainActivity.kt
class MainActivity: ComponentActivity() {
    override fun onCreate(savedInstanceState: Bundle?) {
        super.onCreate(savedInstanceState)
        enableEdgeToEdge()
        val viewModel=ViewModelProvider(this).get(DistrictViewModel::class.java)
                                                                //生成视图模型
        setContent {
            Chapter09Theme {
                Scaffold(modifier=Modifier.fillMaxSize()) {     //渲染界面
                    innerPadding->HomeScreen(modifier =Modifier.padding(innerPadding),
                        viewModel=viewModel)
                }
            }
        }
    }
}
```

(5) 从 Android 11(API 30)开始，在移动应用中使用其他移动应用的外部组件需要在应用配置清单文件 AndroidManifest.xml 中增加 queries 元素，方便检索并调用外部组件，实现与外部组件的交互。本例中调用模块 Ch09_06 已定义并安装的 DistrictProvider 组件，需在模块 Ch09_07 的配置清单文件 AndroidManifest.xml 中配置 queries 元素，代码如下：

```
<!--模块 Ch09_07 的 AndroidManifest.xml -->
<?xml version="1.0" encoding="utf-8"?>
<manifest xmlns:android="http://schemas.android.com/apk/res/android">
    <queries>
        <provider android:authorities="chenyi.book.android.ch09.provider" />
    </queries>
    ...                                       //略
</manifest>
```

运行结果如图 9-21 所示。

观察运行结果，可以发现在日志记录了模块 DistrictProvider 实现对数据库 district.db 的数据表的检索记录的信息。此时，模块 Ch09_07 的省市选择界面将地区的名称在列表选项框中显示，说明模块

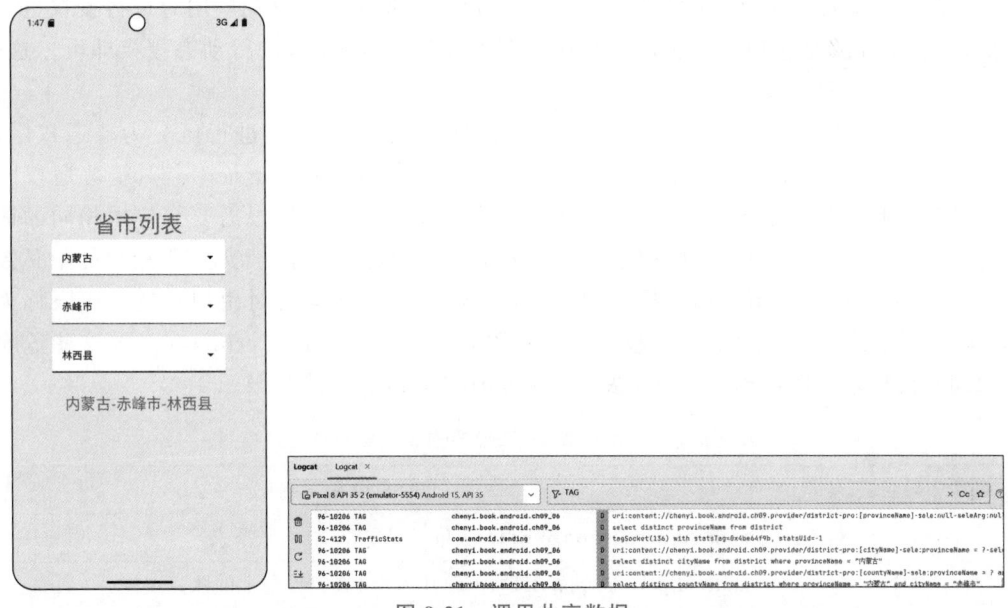

图 9-21 调用共享数据

ch09_07 成功记录了 DistrictProvider 的共享数据。

9.5 调用相机和媒体库

调用移动终端的摄像头拍摄照片或视频是非常常见的。通常情况下,拍摄的照片和视频会保存到 Android 的媒体库中。微信、学习通等很多移动应用中都可进行拍摄照片和录制视频的分享,或者直接调用媒体库中的图片或视频进行分享。本节将介绍如何利用摄像头拍摄图片或视频以及如何分享调用媒体库中的资源。

9.5.1 运行时权限

权限并不是一个陌生概念。前几章已经介绍了如何通过设置,让移动应用具有某些权限。例如,要访问互联网,就需要在应用配置 AndroidManifest.xml 中设置网络访问权限,形式如下：

```
<uses-permission android: name="android.permission.INTERNET" />
```

如果使用前台服务,就需要配置 AndroidManifest.xml 文件设置前台服务许可权限,形式如下：

```
<uses-permission android: name="android.permission.FOREGROUND_SERVICE" />
```

移动应用在使用移动终端中存放的联系人、媒体库、摄像头等敏感资源时,都需要获得用户授权。这是因为,这些资源并不是当前移动应用专属的资源和硬件设备。为了保护用户的隐私,移动应用必须申请相应的使用权限。只有得到用户同意授权,才可以使用这些资源和硬件设备。

Android 6.0 以前的移动终端,所有的授权都是在安装时处理的。如果要安装移动应用,用户就必须许可移动应用的授权申请。一旦安装成功,移动应用就获得了所有的权限。这就导致一些移动应用滥用权限申请,即使不需要的权限,也强行在安装时获取权限。这使得一些移动应用在用户不知情的情况下,具备了收集用户隐私信息的能力,会导致安全隐患。

Android 6.0 以后,提出了运行时权限的概念。用户不需要在安装移动应用时进行授权也可以成功安装移动应用,只要在移动应用为了实现某些功能而需要某些权限时,进行动态授权即可。这样一来,极大地保障了用户的隐私和安全。

Android 中预定义的权限很多。这些权限根据用户隐私信息敏感性进行区分为普通权限(Normal Permission)、签名权限(Signature Permission)和运行时权限(Runtime Permission)。对于普通权限并不会威胁到用户的安全和隐私的权限,系统可以自动授予使用权限。签名权限需要与应用相同的证书进行签名时系统才会授予的权限,但是一些权限严重影响了用户的隐私安全,例如获得用户的隐私数据和设备的控制权等,需要在运行时请求用户授予这些权限。这些必须在运行时由用户授予权限称为运行时权限,因为这些权限涉及用户隐私安全,故又称为危险权限(Dangerous Permission)。这些危险权限必须经过用户同意授权才可以使用。表 9-4 展示了 Android 系统的危险权限。

表 9-4　Android 10 系统的主要危险权限

危险权限组	危险权限名
CONTACTS (通讯录权限组)	android.permission.WRITE_CONTACTS(写入联系人)
	android.permission.GET_ACCOUNTS(查找设备上的账户)
	android.permission.READ_CONTACTS(读取联系人)
ACTIVITY_RECOGNITION (活动识别权限组)	android.permission.ACTIVITY_RECOGNITION(用于需要检测用户的步数或对用户的身体活动)
PHONE (通信权限组)	android.permission.READ_CALL_LOG(读取通话记录)
	android.permission.READ_PHONE_STATE(读取电话状态)
	android.permission.CALL_PHONE(拨打电话)
	android.permission.WRITE_CALL_LOG(修改通话记录)
	android.permission.USE_SIP SIP(视频服务)
	android.permission.PROCESS_OUTGOING_CALLS(修改或放弃拨出电话)
	com.android.voicemail.permission.ADD_VOICEMAIL(加到系统的语音邮件)
CALENDAR (日历权限组)	android.permission.READ_CALENDAR(读取日历)
	android.permission.WRITE_CALENDAR(修改日历)
CAMERA(相机权限组)	android.permission.CAMERA(拍照权限)
SENSORS (传感器权限组)	android.permission.BODY_SENSORS(允许应用程序访问用户用来测量身体内部情况的传感器数据)
LOCATION (位置权限组)	android.permission.ACCESS_COARSE_LOCATION(通过 WiFi 和移动基站获取定位权限)
	android.permission.ACCESS_FINE_LOCATION(通过 GPS 获取定位权限)
	android.permission.ACCESS_BACKGROUND_LOCATION(允许应用程序后台访问位置)

续表

危险权限组	危险权限名
STORAGE （储存权限组）	android.permission.READ_EXTERNAL_STORAGE（读取内存卡）
	android.permission.WRITE_EXTERNAL_STORAGE（写内存卡）
	android.permission.ACCESS_MEDIA_LOCATION（访问媒体位置）
MICROPHONE （麦克风权限组）	android.permission.RECORD_AUDIO（录音权限）
SMS （通信服务权限组）	android.permission.READ_SMS（读取短信）
	android.permission.RECEIVE_WAP_PUSH（接收 WAP PUSH 信息）
	android.permission.RECEIVE_MMS（接收彩信）
	android.permission.RECEIVE_SMS（接收短信）
	permission：android.permission.SEND_SMS（发送短信）
	android.permission.READ_CELL_BROADCASTS（获取小区广播）

在使用蓝牙或照相机等硬件时需要设置应用权限。如果需要使用这些权限，还需要在移动应用的系统配置文件 AndroidManifest.xml 中增加＜uses-feature＞元素的配置，例如：

```
<uses-feature
    android: name=" android.hardware.camera"
    android: required="false" />
```

若属性 android：required 设置为 true，则表示当前的移动应用运行时必须使用这些硬件特征，如果没有则无法工作；若它的设置为 false，则表示应用在运行时需要用到硬件特征；若没有，则应用可能会有一部分功能受到影响，但大部分功能还是可以正常工作。例如，若设置照相机为 false，则表示没有该特征，移动应用会有一部分功能受到限制。

从 Android 11（API30 级）以后，提出了单次授权的概念。每当移动应用请求与地理位置信息、扬声器（俗称麦克风）或相机相关的权限时，面向用户权限的对话框会包含"仅限这一次"的选项。在用户选择这一选项的情况下，Android 会向该移动应用临时授权单次许可，使得在一段时间内可以访问和使用相关资源。当用户下次启动该移动应用时，会提示用户再次授权给移动应用。如果要在移动应用中使用表 9-4 所示的危险权限，那么需要在执行这些权限之前检查是否具有权限，检查权限的代码类似如下：

```
when {
    ContextCompat.checkSelfPermission(CONTEXT, Manifest.permission.REQUESTED_     //在指定的上下文中检查权限
        PERMISSION)==PackageManager.PERMISSION_GRANTED->{    // 如果授予权限成功
        …                                                   //执行操作
    }
    shouldShowRequestPermissionRationale                     //获取显示请求权限的界面
    Manifest.permission.REQUESTED_PERMISSION)->{
        …                                                   //调用显示为什么要调用该权限
                                                            //的界面的提示代码片段
```

```
        }
        else ->{                                              //在指定的上下文中请求权限
            requestPermissions(arrayOf(Manifest.permission.REQUESTED_PERMISSION), REQUEST_CODE)
        }
    }
```

ContextCompat.checkSelfPermission 函数可以用于检查权限,返回 PackageManager.PERMISSION_GRANTED 表示授权成功,如果返回 PackageManager.PERMISSION_DENIED,表示授权失败。

shouldShowRequestPermissionRationale 函数返回一个布尔值,表示是否要求显示请求权限的基本原理,如果为真,则需要定义并调用显示请求权限理由的 GUI 界面进行处理。

requestPermissions 函数用于执行请求权限的操作。

一旦用户响应了请求的权限,就会调用 onRequestPermissionsResult 函数实现权限响应后的处理。处理的代码与如下内容相似:

```
override fun onRequestPermissionsResult(requestCode: Int,permissions: Array<String>,
    grantResults: IntArray) {
    when (requestCode) {                                      //根据请求码进行判断
        PERMISSION_REQUEST_CODE ->{                           //如果授权请求成功,获得请求的结果数组是非空的
            if ((grantResults.isNotEmpty() &&grantResults[0] ==
                PackageManager.PERMISSION_GRANTED)) {
                                                              //授权成功,继续后续的工作流程
                ...
            } else {                                          //授权请求取消获得请求的结果数组为空
                ...                                           //处理提示用户授权失败的操作
            }
            return
        }
        ...                                                   //处理其他的请求
        else ->{
            ...                                               //忽视所有的请求
        }
    }
}
```

例 9-8 在移动应用中内置发送短信功能。

首先,在用用配置清单文件 AndroidManifest.xml 中设置支持发送短信的权限,代码如下:

```
<uses-feature android:name="android.hardware.telephony" android:required="false" />
<uses-permission android:name="android.permission.SEND_SMS" />
```

然后,定义发送短信的界面的可组合函数 HomeScreen,代码如下:

```
//模块 Ch09_08 的发送短信的界面 HomeScreen
@Composable
fun HomeScreen(modifier:Modifier,
    action:(String,String)->Unit){
    var whoTxt by remember{mutableStateOf("")}
    var smsTxt by remember{mutableStateOf("")}
    Column(modifier=modifier.fillMaxSize(),
        horizontalAlignment=Alignment.CenterHorizontally,
```

```
            verticalArrangement=Arrangement.Center){
        Text("短信发送实例",fontSize =32.sp)
        TextField(modifier=Modifier.fillMaxWidth().padding(top=20.dp),
            value=whoTxt, onValueChange={
            it:String->whoTxt=it
        }, leadingIcon={
            Icon(Icons.Filled.Person, contentDescription ="联系人")
        }, label={
            Text("联系人")
        })
        TextField(modifier=Modifier.fillMaxWidth().padding(top=20.dp),
            value=smsTxt, onValueChange={
            it:String->smsTxt=it
            }, leadingIcon={
            Icon(Icons.Default.Info, contentDescription ="SMS")
            },label={
                Text("短信")
            }
        )

        Button(modifier=Modifier.padding(top=50.dp),onClick={
            action.invoke(whoTxt,smsTxt)
        }){
            Text("发送短信")
        }
    }
}
```

MainActivity 调用 HomeScreen，并定义发送短信的动作，并传递给 HomeScreen，以便 HomeScreen 执行权限检查和发送短信，代码如下：

```
//模块 Ch09_08 主活动 MainActivity.kt
class MainActivity: ComponentActivity() {
    private val REQUEST_SEND_SMS_CODE=0x111
    override fun onCreate(savedInstanceState: Bundle?) {
        super.onCreate(savedInstanceState)
        enableEdgeToEdge()
        setContent {
            Chapter09Theme {
                Scaffold(modifier=Modifier.fillMaxSize()) {
                    innerPadding ->HomeScreen(modifier =Modifier.padding
                        (innerPadding),::checkPermission)
                }
            }
        }
    }
    private fun checkPermission(who:String,sms:String) {          //检查权限并请求权限
        when{                                                      //在指定的上下文中检查发送短信权限
            ContextCompat.checkSelfPermission(this,"android.permission.SEND_SMS")==
                PackageManager.PERMISSION_GRANTED ->{
                sendMessage(who,sms)                                //发送消息
```

```kotlin
        }
        //获取显示请求权限的界面
        shouldShowRequestPermissionRationale("android.permission.SEND_SMS")->{
                                        //调用显示为什么要调用该权限的界面的提示代码片段
            showOPMessage("设置权限","发送短信必须设置短信发送权限!")
        }
        else ->{                                        //在指定的上下文中请求权限
            requestPermissions(arrayOf("android.permission.SEND_SMS"),
                REQUEST_SEND_SMS_CODE)
        }
    }
}
private fun sendMessage(who:String,sms:String){            //发送短信
    val smsManager: SmsManager=getSystemService(SmsManager::class.java)
                                                //获取 SMS 管理器
    smsManager.sendTextMessage(who, null, sms, null, null)
    showOPMessage("操作","发送短信成功")
}

private fun showOPMessage(title:String,info:String){       //显示提示权限的对话框
    AlertDialog.Builder(this).apply{
        setTitle(title)
        setMessage(info)
        setPositiveButton("知道了",null)
        create()
    }.show()
}
}
```

运行结果如图 9-22 所示。

(a) 运行界面　　　　(b) 运行时申请权限　　　　(c) 发送短信成功

图 9-22　在移动终端上发送短信的运行界面

图 9-22 中展示了在移动终端上发送短信的界面。注意,在测试时会产生短信服务费。

9.5.2 拍照和显示媒体库的图片

使用照相机拍照并把照片保存到媒体库,或者从媒体库中调用图片,是移动应用常用的功能。在本节中将介绍如何实现拍照功能以及调用媒体库中的图片。

1. 权限处理

因为拍照时不但需要使用照相机,并将拍下的图片保存到媒体库中,而且还需要从媒体库中调出图片进行查看,因此在创建的项目中的应用配置清单文件 AndroidManifest.xml 中增加如下权限设置:

```xml
<!--设置使用照相机许可权限-->
<uses-permission android: name="android.permission.CAMERA"/>
<uses-feature android: name="android.hardware.camera"/>
<!--设置读取外部存储的访问权限 -->
<uses-permission android: name="android.permission.READ_EXTERNAL_STORAGE" />
<!--设置写入存储的访问权限-->
<uses-permission android: name="android.permission.WRITE_EXTERNAL_STORAGE" />
```

MainActivity 可以调用不同的组合函数实现界面的切换。此外,即使在 AndroidManifest.xml 申明要使用这些权限,仍需要在移动应用的运行时需要申请使用权限。因此,在本项目的 MainActivity 中定义权限的申请,代码如下:

```kotlin
//模块 Ch09_09 主活动 MainActivity.kt
class MainActivity: ComponentActivity() {
    private val REQUEST_MEDIASTORE=0x111
    private lateinit var launcher: ActivityResultLauncher<Intent>
    override fun onCreate(savedInstanceState: Bundle?) {
        super.onCreate(savedInstanceState)
        enableEdgeToEdge()
        checkPermission()                                           //检查权限
        val cameraViewModel=ViewModelProvider(this).get(CameraViewModel::class.java)
        launcher=registerForActivityResult(ActivityResultContracts.
            StartActivityForResult()) {
            cameraViewModel.getImagesFrmMedias(this)
        }
        setContent {
            Chapter09Theme {
                Scaffold(modifier=Modifier.fillMaxSize()) {
                    innerPadding->cameraViewModel.getImagesFrmMedias(MainActivity@this)
                    CameraScreen(modifier=Modifier.padding(innerPadding),
                        cameraViewModel=cameraViewModel,
                        takePicAction=::takePictures)               //生成界面
                }
            }
        }
    }
    private fun checkPermission() {                                 //检查权限
        if (ContextCompat.checkSelfPermission(this,
            android.Manifest.permission.READ_EXTERNAL_STORAGE)
```

```kotlin
            ==PackageManager.PERMISSION_GRANTED &&ContextCompat.checkSelfPermission
            (this,android.Manifest.permission.CAMERA)
            ==PackageManager.PERMISSION_GRANTED&&ContextCompat.checkSelfPermission(this,
            android.Manifest.permission.
                WRITE_EXTERNAL_STORAGE)==PackageManager.PERMISSION_GRANTED){
                                                                        //检查权限
        } else {
            requestPermissions(arrayOf(android.Manifest.permission.READ_EXTERNAL_
                STORAGE,android.Manifest.permission.CAMERA,
                android.Manifest.permission.WRITE_EXTERNAL_STORAGE),
                REQUEST_MEDIASTORE)                                     //请求权限
        }
    }
    private fun takePictures(){                                         //拍照
        val values=ContentValues()
        val photoUri=contentResolver?.insert(MediaStore.Images.Media.
            EXTERNAL_CONTENT_URI, values)
        val intent=Intent(MediaStore.ACTION_IMAGE_CAPTURE)
        intent.putExtra(MediaStore.EXTRA_OUTPUT, photoUri)
        launcher.launch(intent)
    }
}
```

本移动应用中需要申请的运行时权限如下。

（1）android.Manifest.permission.READ_EXTERNAL_STORAGE：读取应用外部存储权限。

（2）android.Manifest.permission.CAMERA：拍照权限。

（3）android.Manifest.permission.WRITE_EXTERNAL_STORAGE：写入应用外部存储权限。

通过设置这些权限，实现拍照、将图片保存到媒体库和浏览媒体库。

2. 定义视图模型 CameraViewModel

通过定义视图模型 CameraViewModel，获取媒体库所有的照片，并将照片保存到 MutableLiveData 中，便于后续观察媒体库的数据是否发生变化，代码如下：

```kotlin
//模块 Ch09_09 的 CameraViewModel.kt
class CameraViewModel: ViewModel() {
    val mediaStore:MutableLiveData<List<MediaEntity>>=MutableLiveData()
    fun getImagesFrmMedias(context: Context): MutableLiveData<List<MediaEntity>>{
                                                            //从媒体库中获得所有图片
        val projections=arrayOf(
            MediaStore.Images.ImageColumns._ID,
            MediaStore.Images.ImageColumns.DISPLAY_NAME)
        val contentResolver=context.contentResolver         //创建 contentResolver 对象
        val cursor=contentResolver.query(MediaStore.Images.Media.EXTERNAL_CONTENT_
            URI,projections,null,null,MediaStore.Images.ImageColumns.DATE_MODIFIED +"
            desc")                                          //检索媒体库的图片
```

```kotlin
        val imageList=mutableListOf<MediaEntity>()
        cursor?.apply{
        while(moveToNext()){
            val imageId=getLong(getColumnIndexOrThrow(MediaStore.Images.
                ImageColumns._ID))
            val imageName =getString(getColumnIndexOrThrow(MediaStore.Images.
                ImageColumns.DISPLAY_NAME))
            val imageUri =ContentUris.withAppendedId(MediaStore.Images.Media.
                EXTERNAL_CONTENT_URI,imageId)
            imageList.add(MediaEntity(imageId,imageName,imageUri))
                                                     //将检索结果保存到列表中
        }
    }
    mediaStore.value=imageList
    return mediaStore
}
```

3. 定义界面

为了便于照片库的显示，需要定义对应图片基本信息的实体类 MediaEntity，代码如下：

```kotlin
//模块 Ch09_09 的 MediaEntity.kt
data class MediaEntity(val imageId:Long,val imageName:String,val uri:Uri)
```

界面由两个可组合函数构成：ImageCard 和 CameraScreen。可组合函数 ImageCard 用来显示单张照片和照片文件名，代码如下：

```kotlin
//模块 Ch09_09 可组合函数 ImageCard
@Composable
fun ImageCard(mediaEntity: MediaEntity) {
    Card(elevation=CardDefaults.cardElevation(defaultElevation=3.dp),
        colors=CardDefaults.cardColors(containerColor=Color.Gray),
        shape=RoundedCornerShape(5.dp),) {
        Column(modifier=Modifier
            .wrapContentSize()
            .padding(bottom=10.dp),
            horizontalAlignment=Alignment.CenterHorizontally) {
            AsyncImage(modifier=Modifier.size(120.dp,150.dp),
                model=mediaEntity.uri,contentDescription=null)
            Text(text=mediaEntity.imageName,fontSize=20.sp,color=Color.White)
        }
    }
}
```

定义的 CameraScreen 可组合函数实际承担了拍照和显示媒体库所有照片的功能。为了能将媒体库的照片有序排列，因此调用 ImageCard 来布局照片库的每张照片。此外，通过调用传递的 takePicAction 函数对象，执行实际的拍照功能，代码如下：

```kotlin
//模块 Ch09_09 可组合函数 CameraScreen
@Composable
fun CameraScreen(modifier:Modifier,cameraViewModel: CameraViewModel,
    takePicAction:()->Unit){
    val pictures=cameraViewModel.mediaStore
    val context=LocalContext.current as MainActivity
    val pictureLst=SnapshotStateList<MediaEntity>()
    pictures.observe(context) {                              //观察数据是否变化
        it:List<MediaEntity>->pictureLst.clear()
        pictureLst.addAll(it)
    }
    Box(modifier=modifier.fillMaxSize(),
        contentAlignment=Alignment.TopCenter) {
        Column(horizontalAlignment=Alignment.CenterHorizontally) {
            Text("照片库",fontSize=24.sp)
            LazyVerticalGrid(columns=GridCells.Adaptive(minSize=180.dp)) {
                items(pictureLst){
                    ImageCard(it)
                }
            }
            TextButton(onClick={
                takePicAction.invoke()                       //执行拍照
            }){
                Text("拍照",fontSize=20.sp)
                Icon(Icons.Filled.ThumbUp,contentDescription=null)
            }
        }
    }
}
```

运行结果如图 9-23 所示。

(a) 运行时权限申请　　(b) 浏览媒体库中的图片

图 9-23　拍照和浏览媒体库中的图片

9.5.3 访问媒体库中的视频

访问媒体库中视频的操作流程与访问媒体库中图片的操作流程一样,也需要设置访问权限。一方面,在系统应用配置 AndroidManifest.xml 文件中;另一方面,需要在程序代码中设置运行时访问权限,具体的设置同 9.5.2 节。

媒体库中的文件主要分为 3 种类型。

(1) MediaStore.Images:图片集合。
(2) MediaStore.Video:视频集合。
(3) MediaStore.Audio:音频集合。

可以通过内容 URI 来分别访问它们。

图片集合的图片文件通过 MediaStore.Images.Media.EXTERNAL_CONTENT_URI。
视频集合的视频文件通过 MediaStore.Video.Media.EXTERNAL_CONTENT_URI。
音频集合中音频文件通过 MediaStore.Audio.Media.EXTERNAL_CONTENT_URI。

在访问媒体库的视频时,也需要通过 ContentResolver 对象对媒体库的视频进行检索以及查看,代码如下:

```
val projection=arrayOf(MediaStore.Video.Media._ID,         //视频的编号
    MediaStore.Video.Media.DISPLAY_NAME,                   //视频的名称
    MediaStore.Video.Media.DURATION,                       //视频的播放时间
    MediaStore.Video.Media.SIZE)                           //视频文件的大小
```

project 直接量表示映射存储集合对象的字段或者对应为表的列名,代码如下:

```
val cursor=ContentResolver.query(
    MediaStore.Video.Media.EXTERNAL_CONTENT_URI,    //定义媒体库的视频位置
    projection,                                     //定义媒体库中视频的投影字段
    selection,                                      //selection 定义选择条件
    selectionArgs,                                  //定义选择参数
    sortOrder)                                      //定义排序的顺序
```

Cursor 可通过 ContentResolver 对象的 query 检索方法获得。在 query 函数中定义的 MediaStore.Video.Media.EXTERNAL_CONTENT_URI 表示定义媒体库中视频的位置,project 执行投影操作,映射到特定的字段(表的字段),selection 表示选择条件,通常使用"?"表示占位符,接收选择 selectionArgs 传递的参数。例如:

```
selection="MediaStore.Video.Media._ID =? ",selectionArgs=1234L
```

表示检索视频编号 MediaStore.Video.Media._ID 为 1234 的视频。

sortOrder 表示检索的结果按照特定的顺序进行,例如:

```
sortOrder="${MediaStore.Video.Media.DURATION } ASC"
```

表示按照视频播放时间从小到大进行排序。

对检索的结果集通过游标 Cursor 对象依次进行遍历来获得,代码如下:

```
cursor?.use {
    c->while (c.moveToNext()) {
        val id=cursor.getLong(c.getColumnIndex (MediaStore.Video.Media._ID))
                                                        //获得视频编号
        val name=cursor.getString(c.getColumnIndex (MediaStore.Video.Media.
            DISPLAY_NAME))                              //获得视频名称
        val duration=cursor.getInt(c.getColumnIndex (MediaStore.Video.Media.
            DURATION))                                  //获得视频的播放时间
        val size=cursor.getInt(c.getColumnIndex (MediaStore.Video.Media.
            SIZE))                                      //获得视频的大小
        val uri: Uri =ContentUris.withAppendedId(MediaStore.Video.Media.EXTERNAL_
            CONTENT_URI, id)                            //将内容URI与id编号连接在一起生
                                                        //成新的内容uri
    }
}
```

例 9-9 创建移动应用浏览并选择播放媒体库中的视频。

在配置清单文件 AndroidManifest.xml 中设置读取外部存储的访问权限,代码如下:

```
<!--设置读取外部存储的访问权限 -->
<uses-permission android:name="android.permission.READ_EXTERNAL_STORAGE" />
```

定义实体类 VideoEntity,表示媒体库的视频基本信息,代码如下:

```
//模块 Ch09_10 的 VideoEntity.kt
/* * 视频实体类
 * @property uri Uri: 视频 uri 播放地址
 * @property thumnail ImageBitmap: 缩略图
 * @property name String: 视频文件名
 * @property duration Int: 视频的时长
 * @property size Int: 视频大小 */
data class VideoEntity(val uri:Uri, val thumnail: ImageBitmap,val name:String,val
    duration:Int,val size:Int)
```

定义视图模型 VideoViewModel,执行实际上的业务加载媒体库中的所有视频,以及修改当前要显示的视频。通过修改界面的状态,使得界面刷新,代码如下:

```
//模块 Ch09_10 的 VideoViewModel.kt
class VideoViewModel: ViewModel() {
    var _mediaStore: MutableStateFlow<List<VideoEntity>>=
        MutableStateFlow<List<VideoEntity>>(listOf())
    val mediaStore: StateFlow<List<VideoEntity>>=
        _mediaStore.asStateFlow()
    var _current_Video:MutableStateFlow<VideoEntity?>=MutableStateFlow(null)
    val currentVideo=_current_Video.asStateFlow()
    @SuppressLint("Range")
    fun loadVideosFrmMedias(context: Context){           //从媒体库中获得所有的视频
        val videoList=mutableListOf<VideoEntity>()
        val project=arrayOf(                             //投影
            MediaStore.Video.Media._ID,                  //视频编号
```

```
            MediaStore.Video.Media.DISPLAY_NAME,          //视频显示名称
            MediaStore.Video.Media.DURATION,              //视频时长
            MediaStore.Video.Media.SIZE                   //视频大小
        )
        val contentResolver=context.contentResolver       //创建 contentResolver 对象
        val cursor=contentResolver?.query(                //检索媒体库的视频
            MediaStore.Video.Media.EXTERNAL_CONTENT_URI,
            project, null, null, null )
        cursor?.use {                                     //将检索的结果保存到列表中
            it: Cursor->while (it.moveToNext()) {
                val id=it.getLong(it.getColumnIndex(MediaStore.Video.Media._ID))
                                                          //获得视频编号
                val name=
                    it.getString(it.getColumnIndex(MediaStore.Video.Media.DISPLAY_NAME))
                                                          //获得视频名称
                val duration=
                    it.getInt(it.getColumnIndex(MediaStore.Video.Media.DURATION))
                                                          //获得视频的播放时间
                val size=it.getInt(it.getColumnIndex(MediaStore.Video.Media.SIZE))
                                                          //获得视频的大小
                val uri=ContentUris.
                    withAppendedId(MediaStore.Video.Media.EXTERNAL_CONTENT_URI, id)
                                                          //获得视频的内容 uri
                val thumnail=contentResolver?.loadThumbnail(uri,
                    Size(200, 160), null)?.asImageBitmap()
                                                          //获得视频的缩略图
                videoList.add(VideoEntity(uri, thumnail!!, name, duration, size))
            }
        }
        _mediaStore.value=videoList
    }
    fun updateVideo(videoEntity:VideoEntity){             //变更当前视频对象
        _current_Video.value=videoEntity
    }
}
```

在本例中需要显示两个界面：显示媒体库视频列表界面和显示播放视频的界面。这两个界面之间会相互切换。视频列表界面可以选择视频跳转到播放视频的界面，播放视频的界面也可以返回视频列表界面。因此，基于 Compose 的 Navigation 组件定义导航图界面 NavigationGraphScreen，实现不同界面的切换，代码如下：

```
//模块 Ch09_10 的 NavigationGraphScreen 可组合函数
@Composable
fun NavigationGraphScreen(modifier: Modifier,videoViewModel: VideoViewModel){
    val navController: NavHostController=rememberNavController()
    NavHost(navController=navController, startDestination="videoList" ) {
        composable(route="videoList"){
            VideoScreen(modifier=modifier, navController=navController,
                videoViewModel=videoViewModel )
```

```
        }
        composable(route="displayVideo"){
            DisplayVideoScreen(modifier=modifier,videoViewModel=videoViewModel,
                navController=navController )
        }
    }
}
```

定义的媒体库视频列表界面,还需要定义视频列表中单项的界面 ImageCard,具体包括视频的缩略图、视频名称、视频播放的时间和视频文件的大小对应组件的定义,代码如下:

```
//模块 Ch09_10 的可组合函数 ImageCard
@Composable
fun ImageCard(videoEntity: VideoEntity,videoViewModel:VideoViewModel,
    navController: NavController){
    Card(elevation=CardDefaults.cardElevation(defaultElevation =3.dp),
        colors=CardDefaults.cardColors(containerColor =Color.Gray),
        shape=RoundedCornerShape(5.dp),
    ){
        Column(modifier=Modifier.wrapContentSize().padding(bottom=10.dp),
            horizontalAlignment=Alignment.CenterHorizontally){
            Image(modifier=Modifier
                .size(120.dp, 150.dp)
                .clickable {
                    videoViewModel.updateVideo(videoEntity)          //变更当前视频
                    navController.navigate("displayVideo")            //跳转到视频播放界面
                }, bitmap=videoEntity.thumnail, contentDescription="缩略图" )
            Text(text=videoEntity.name,fontSize=12.sp,color=Color.White)
            Text(text="大小: ${videoEntity.size}",fontSize=12.sp,color=Color.White)
            Text("时长: ${videoEntity.duration}",fontSize=12.sp,color=Color.White)
        }
    }
}
```

媒体库视频列表界面由可组合函数 VideoScreen 来完成,它在 LazyVerticalGrid 可组合项中多次调用 ImageCard,生成视频网格列表,代码如下:

```
//模块 Ch09_10 的可组合函数 VideoScreen
@Composable
fun VideoScreen(modifier: Modifier,navController:NavController,
    videoViewModel: VideoViewModel) {
    val videos=videoViewModel.mediaStore.collectAsState()
    val context=LocalContext.current
    LaunchedEffect(videos) {
        videoViewModel.loadVideosFrmMedias(context)                  //加载媒体库中的视频
    }
    Box(modifier=modifier.fillMaxSize(),
        contentAlignment=Alignment.TopCenter) {
```

```
        Column(horizontalAlignment=Alignment.CenterHorizontally) {
            Text("视频库",fontSize=24.sp)
            LazyVerticalGrid(columns=GridCells.Adaptive(minSize =180.dp)) {
                items(videos.value){
                    it:VideoEntity->ImageCard(it,videoViewModel,navController)
                }
            }
        }
    }
}
```

可组合函数 DisplayVideoScreen 定义了视频播放的界面，它结合传统的 VideoView 实现视频的播放，代码如下：

```
//模块 Ch09_10 DisplayVideoScreen 可组合函数
@Composable
fun DisplayVideoScreen(modifier: Modifier, videoViewModel: VideoViewModel,
    navController: NavController) {
    val currentVideo=videoViewModel.currentVideo.collectAsState()
    val context=LocalContext.current
    Column(modifier=modifier.fillMaxSize(),
        horizontalAlignment=Alignment.CenterHorizontally,
        verticalArrangement=Arrangement.Center) {
        currentVideo.value?.run {
            val video=this
            Text("正在播放$name", fontSize=20.sp)
            AndroidView(factory={
                VideoView(it).apply{                                    //创建 VideoView
                    setMediaController(MediaController(context))        //媒体控制
                    setVideoURI(video.uri)                              //媒体播放地址
                    start()                                             //播放媒体
                }
            })
        }
        Button(onClick={
            navController.navigate("videoList")                         //返回媒体列表界面
        }) { Text("返回",fontSize =20.sp) }
    }
}
```

最后在 MainActivity 中处理权限的访问，还调用导航图 NavigationGraphScreen，实现视频列表界面和视频播放界面的切换，代码如下：

```
//模块 Ch09_10 MainActivity.kt
class MainActivity: ComponentActivity() {
    private val REQUEST_MEDIASTORE=0x111
    override fun onCreate(savedInstanceState: Bundle?) {
        super.onCreate(savedInstanceState)
        enableEdgeToEdge()
```

```
            checkPermission()                                              //检查权限
            val videoViewModel=ViewModelProvider(this).get(VideoViewModel::class.java)
                                                                           //创建视图模型
            setContent {
                Chapter09Theme {
                    Scaffold(modifier=Modifier.fillMaxSize()) {
                        innerPadding->NavigationGraphScreen(modifier=
                            Modifier.padding(innerPadding), videoViewModel=videoViewModel)
                                                                           //定义导航图
                    }
                }
            }
        }
        private fun checkPermission(){                                     //检查权限
            if (ContextCompat.checkSelfPermission(this,
                android.Manifest.permission.READ_EXTERNAL_STORAGE)
                ==PackageManager.PERMISSION_GRANTED &&
                ContextCompat.checkSelfPermission(this,android.Manifest.permission.CAMERA)
                ==PackageManager.PERMISSION_GRANTED&&
                ContextCompat.checkSelfPermission(this,
                    android.Manifest.permission.WRITE_EXTERNAL_STORAGE)
                ==PackageManager.PERMISSION_GRANTED) {                     //检查权限
            } else {
                requestPermissions(arrayOf(android.Manifest.permission.READ_EXTERNAL_
                    STORAGE,android.Manifest.permission.CAMERA,
                    android.Manifest.permission.WRITE_EXTERNAL_STORAGE),
                    REQUEST_MEDIASTORE)                                    //请求权限
            }
        }
    }
```

习 题 9

一、选择题

1. 通过 Context 的 val Context.dataStore：DataStore＜Preferences＞ by preferencesDataStore(name="settings")创建 Preferences DataStore 对象。在这个函数中，name 参数表示_____。

  A. Preferences 对象名   B. Preferences 文件名

  C. Preferences 键值对   D. Preferences 类名

2. 根据 res/raw 目录下的资源文件 file.csv，创建读取文件的字节输入流操作正确的是_____。

  A. val in=getResources().openResource(R.raw.file)

  B. val in=resources.openRawResource(R.raw.file)

  C. val in=getResources().openAssetsResource(R.raw.file)

  D. 以上答案均不正确

3. 根据 assets 目录下的资源文件 file.csv，创建读取文件的字节输入流操作正确的是_____。

  A. val in=resources.getAssets().openAssetsFile("file.csv")

  B. val in=resources.assets.openAssetsFile(R.assets.file)

C. val in＝resources.assets.open("file.csv")

D. 以上答案均不正确

4. 已知自定义 ContentProvider 组件为 MyProvider，如果要让其他的应用可以调用 MyProvider，需要在 AndroidManifest.xml 文件配置 MyProvider 并设置属性_____为 true。

    A. android：name   B. android：enabled

    C. android：exported   D. android：authorities

5. 结合 Room 组件使用_____注解创建对象访问对象，定义访问数据库的常见的操作（例如插入、删除、修改和检索等），以达到实现创建映射数据表的实体类对象，以及对该实体类的对象实例的属性值进行设置和获取的目的。

    A. @Dao   B. @Entity   C. @Parcelize   D. @Database

6. Room 组件提供了_____实现数据库的增量迁移。

    A. Database   B. Migration   C. migrate()   D. Dao

7. 自定义的 ContentProvider 可以结合 Room 组件调用 SQLite 数据库，可以调用 RoomDatabase 对象的_____获得可写的 SupportSQLiteDatabase 对象将 ContentValue 数据库的读写操作。

    A. OpenHelper.readableDatabase   B. OpenHelper.writableDatabase

    C. OpenHelper.writableDatabase   D. openHelper.writableDatabase

8. 已知要修改的记录保存在 ContentValues 对象，可以调用 SupportSQLiteDatabase 的_____函数利用这个 ContentValues 执行修改记录操作。

    A. updateRecord   B. updateValues

    C. update   D. execSQL

9. 可以通过调用 Context 对象的_____函数获得 ContentResolver 对象。

    A. content   B. getResolver

    C. getContent   D. getContentResovler

10. 已知 android.permission.WRITE_CONTACTS 用于设置写入联系人权限，它属于_____。

    A. 普通权限   B. 签名权限

    C. 危险权限   D. 以上说法均不正确

二、问答题

1. 什么是 Proto DataStore？它是如何实现数据的保存的？
2. 说明如何结合文件输入输出流实现对文件的持久化处理。
3. 结合实例说明 Room 组件如何升级 SQLite 数据库。
4. 结合实例说明 Room 组件如何对 SQLite 数据库执行检索操作。
5. 解释说明可运行时权限。

三、上机实践题

1. 编写一个移动应用，用于浏览和选择媒体库中的视频。
2. 编写一个移动应用，用于浏览和选择媒体库中的音频。
3. 编写一个移动应用，用于拍照、将照片保存到媒体库，以及浏览媒体库的图片。
4. 编写一个移动应用，用于写日记、将所写的日记保存到数据库中，以及按照日期检查和阅读数据库中当日的日记。

第 10 章　Paging 组件

Android JetPack 是 2018 年由谷歌公司推出的开发移动应用的套件，由多个库构成。Android JetPack 帮助开发者遵循最佳的开发方法和规范，取消样本代码，确保各种库在不同的版本中运行方式一致，达到更容易构建移动应用的目的。在前面几章中，陆续介绍了 Compose 组件、Navigation 组件、ViewModel 组件、WorkManager 组件、DataStore 组件、Room 组件等。

本章将介绍 Paging 组件。Paging 组件是 Android JetPack 开发套件的分页加载库。Paging 组件非常强大，它可用于加载和显示来自本地存储或网络中更大的数据集中的数据页面，可让移动应用更高效地利用网络带宽和系统资源。

10.1　分页组件概述

移动应用往往需要存在大量的数据进行交互。数据可以来源于移动应用的数据库、互联网或者其他的数据来源。特别对于基于 Internet 的移动应用，它们往往需要从网络中获取数据实现交互。如果一次将所有的网络所有的记录在列表组件全部展示，一方面会造成网络带宽负载过大，另一方面造成系统资源的浪费。更何况，移动终端的屏幕受限，没有必要一次将所有的数据全部展示，减低应用的性能，导致用户体验差。

Android JetPack 推出的 Paging 组件可以实现分页处理。界面的列表中显示异步加载部分数据，滚动列表可以增量加载更多的数据以方便显示。这种增量提取信息在界面显示部分数据的过程为分页，每个分页对应一个要提取数据块。这种分页方式减少系统资源和网络带宽的使用。

Paging 组件主要支持 3 种数据来源，如图 10-1 所示。

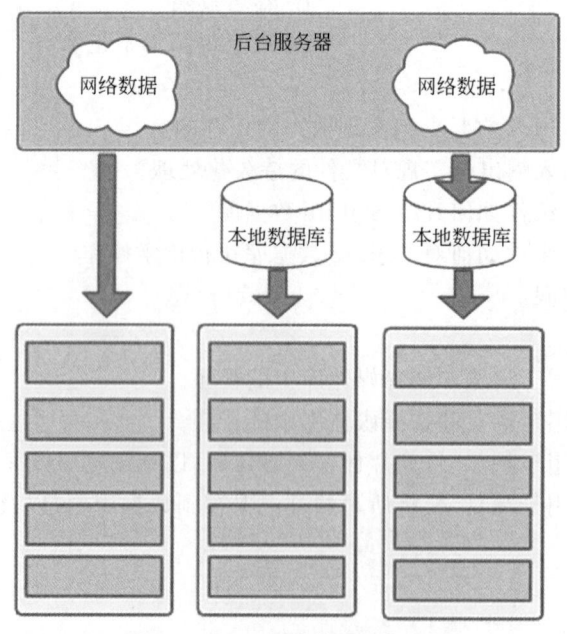

图 10-1　Paging 支持的数据来源

(1) 数据从后台服务器直接获取并应用到移动应用,并结合 Paging 组件和传统的 RecyclerView 视图组件或 Compose 的列表或网格可组合项进行列表展示。

(2) 从存储在本地设备上的数据库获得结构化数据,并结合 Paging 组件和传统的 RecyclerView 视图组件或 Compose 的列表或网格可组合项进行列表展示。

(3) 使用设备上的数据库作为缓存的其他来源组合,例如从互联网的数据缓存到本地设备的数据库中,然后再从数据库中获取数据。并结合 Paging 组件和传统的 RecyclerView 视图组件或 Compose 的列表或网格可组合项进行列表展示。

当前的 Paging 组件是 3.3.2 版本,简称 Paging 3 组件。要使用 Paging3 组件需要在项目的 libs.versions.toml 文件中配置相应的版本信息,代码如下:

```
[versions]
paging="3.3.2"
[libraires]
androidx-paging-runtime={
    group="androidx.paging",name="paging-runtime",version.ref="paging"
}
androidx-paging-rxjava3={
    group="androidx.paging",name="paging-rxjava3",version.ref="paging"
}
androidx-paging-compose={
    group="androidx.paging",name="paging-compose",version.ref="paging"
}
androidx-room-paging={
    group="androidx.room", name="room-paging",version.ref="room"
}
```

在使用 Paging3 组件的模块的 build.gradle.kts 中需要配置 Paging 3 依赖库,代码如下:

```
dependencies {
    implementation(libs.androidx.paging.runtime)
    implementation(libs.androidx.paging.compose)
    implementation(libs.androidx.paging.rxjava3)
    ...                   //略
}
```

Paging 3 库采用 Kotlin 的协程 Coroutines 实现。Paging 3 实现从大数据集自动加载分页数据,支持 Kotlin Coroutines,Flows 或 RxJava,有利于异步数据处理,而且便于展示从数据集中映射 map 或过滤 filter 特定的数据。Paging 3 库内置错误处理,如果错误发生可以刷新数据或重试请求。

Paging 3 直接集成到推荐的 Android 应用架构中,如图 10-2 所示。Android 应用架构包括 3 个层次:Repository 层:仓库层,用于数据的持久化处理;ViewModel 层:视图模型层,为视图组件保存数据;UI 界面层,用于界面的展示和处理。

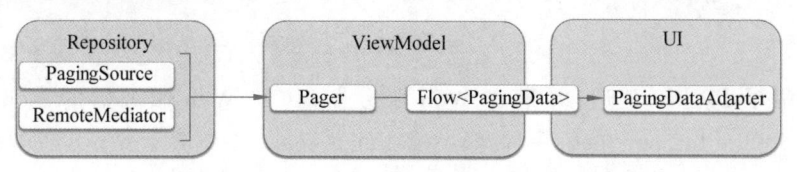

图 10-2 Android 应用架构中的 Paging 组件

1. Repository 中的 Paging 组件

在 Repository 仓库层中,包括了两个非常重要的组件:PagingSource 和 RemoteMediator。

(1) PagingSource:定义数据源以及提供了检索数据的处理方式。PagingSource 从本地数据库或互联网中获得数据。

如果数据源通过 Room 访问,则需要在模块的 build.gradle.kts 增加 Room 组件的配置,参考 9.3 节。

(2) RemoteMediator:是高级分页操作,用于处理从多层数据源中分页处理。例如具有本地数据库缓存的网络数据源,对这些不同层次的数据进行分页处理。

2. ViewModel 层的 Paging 组件的处理

ViewModel 可为 View 组件提供数据。在本层中包括了 Pager、PageingData、PagingConfig 和 Flow。

Pager 提供一套公共的 API。Pager 根据 PagingSource 提供的数据源和 PagingConfig 配置的分页参数,构建 PagingData 的实例,并将 PagingData 会提供给反应流 Flow,形成 Flow<PagingData>实例。在图 10-2 中,Flow<PagingData>介于 ViewModel 和 UI 之间,这是因为 Flow<PagingData>可以实现数据流的共享。

PagingData 是分页数据的容器,将 ViewModel 层与 UI 层连接在一起。在 PagingData 中包含了要分页的数据快照。它用于实现对 PagingSource 数据源的数据检索,并保存检索的结果。

3. UI 界面层的 Paging 组件

界面层通过结合 Compose 组件的 LazyColumn 等列表或网格方式显示分页加载的数据。

10.2 分 页 处 理

为了更好地解释 Paging 3 组件实现分页的功能。笔者从网络爬取了一些电影视频的相关数据,将这些数据生成 JSON 数据。将保存电影数据的 json 文件放置到 Web 服务器中。Web 服务器可以是 Tomcat 服务器或 Flask 服务器或 Nginx 服务器等。本节中,结合 Python+Flask,生成本地服务器。JSON 文件定义了 JSON 数组,数据结构的格式如下所示:

```
[
    {"actors":"演员",
    "directors":"导演",
    "intro":"电影简介",

    "poster":"http://localhost:5000/photo/s_ratio_poster/public/p2626067725.jpg",
    "region":"地区",
    "release":"发布年份",
    "trailer_url":"https://localhost:5000/trailer/268661/# content",

    "video_url":"https://localhost:5000/d04d3c0d2132a29410dceaeefa97/view/movie/M/
        402680661.mp4"},
    ...]
```

在本节中采用了 Python+Flask 搭建 Web 服务器,访问 Web 服务器的 json 文件的 URL 为 http://localhost:5000/film.json?page=1&size=5。此处,page 参数表示要显示的页面,参数 size 表示每页显示的记录数。其中 page 和 size 可以传递不同的参数。

例 10-1 结合 PagingSource 访问网络单一数据源的应用实例。

如图 10-3 所示,将访问网络获取数据封装成 PagingSource 分页数据源,然后通过 Pager 将

PagingSource 进行配置并生成 PagingData 实例,通过异步流将数据发射给 UI 界面的列表如 LazyColumn 中显示。

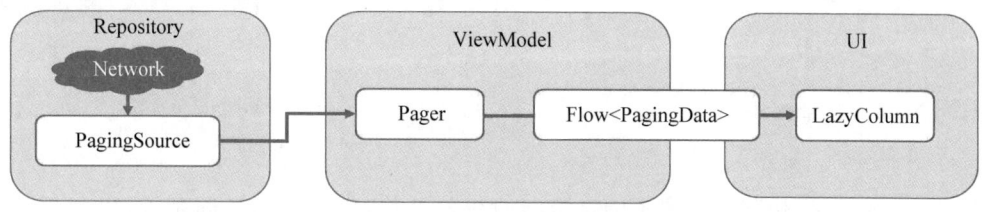

图 10-3 单一数据源访问架构

1. 定义实体类 Film

因为需要将在线的 JSON 数据映射成对象。因此根据 JSON 数组包含的每个 JSON 对象,定义实体类 Film,代码如下:

```kotlin
//模块 Ch10_01 实体类 Film.kt
data class Film(
    @SerializedName("name") val name:String,
    @SerializedName("release") val release:String,
    @SerializedName("region") val region:String,
    @SerializedName("directors") val directors:String,
    @SerializedName("actors") val actors:String,
    @SerializedName("intro") val intro:String,
    @SerializedName("poster") val poster:String,
    @SerializedName("trailer_url") val trailer:String,
    @SerializedName("video_url") val video:String
)
```

注解@SerializedName 包含的名称表示与 JSON 数据的字段名表示一致。

2. 网络访问处理

在本应用中需要访问在线资源 http://127.0.0.1/film.json 并通过传递参数 page 和参数 size 分别表示访问资源的页面和每个页面资源记录的个数。

结合 Retrofit2 访问网络资源。在此前提下,定义网络服务访问接口 FilmApiService,代码如下:

```kotlin
//模块 Ch10_01 网络访问服务接口 FilmApiService.kt
interface FilmApiService {
    @GET("film.json")
    suspend fun getData(
        @Query("page") page:Int,
        @Query("size") size:Int
    ):List<Film>
}
```

上述代码中,page 属性对应访问 URL 中的 page 参数,表示页,size 表示每页的记录数,与 URL 中传递的参数 size 一致。

利用 Retrofit2 构建网络访问服务,定义 Retrofit 构建器 RetrofitBuilder,代码如下:

```kotlin
//模块 Ch10_01 构建网络访问 RetrofitBuilder.kt
object RetrofitBuilder {
    private const val BASE_URL="http://10.0.2.2:5000/"
    private fun getRetrofit(): Retrofit {
```

```
        return Retrofit.Builder()
            .baseUrl(BASE_URL)
            .addConverterFactory(GsonConverterFactory.create())
            .build()
    }
    val apiService:FilmApiService =getRetrofit().create(FilmApiService::class.java)
}
```

通过 Retrofit.Builder 构建器配置了网络访问的相关参数,指定网络访问基址,以及指定了数据序列化和反序列化的转换器为 GsonConverterFactory.create(),即通过调用可以将网络的 JSON 数据转换成对应的对象,以及创建一个 Retrofit 网络访问服务对象。

3. 定义 Repository 仓库层

首先定义仓库层需要执行操作的接口。此处定义接口 FilmRepository,表示需要要根据提供的 page 参数和 size 参数获取数据的接口,代码如下:

```
//模块 Ch10_01 的 FilmRepository.kt
interface FilmRepository {
    /**获取指定 page 页面的信息
     * @param page Int
     * @return List<Film> */
    suspend fun getFilms(page:Int,limit:Int=5):List<Film>
}
```

然后,定义 FilmRepositoryImp 类实现 FilmRepository 接口,通过调用网络访问服务接口 FilmApiService 完成实际从网络获取资源的操作,代码如下:

```
//模块 Ch10_01 的 FilmRepositoryImp.kt
class FilmRepositoryImp:FilmRepository {
    private val apiService: FilmApiService=RetrofitBuilder.apiService
    override suspend fun getFilms(page: Int,limit:Int): List<Film>
        =apiService.getData(page, limit)
}
```

4. 定义数据源 FilmSource

将结合 Retrofit2 获取的网络资源封装成 PageSource,为后续的分页处理提供支持,代码如下:

```
//模块 Ch10_01 的 FilmSource.kt
class FilmSource(private val filmRepository: FilmRepository):
PagingSource<Int, Film>() {
    override suspend fun load(params: LoadParams<Int>):LoadResult<Int, Film>{
        return try {
            val currentPage=params.key ?:1
            val filmResponse=filmRepository.getFilms(currentPage)    //请求当前页面
            val prevKey=if (currentPage==1) null else currentPage-1  //修改上一页
            val nextKey=if (filmResponse.isEmpty()) null else currentPage+1
                                                                     //修改下一页
```

```
                LoadResult.Page(                                    //加载数据成功
                    data=filmResponse,
                    prevKey=prevKey,
                    nextKey=nextKey
                )
            } catch(e:Exception){
                LoadResult.Error(throwable=e)                       //加载数据失败
            }
        }
        override fun getRefreshKey(state: PagingState<Int, Film>): Int? {
            return state.anchorPosition                             //返回锚点
        }
    }
```

在 FilmSource 的 load 函数的定义中,通过 prevKey 和 nextKey 分别表示修改当前的页面为上一页和下一页。当前页面发生变化时,通过调用 FilmRepository 对象获取的数据也要求重新加载。最后将获取的数据和页面的信息封装成 LoadResult.Page 作为成功返回的结果。如果操作发生错误,则抛出异常。

5. 定义视图模型 FilmViewModel

定义视图模型 FilmViewModel,执行实际的获取数据源的业务,代码如下:

```
//模块 Ch10_01 的 FilmViewModel.kt
class FilmViewModel: ViewModel() {
    private val filmRepository: FilmRepository =FilmRepositoryImp()
    fun getFilms(): Flow<PagingData<Film>>=Pager(PagingConfig(pageSize =5)){
        FilmSource(filmRepository)
    }.flow
}
```

上述代码的 Pager 是分页的主入口,这里分页配置每页的记录数 pageSize 为 5,最后获取 PagingData 的响应式流。这个响应式流,当 PagingSource 对象失效,flow 流就会刷新并发射出新的 PageData,为后续界面的刷新提供新的数据源。

6. 界面的定义

为了更好地实现列表的显示,首先定义列表的单项可组合函数 FilmCard,代码如下:

```
//模块 Ch10_01 的 FilmCard 可组合函数
@Composable
fun FilmCard(film: Film?) {
    Card(modifier=Modifier.fillMaxSize().padding(2.dp),
        elevation=CardDefaults.cardElevation(5.dp),
        colors=CardDefaults.cardColors(containerColor=Color.Black)){
        Column{
            Row(modifier=Modifier.fillMaxSize()){
                AsyncImage(model="${film?.poster}",contentDescription="${film?.name}")
                Column{
                    Text("${film?.name}",fontSize=18.sp,color=Color.White)
                    Text("导演: ${film?.directors}",fontSize=14.sp,color=Color.Green)
```

```kotlin
                Text("演员: ${film?.actors}", fontSize=14.sp,color=Color.White)
            }
        }
        Text("${film?.intro?.subSequence(0,50)}...",fontSize =14.sp,color=Color.
            Green)
        Row(horizontalArrangement=Arrangement.End,
            modifier=Modifier.fillMaxSize()){
            Text("More",fontSize=12.sp)
            IconButton(onClick={}){
                Icon(imageVector=Icons.Default.MoreVert,
                    tint=Color.Green,contentDescription ="更多...")
            }
        }
    }
}
```

然后在可组合函数 FilmScreen 中将获取的电影数据源结合 LazyColumn 列表一次遍历的过程中，对每个 Film 对象按照 FilmCard 进行渲染，代码如下：

```kotlin
//模块 Ch10_01 的 FilmScreen 可组合函数
@Composable
fun FilmScreen(modifier:Modifier,filmViewModel: FilmViewModel) {
    val films=filmViewModel.getFilms().collectAsLazyPagingItems()
    Column(horizontalAlignment=Alignment.CenterHorizontally,
        modifier=Modifier.background(Color.White)){
        LazyColumn{
            items(films.itemCount){
                FilmCard(films[it])
            }
        }
    }
}
```

FilmScreen 中通过语句 val films = filmViewModel.getFilms().collectAsLazyPagingItems() 从 PagingData 的流中收据数据，保存在 LazyPageItems＜Film＞对象，并赋值给 films。通过 films.itemCount 获取 LazyPageItems 的数据总个数。通过 items 函数对 films 按照元素的索引值依次对每个元素生成列表的单项，继而生成列表界面。

最后在 MainActivity 中调用 FilmScreen，生成主界面。并结合 FilmViewModel 完成业务逻辑的处理，使得数据变更，界面刷新，代码如下：

```kotlin
//模块 Ch10_01 的 MainActivity.kt
class MainActivity: ComponentActivity() {
    override fun onCreate(savedInstanceState: Bundle?) {
        super.onCreate(savedInstanceState)
        enableEdgeToEdge()
        val filmViewModel=ViewModelProvider(this).get(FilmViewModel::class.java)
        setContent {
```

```
            Chapter10Theme {
                Scaffold(modifier=Modifier.fillMaxSize()) {
                    innerPadding ->FilmScreen(modifier=Modifier.padding(innerPadding),
                       filmViewModel=filmViewModel)
                }
            }
        }
    }
}
```

上述的例题说明,从网络获取数据并在 Compose 组件构成的界面以列表方式展示。但是一旦处于断网状态,上述的移动应用将无法正常使用。理想状态是,首先在移动应用本地获取数据,如果没有数据或有新数据就从网络中获取数据,并保存到移动终端的 SQLite 数据库。这样,即使处于断网离线情况下,移动应用仍可以通过本地的 SQLite 数据库获取数据并展示数据。Paging 3 支持结合网络和本地数据库的种多层次的数据访问。

例 10-2 结合 PagingSource 访问如图 10-4 所示多层次数据源(网络和本地数据库)的应用实例。

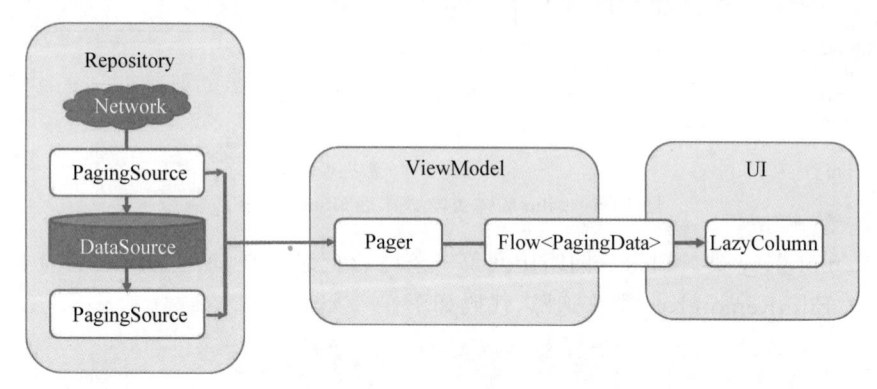

图 10-4 多层次数据源访问的应用架构

与单一数据源结构不同在于增加了 RemoteMediator。当应用的已缓存数据用尽时,RemoteMediator 会充当来自 Paging 库的中介信号。可以使用此信号从网络增量加载更多的数据并将其存储在本地数据库中,PagingSource 可以从本地数据库加载这些数据并将其提供给界面进行显示。

当需要更多数据时,Paging 库从 RemoteMediator 实现调用挂起 load 函数。它在协程中调用,因此可以放心地执行长时间运行的工作。此功能通常从网络源提取新数据并将其保存到本地存储空间。

这一过程会处理新数据,但长期存储在数据库中的数据需要进行失效处理(例如,当用户手动触发刷新时)。这由传递到 load 方法的 LoadType 属性表示。LoadType 会通知 RemoteMediator 是需要刷新现有数据,还是提取需要附加或前置到现有列表的更多数据。

通过这种方式,RemoteMediator 可确保应用以适当的顺序加载用户要查看的数据。

1. 定义实体类

定义实体类 Film,它可以与 JSON 数据映射,将网络获取的电影 JSON 数据转换成 Film 对象,代码如下:

```
//模块 Ch10_02 Film.kt
@Entity(tableName="films")
data class Film(
```

```kotlin
    @PrimaryKey(autoGenerate=false) @SerializedName("name") val name:String,
    @SerializedName("release") val release:String,
    @SerializedName("region") val region:String,
    @SerializedName("directors") val directors:String,
    @SerializedName("actors") val actors:String,
    @SerializedName("intro") val intro:String,
    @SerializedName("poster") val poster:String,
    @SerializedName("trailer_url") val trailer:String,
    @SerializedName("video_url") val video:String
)
```

因为是多层次的数据处理,需要将获取的网络数据保存到本地 SQLite 数据库的数据表 films 中,实体类 Film 与数据表 films 的映射如图 10-5 所示。

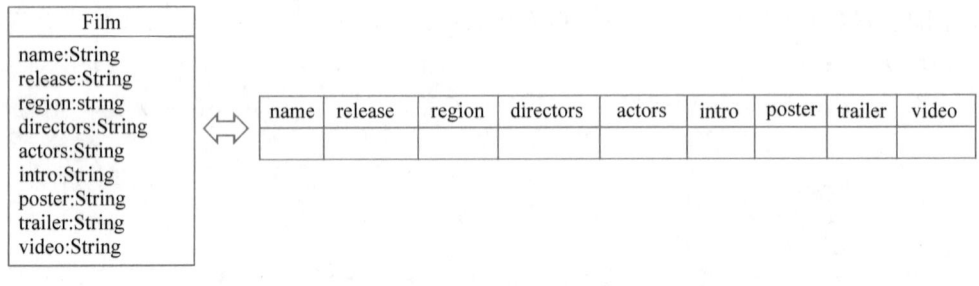

图 10-5 Film 实体类与数据表 films 的映射

定义 FilmRemoteKey 类。因为从网络访问每一条电影记录需要知道这条记录的上一页和下一页的内容,因此定义 FilmRemoteKey 类来追踪,代码如下:

```kotlin
//模块 Ch10_02 的 FilmRemoteKey.kt
@Entity(tableName="filmRemoteKeys")
data class FilmRemoteKey(
    @PrimaryKey(autoGenerate=false)
    val name:String,
    val prePage:Int?,
    val nextPage:Int?
)
```

图 10-6 展示了实体类 FilmRemoteKey 和数据表 filmRemoteKeys 的映射关系。

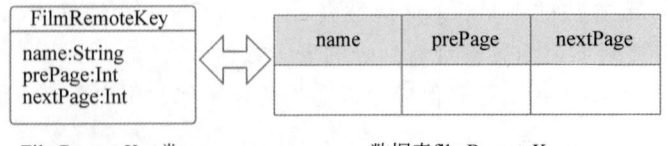

图 10-6 实体类 FilmRemoteKey 和数据表 filmRemoteKeys 的映射

filmRemoteKeys 的 name 字段表示电影名,prePage 字段表示记录所在页的上一页的页码,nextPage 字段表示记录所在页的下一页的页码。filmRemoteKeys 的数据表保存的记录示例如表 10-1 所示。

表 10-1　filmRemoteKeys 数据表样例

name	prePage	nextPage	name	prePage	nextPage
电影名 1		2	电影名 6	1	3
电影名 2		2	电影名 7	1	3
电影名 3		2	电影名 8	1	3
电影名 4		2	电影名 9	1	3
电影名 5		2	电影名 10	1	3

假设网络 http://127.0.0.1:5000/film.json? page=1&size=5，每页共 5 条电影记录，保存到本地 SQLite 数据库。从电影名 1~电影名 5 是第一页的记录，当前页码是 1，因此它们的 prePage 的值为空，而 nextPage 是当前页码的下一页，因此这 5 条记录的下一页的页码为 3。同理，网络资源 http://127.0.0.1:5000/film.json? page=2&size=5 的 5 条记录，对应电影名 6~电影名 10 这 5 条记录。这 5 条记录的当前页码为 2，它们的 prePage 为前一页，页码为 1，而 nextPage 为后一页，则这 5 条记录的 nextPage 的值为 3。

2. Retrofit2 构建网络访问服务

本例仍是采用 Retrofit2 库构建网络访问，定义的 FilmApiService 网络访问接口和实现创建 Retrofit 网络访问构建器 RetrofitBuilder 类，代码同例 10-1，此处不再介绍。

3. 定义数据库的访问

因为需要将网络数据保存到移动终端的本地 SQLite3 数据库。因此结合 Room 组件来完成。
首先定义数据库访问接口 FilmDao。定义对数据表 films 操作插入、检索和删除操作，代码如下：

```kotlin
//模块 Ch10_02 的 FilmDao.kt
    @Dao
interface FilmDao {
    @Insert(onConflict=OnConflictStrategy.REPLACE)
    suspend fun insertAll(films: List<Film>)              //插入数据列表
    @Query("select * from films")
    fun queryAll(): PagingSource<Int, Film>               //检索所有的 Film 记录
    @Query("DELETE FROM films")
    suspend fun deleteAll()                               //删除表 films 中所有记录
}
```

注意：在访问电影的数据库访问接口中 queryAll 函数返回的是 PageSource<Int,Film>对象。其中，Int 表示页码，Film 表示该页面的所有 Film 对象。

定义电影页码数据访问接口 FilmRemoteKeyDao。对数据表 filmRemoteKeys 的检索、插入和删除记录，代码如下：

```kotlin
//模块 Ch10_02 的 FilmRemoteKeyDao.kt
@Dao
interface FilmRemoteKeyDao {
    @Query("SELECT * FROM filmRemoteKeys WHERE name=:name")
    suspend fun findByName(name:String): FilmRemoteKey
    @Insert(onConflict=OnConflictStrategy.REPLACE)
    suspend fun insertAllKeys(remoteKeys:List<FilmRemoteKey>)
    @Query("DELETE FROM filmRemoteKeys")
```

```
        suspend fun deleteAllKeys()
}
```

定义 FilmDatabase 类来实现创建数据库 filmDB.db 以及访问 filmDB.db 数据库的唯一实体对象，代码如下：

```
//模块 Ch10_02 的 FilmDatabase.kt
@Database(entities=[Film::class, FilmRemoteKey::class], version=1)
    abstract class FilmDatabase: RoomDatabase() {
    abstract fun filmDao(): FilmDao                      //访问数据表 films 接口
    abstract fun filmRemoteKeyDao():FilmRemoteKeyDao//访问数据表 filmRemoteKeys 接口
    companion object{
        private var instance: FilmDatabase? =null
        @Synchronized
        fun getInstance(context:Context=FilmApp.context): FilmDatabase {
            instance?.let{
                return it
            }
            return Room.databaseBuilder(
                context,
                FilmDatabase::class.java,
                "filmDB.db"
            ).build()
                                                         //单例模式创建为一个 FilmDB 对象示例
        }
    }
}
```

4. 定义仓库层

定义的 FilmRemoteMediator 类是一个 RemoteMediator 类的子类，它起到非常重要作用。它的 load 函数可以从数据库加载历史记录，如果需要更新，则访问网络资源，并将更新的数据保存到数据库中，代码如下：

```
//模块 Ch10_02 的 FilmRemoteMediator.kt
@OptIn(ExperimentalPagingApi::class)
class FilmRemoteMediator(
    private val database: FilmDatabase,              //指定数据库
    private val networkService: FilmApiService       //指定网络访问
): RemoteMediator<Int, Film>() {
    private val filmDao=database.filmDao()
    private val filmRemoteKeyDao=database.filmRemoteKeyDao()
    override suspend fun load(loadType: LoadType,
        state: PagingState<Int, Film>): MediatorResult {
        return try{                                  //从数据库获取缓存的当前页面
            val currentPage:Int=when(loadType){      //根据加载类型判断
                LoadType.REFRESH ->{                 //UI 初始化刷新
                    val remoteKey: FilmRemoteKey? =getRemoteKeyToCurrentPosition(state)
                    remoteKey?.nextPage?.minus(1)?:1 //计算当前页码
                }
                LoadType.PREPEND ->{                 //在当前列表头添加数据使用
```

```kotlin
            val remoteKey=getRemoteKeyForTop(state)
                val prevPage = remoteKey?.prePage?: return MediatorResult.Success
                (remoteKey!=null)
            prevPage
        }
        LoadType.APPEND ->{                              //尾部加载更多的记录
            val remoteKey=getRemoteKeyForTail(state)
                val nextPage = remoteKey?.nextPage?: return MediatorResult.Success
                (remoteKey!=null)
            nextPage
        }
    }
    /**
        * 联网状态下的处理获取网络资源
        * response
        */
    val response=networkService.getData(currentPage,5)   //获取网络数据
    val endOfPaginationReached=response.isEmpty()         //网络响应是否为空
    val prePage=if (currentPage==1) null else currentPage-1  //计算前一页
    val nextPage=if (endOfPaginationReached) null else currentPage+1
                                                          //到达本页尾
    database.withTransaction{
        if (loadType==LoadType.REFRESH) {                 //刷新记录要删除原有的记录
            filmDao.deleteAll()
            filmRemoteKeyDao.deleteAllKeys()
        }
        val keys:List<FilmRemoteKey>=response.map{
                                                          //获取的记录映射成对应的索引记录
            film:Film->FilmRemoteKey(film.name,prePage,nextPage)
        }
        filmRemoteKeyDao.insertAllKeys(keys)
        filmDao.insertAll(response)
    }
    MediatorResult.Success(endOfPaginationReached)
}catch(e: IOException){
    MediatorResult.Error(e)
}catch(e: HttpException){
    MediatorResult.Error(e)
}
}
/** * 获取当前位置对应的 FilmRemoteKey
    * @param state PagingState<Int, Film>
    * @return FilmRemoteKey? */
private suspend fun getRemoteKeyToCurrentPosition(state:PagingState<Int,Film>)
    :FilmRemoteKey?=state.anchorPosition?.let{
    position:Int->state.closestItemToPosition(position)?.name?.let {
        name:String->filmRemoteKeyDao.findByName(name)
        }
    }
/** * 获取当前页面从头部第一个位置对应的 FilmRemoteKey
```

```
 * @param state PagingState<Int, Film>
 * @return FilmRemoteKey? */
private suspend fun getRemoteKeyForTop(state:PagingState<Int,Film>)
    :FilmRemoteKey?=state.pages.firstOrNull{
        it: PagingSource.LoadResult.Page<Int,Film>->it.data.isNotEmpty()
    }?.data?.firstOrNull()?.let{
        film:Film ->filmRemoteKeyDao.findByName(film.name)
    }
/** 获取当前尾部最后一个位置对应的 FilmRemoteKey
 * @param state PagingState<Int, Film>
 * @return FilmRemoteKey? */
private suspend fun getRemoteKeyForTail(state:PagingState<Int,Film>)
    :FilmRemoteKey?=state.pages.lastOrNull{
        it:PagingSource.LoadResult.Page<Int,Film>->it.data.isNotEmpty()
    }?.data?.lastOrNull()?.let{film:Film ->filmRemoteKeyDao.findByName(film.name)
    }
}
```

load 函数包含两种参数,即 loadType 和 state。loadType 是加载 PagingData 数据触发执行 PagingSources 的加载类别 LoadType。共有 3 中加载类别。

（1）LoadType.REFRESH：正在刷新的 PagingData 内容,这可能是由于 PagingSource 失效可能包含内容刷新或初始加载。在上述代码中,通过访问数据库的 remoteFilmKeys 数据表获取当前访问页码。

（2）LoadType.PREPEND：开始加载 PagingData。从 state 中获取当前位置对应的当前页码判断是否是第一页或者是 null,如果是从数据表 remoteFilmKeys 中根据电影名获取当前页码。

（3）LoadType.APPEND：结束加载 PagingData。从 state 中获取当前位置对应的当前页码是否是最后一页或已经是页尾,如果是从数据表 remoteFilmKeys 中根据电影名获取当前页码。

而 state 参数是 LoadState 类型对象。state 表示可观察 LoadType 变换的加载状态。

根据获取的当前页码,从网络中获取电影数据,并计算当前页、前一页和后一页的页码。然后数据库 database 根据加载类型做出处理,如果是 LoadType.REFRESH,表示数据更新,则将数据库原有记录删除,在将网络获取的数据插入数据库中。

定义 FilmRepository 获取 PagingSource 数据源,代码如下：

```
//模块 Ch10_02 的 FilmRepository.kt
@OptIn(ExperimentalPagingApi::class)
class FilmRepository(
    private val filmApiService: FilmApiService,
    private val filmDatabase: FilmDatabase
) {
    fun getAllFilms(): Flow<PagingData<Film>> {
        val pagingSourceFactory:()->PagingSource<Int, Film>={
            filmDatabase.filmDao().queryAll()
        }
        return Pager (
            config=PagingConfig(pageSize=5),
```

```
            initialKey=null,
            remoteMediator=FilmRemoteMediator(filmDatabase,filmApiService),
            pagingSourceFactory=pagingSourceFactory
        ).flow
    }
}
```

5. 定义视图模型 FilmViewModel

它承担实际的业务逻辑,获取电影记录,代码如下:

```
//模块 Ch10_02 的 FilmViewModel.kt
class FilmViewModel: ViewModel() {
    val filmRepository: FilmRepository=FilmRepository(RetrofitBuilder.apiService,
    FilmDatabase.getInstance())
    fun getFilms()=filmRepository.getAllFilms()
}
```

6. 定义界面

FilmCard 可组合函数定义显示电影单项的界面,代码如下:

```
//模块 Ch10_02 的 FilmCard 可组合函数
@Composable
fun FilmCard(film: Film?) {
    Card(modifier=Modifier
        .fillMaxSize()
        .padding(2.dp),
        elevation=CardDefaults.cardElevation(5.dp),
        colors=CardDefaults.cardColors(containerColor =Color.DarkGray)) {
            Column{
            Row(modifier=Modifier.fillMaxSize()){
                AsyncImage(
                    modifier=Modifier.width(180.dp).height(240.dp),
                    model="${film?.poster}",
                    contentDescription ="${film?.name}")
                Column{
                    Text("${film?.name}",fontSize=18.sp,color=Color.Green)
                    Text("导演：${film?.directors}",fontSize=14.sp,
                    color=Color.White)
                    Text("演员：${film?.actors}", fontSize=14.sp,color=Color.Green)
                }
            }
            Text("${film?.intro?.subSequence(0,60)} ...",fontSize=14.sp,
                color=Color.White)
            Row(horizontalArrangement=Arrangement.End,
                modifier=Modifier.fillMaxSize()) {
                Text("More",fontSize=12.sp)
                IconButton(onClick={
                }){Icon(imageVector=Icons.Default.MoreVert,
                    tint=Color.Green,contentDescription="更多...")
                }
            }
        }
```

```
            }
        }
}
```

FilmScreen 可组合函数定义电影列表界面,代码如下:

```
//模块 Ch10_02 的 FilmScreen 可组合函数
@Composable
fun FilmScreen(modifier:Modifier,filmViewModel: FilmViewModel) {
    val films=filmViewModel.getFilms().collectAsLazyPagingItems()
    val TAG="加载状态"
    Column(horizontalAlignment=Alignment.CenterHorizontally,
        modifier=modifier.background(Color.White)){
        LazyColumn{
            items(films.itemCount) {
                FilmCard(films[it])
            }

            films.apply{
                when{
                    loadState.refresh is LoadState.Loading ->item {
                    }
                    loadState.refresh is LoadState.Error ->item{
                        Log.d(TAG,"加载失败")
                    }
                    loadState.append is LoadState.Loading ->item {
                        Log.d(TAG,"尾部加载成功")
                    }
                    loadState.append is LoadState.Error ->{
                        Log.d(TAG,"尾部添加错误!")
                    }
                }
            }
        }
    }
}
```

定义 MainActivity,调用 FilmScreen,加载主界面,显示电影列表,代码如下:

```
//模块 Ch10_02 的 MainActivity.kt
class MainActivity: ComponentActivity() {
    override fun onCreate(savedInstanceState: Bundle?) {
        super.onCreate(savedInstanceState)
        enableEdgeToEdge()
        val filmViewModel=ViewModelProvider(this).get(FilmViewModel::class.java)
        setContent {
            Chapter10Theme {
                Scaffold(modifier=Modifier.fillMaxSize()) {
                    innerPadding ->FilmScreen(modifier=Modifier.padding
                    (innerPadding),filmViewModel= filmViewModel)
```

```
                }
            }
        }
    }
}
```

习 题 10

1. 结合 Android JetPack 架构组件实现货币兑换的移动应用。

2. 结合 Paging 3 组件实现智能聊天的移动应用。在线智能聊天可以采用青云客网站。

3. 结合 Paging 3 组件实现在线电影预告片观看的移动应用。在线电影数据相关(电影名、电影预告片网址等数据)可以通过网络爬虫来获取。

4. 结合 Paging 3 组件实现天气预报的移动应用。天气预报数据可以来自天气预报 API 接口,如 https://yiketianqi.com/。

5. 结合 Paging 3 组件实现在线新闻的移动应用。

6. 结合 Paging 3 组件实现在线音乐播放器的移动应用,功能类似 QQ 音乐或网易云音乐等 App。

参考文献

[1] 郭霖. 第一行代码 Android[M]. 3版. 北京：人民邮电出版社，2020.
[2] 李刚. 疯狂 Android 讲义[M]. 4版. 北京：电子工业出版社，2018.
[3] JEMEROV D,ISAKOVA S. Kotlin 实战[M]. 覃宇，等译. 北京：电子工业出版社，2017.
[4] 叶坤. Android JetPack 应用指南[M]. 北京：电子工业出版社，2020.
[5] SMYTH N. Android Studio 3.3 Development Essentials[M]. Middle Town：Payload Media Inc,2019.
[6] 夏辉,杨伟吉,张瑾. Android 移动应用开发技术与实践[M]. 北京：机械工业出版社，2021.
[7] DARWIN I F. Android 应用开发实战[M]. 胡训强，夏红梅，张文婧，译. 北京：机械工业出版社，2018.
[8] 林学森. 深入理解 Android 内核设计思想[M]. 2版. 北京：人民邮电出版社，2017.